OIL SPACES

Exploring the Global Petroleumscape

Edited by
Carola Hein

Routledge
Taylor & Francis Group
NEW YORK AND LONDON

First published 2022
by Routledge
605 Third Avenue, New York, NY 10158

and by Routledge
2 Park Square, Milton Park, Abingdon, Oxon, OX14 4RN

Routledge is an imprint of the Taylor & Francis Group, an informa business

Library of Congress Cataloging-in-Publication Data
Names: Hein, Carola, editor.
Title: Oil spaces : exploring the global petroleumscape / edited by Carola Hein.
Description: 1 Edition. | New York, NY : Routledge, 2022. | Includes bibliographical references and index. |
Identifiers: LCCN 2021014469 (print) | LCCN 2021014470 (ebook) | ISBN 9780367417512 (hardback) | ISBN 9780367417499 (paperback) | ISBN 9780367816049 (ebook)
Subjects: LCSH: Petroleum industry and trade—History. | Infrastructure (Economics)—History.
Classification: LCC HD9560.5 .O3689 2022 (print) | LCC HD9560.5 (ebook) | DDC 338.2/7282—dc23
LC record available at https://lccn.loc.gov/2021014469
LC ebook record available at https://lccn.loc.gov/2021014470

ISBN: 978-0-367-41751-2 (hbk)
ISBN: 978-0-367-41749-9 (pbk)
ISBN: 978-0-367-81604-9 (ebk)

DOI: 10.4324/9780367816049

Typeset in Bembo
by codeMantra

OIL SPACES

Oil Spaces traces petroleum's impact through a range of territories from across the world, showing how industrially drilled petroleum and its refined products have played a major role in transforming the built environment in ways that are often not visible or recognized. Over the past century and a half, industrially drilled petroleum has powered factories, built cities, and sustained nation-states. It has fueled ways of life and visions of progress, modernity, and disaster.

In detailed international case studies, the contributors consider petroleum's role in the built environment and the imagination. They study how petroleum and its infrastructure have served as a source of military conflict and political and economic power, inspiring efforts to create territories and reshape geographies and national boundaries. The authors trace ruptures and continuities between colonial and postcolonial frameworks, in locations as diverse as Sumatra, northeast China, Brazil, Nigeria, Tanzania, and Kuwait as well as heritage sites including former power stations in Italy and the port of Dunkirk, once a prime gateway through which petroleum entered Europe.

By revealing petroleum's role in organizing and imagining space globally, this book takes up a key task in imagining the possibilities of a post-oil future. It will be invaluable reading to scholars and students of architectural and urban history, planning, and geography of sustainable urban environments.

Carola Hein is professor of history of architecture and urban planning at Delft University of Technology. Her authored and (co-) edited books include *Adaptive Strategies for Water Heritage* (2019), *The Routledge Handbook of Planning History* (2018), *Port Cities* (2011), *Cities, Autonomy and Decentralisation in Japan* (2006), *The Capital of Europe* (2004), and *Rebuilding Urban Japan after 1945* (2003).

CONTENTS

PREFACE AND ACKNOWLEDGMENTS

Carola Hein

Lecturing on the impact of petroleum flows on the built environment, I often tell my audience "Drop your oil!" a command that I also used as the title of a short piece.[i] Once people start investigating the composition of their clothes and accessories, they also start looking more carefully at their surroundings, from furniture to building materials. Zooming out to a larger scale, they come to realize the impact of oil as material and financial flow, on lifestyles, and on the formation of space and its imaginary, what I call the global petroleumscape. These spaces of oil differ from each other according to geographic, political, economic, social, and cultural local context, while still being intimately connected. Exploring the particularities in their local detail is more than a single person can research.

Research on global landscapes of oil has been part of my studies since the 1990s, when I studied the urban and architectural history of the City Nord in Hamburg, a business district developed in the 1950s and 1960s. Whatever research topic I took on after that, whether the architecture and urban form of capital cities, global flows of urban planning ideas, or the development of port cities, the impact of petroleum on urban form was included. Over the years, as I have developed the petroleumscape concept, notably in "Landscapes of Oil" in *New Geographies* and "Oil Spaces" in the *Journal of Urban History*,[ii] I have been able to share my reflections with others and benefit from their research approaches. While taking this approach further and investigating the role of petroleum in urban and rural spaces in a forthcoming monograph, I have been able to refine my own work and connect with colleagues and students through conference sessions and workshops, relating their research to the petroleumscape concept. I have worked with master's and PhD students who have taken on the concept for their own research.

The current book is the outcome of long-standing efforts, collaborations, and grants and would not have been possible without the contributions of many people. A Guggenheim

i Carola Hein, "Editorial: Drop Your Oil!," The Beam 7 (2018).

ii "Global Landscapes of Oil," New Geographies 2 (2010); "Oil Spaces: The Global Petroleumscape in the Rotterdam/the Hague Area," Journal of Urban History 44, no. 5 (2018).

Fellowship in 2007–2008 was the springboard to seriously start the research on the global petroleumscape. An NWO Creative Industries KIEM grant in 2016–2017 for the project Petroleumscapes along the Maas: Visualizing Oil's Impact and Promoting Citizen Science at Museum Rotterdam allowed me to refine the concept and the spatial mapping. The opening event occurred during the 2016 International Planning History Society (IPHS) meeting in Delft. During the exhibition and the conference, we organized several subsequent events and meetings on the topic. A grant by the KNAW provided additional funding for a workshop on the global petroleumscape held in 2017 that brought together many of the contributors to this volume. Additional opportunities for presentations of the evolving concept included the SACRPH conference in Cleveland 2017 and the IPHS meeting in Yokohama 2018. Each of these meetings provided new inspirations. An NWO grant made this open access publication possible. Many of the authors in this volume presented their work at the 2017 workshop in Delft. I am grateful to everyone who participated in these meetings as a speaker or an attendee. While I cannot list all the participants in the diverse events, I would like to thank all of them for inspiring discussions. Several of the book authors have since become involved in the Leiden-Delft-Erasmus PortCityFutures initiative, which I lead. Because ships transport oil around the world, port city regions function as key petroleum hubs. They present particular challenges for sustainable post-oil future development, and they have become a focus of attention for policy makers.

Special thanks go to Molly Mullin for her excellent editorial support and to Rosemary Wakeman for her careful review of the final text.

As always, the project would not have been possible without the patience and support of my family. With love to Patrick, Caya, Aliya, Jolan, Joris, and my parents Wuppi and late Walter.

Introduction

1

SPACE, TIME, AND OIL

The Global Petroleumscape

Carola Hein

Industrially drilled petroleum and its refined products have transformed life and landscapes around the world in ways that are often not visible or recognized. It is difficult to see petroleum's shaping force when it forms part of so many of the objects and materials that surround us: the fabric of our clothes, the toys of our children, the asphalt on which we drive, even the food we eat. Petroleum's influence on daily life is often indirect, such as when its availability influences choices about where one should live or how one should spend one's leisure time. We take its presence and what it makes possible for granted and have only recently started to track its impact through CO_2 emissions, plastic waste, and petrochemical fertilizers. Yet, petroleum and its derivatives have been at work for a century and a half, powering factories, building cities, shaping landscapes, and sustaining nation-states. The diverse spaces of petroleum—oil fields, refineries, company headquarters, gas stations, housing, heritage sites, and even philanthropic structures—are dispersed around the world in uneven ways, facilitating everyday life for some and placing burdens on others. The spatial realities of petroleum and its representation have been game changers in the history of humans. Space is particularly important in the permeation of petroleum use; once established, physical infrastructures and spaces can shape future developments. Analyzing the oil revolution and the way it shaped and used buildings, cities, and landscapes, from underground to atmosphere, is important as we face the need to transition to a new energy landscape and imagine the possibilities and challenges of a post-oil future.

The investigations in this volume are based on the concept of a *global palimpsestic petroleumscape*[1]: a layered physical and social landscape that reinforces itself over time through human action and connects urban and rural spaces, culture and nature, materials and intangible practices. An interest in space and spatial practices in the built environment has been a constant in my research in architectural and urban history. Early in my career, as I explored topics as diverse as the construction of business districts in the 1950s and 1960s, the question of a capital for Europe, and the works of the French planner Maurice Rotival, petroleum continually emerged as an agent of change. Researching port cities and the impact of commodity flows on the built environment, I started to address the question of petroleum more

DOI: 10.4324/9780367816049-2

directly. With a Guggenheim fellowship in 2007–2008, I began to theorize the concept of the petroleumscape. With this concept in mind, I conducted research on the spaces of petroleum in various cities, including Rotterdam, Dunkirk, and Philadelphia; I collaborated with colleagues and PhD students; I organized conference sessions and workshops. The contributors to this volume, many of whom attended the conference on the Global Petroleumscape in Delft in 2017, engage with the petroleumscape concept by revealing through case studies petroleum's role in organizing and imagining spaces globally.

Following a brief discussion of the materiality and agency of petroleum, this chapter introduces the concept of the petroleumscape as an attempt to theorize the role of petroleum in the built environment through the lenses of the spatial and the representational petroleumscape. It concludes by briefly introducing the chapters that form the three parts of the book.

Petroleum as Material and Agent over Five Generations of the Petroleumscape

The concept of the petroleumscape starts from the understanding that petroleum is a physical material with a pervasive impact on physical space in terms of architecture, cities, and landscapes, and is not a magic fluid that fuels economies without a spatial imprint. Petroleum, or "rock oil," is a natural product that bubbles up from the ground, and humans have long used it for specialized purposes in ways that foreshadow our uses today. Incendiary weapons such as "Greek fire" anticipated our current use of oil for warfare, lighting, and warming. Bitumen, once used to make pools and basins watertight in Mesopotamia and to mummify bodies in Egypt, today appears in asphalt street cover, roofing materials, and waterproofing. And the historic use of petroleum as a medicine precedes more recent pharmaceutical and cosmetic uses; for some 150 years, people have been applying Vaseline petroleum jelly to dry skin and minor wounds. This materiality of petroleum has shaped physical space and people's lifestyles around the world.

Oil extraction, production, and resale are all spatially anchored through extraction, transformation, transportation, storage, and consumption. Petroleum interests have created and promoted geopolitical conflict. The drive to access petroleum sources has led to numerous wars and reshaped the boundaries and structure of nations. Transnational companies have pursued their interests, often forcing the hands of nation-states, influencing internal politics and policies. Conflicting interests within nations have spurred internal conflicts, visible in ongoing discussions around the future energyscape. Although petroleum has been at the origin of numerous conflicts and although its actors are multiple and have competing interests, the history of petroleum as an agent of spatial transformation has been one of seemingly unhindered and continual growth and expansion.

With industrial petroleum's end in sight, its past merits scrutiny. The petroleumscape did not emerge as if squeezed from a tube; rather, it was built incrementally, with one generation supporting the next. Industrial drilling provided the initial foundation for the development of the modern petroleumscape. In 1859, on Oil Creek in Pennsylvania (later known as Titusville), Edwin Laurentine Drake developed the technology of drilling wells, providing access to deeper oil fields and giving rise to the first generation of the petroleumscape (1859–1910). The industry at this point revolved mostly around kerosene, a petroleum distillate, which eventually replaced whale oil for lighting, and provided select countries with a head start for building new infrastructures and developing petroleum-related knowledge. Industrialized drilling made petroleum available around the world for a huge number of purposes, but there was no large demand for oil before the arrival of advanced refining techniques and other technological inventions. In particular, gasoline-powered automobiles and their later mass production along assembly lines led to a new demand for petroleum as an engine lubricant, and consumers and companies alike pushed for further oil drilling.

The second generation (1910–1940) saw the expansion of existing installations to other parts of the world, notably the Middle East, often through colonial networks. Access to petroleum sources, knowledge of its physical qualities, creativity, and innovation in developing new products and increased capacity of distribution have led to the growth of huge corporations and have shaped the form and function of entire nation-states. The need to find new applications and consumers led to the third generation (1940–1970), which included the extensive introduction of plastics into architecture and everyday life. Petroleum became the most important fossil fuel of the twentieth century. The emergence of critical voices brought about the fourth generation (1970–2000), at a time when spaces on sea and land had become saturated with petroleum-related structures. The fifth generation (2000–) has seen multiple attempts to overcome the petroleumscape, yet with only partial success as the increased demand for a growing number of petroleum products spurred the development of new technologies for finding oil and extracting it from more and more difficult sites on land and sea, in ice, rainforests, or from shale formations.

The oil industry has produced numerous powerful global players, who control interconnected oil spaces around the world. Well-capitalized operators and politically powerful nations have managed the consolidation of the sector. They have been capable of investing in technical innovation and professional knowledge and able to endure failures, disasters, wars, and price busts, and able to balance income and investment across their global supply chains. The petroleum industry spans multiple areas of intervention, the so-called upstream, midstream, and downstream activities. The *upstream* sector includes oil prospecting and extraction. Its sites are often in rural areas or places that are difficult to access. The *midstream* sector includes road, rail, and pipeline infrastructures that transport petroleum and its products to and from refineries, often on a global scale. The *downstream* sector includes oil refining and processing as well as marketing and some midstream activities, all of which are important in distributing oil products to consumers. Each of these sectors has similar functions and typologies around the world, with its own location, typology, built form, and function. They interconnect to form a single landscape while engaging in unique ways with local needs and practices.

Among the best-known and largest contemporary oil companies is Exxon Mobil. Its roots go back to 1863, when John D. Rockefeller joined a Cleveland-based oil-refining business that in 1870 became the Standard Oil Company. By 1880, it controlled the refining of 90–95 percent of all oil in the US. Nine trustees, including Rockefeller, created the Standard Oil Trust in 1882 from various companies engaged in producing, refining, and marketing oil. The company became the image of economic concentration and control and successfully expanded beyond national borders, making petroleum the primary source of energy around the world. In 1911, enforcement of federal anti-trust laws led to its dissolution. The pieces of the complex themselves became global giants, including Exxon, Mobil, Amoco, and others (some have reunited in subsequent years). The history of Exxon and other giant oil companies, including Shell, BP, and TOTAL, is the object of important historical studies and corporate publications, but these studies do not engage in a close exploration of the role of space as a system.[2]

Petroleum practices depend on small and large businesses as well as national governments with differing degrees of control over land, funds, and laws. While oil companies are active on a global scale, they need governments to provide access to local spaces and they depend on national infrastructures, policies, and legal frameworks. Their diverse interests create multiple transnational linkages and friction points. Oil companies and governments have closely collaborated at home and in colonial settings, leading, for example, to British government support for the Anglo-Persian Oil Company shortly before World War I under the influence of Winston Churchill, and the Compagnie Française des Pétroles founded in 1924 and today known as TotalFinaElf. Oil companies and national governments have also come into conflict; numerous nationalization movements around the world have brought refineries and oil exploitation originally developed by private companies into the hands of new nations. The expropriations of petroleum reserves, oil facilities, and companies in Mexico, Suez, and Abadan are examples.

The discovery and ownership of petroleum, the specialized knowledge and equipment required, the enormous amounts of funding generated and consumed have brought about new nations and changed global power balances. The global leadership of oil business in Europe and the US peaked in the nineteenth and early twentieth centuries. Other countries, like China, discovered oil later, but have already largely depleted their national sources, making them important agents in global petroleum extraction. The discovery of large oil fields in the Middle East, Venezuela, Nigeria, and Russia created new global players with national economies largely dependent on petroleum income. In 1960, some of them, including Iran, Iraq, Kuwait, Saudi Arabia, and Venezuela, formed a cartel, the Organization of the Petroleum Exporting Countries (OPEC), to control oil prices and flow. The discovery of oil in the North Sea further changed the global spaces of oil, giving countries like Norway, the UK, and the Netherlands (through gas) new riches, and effectively leading to a redefinition of maritime territories and laws.

The exploitation, transformation, transportation, use, and even the cleanup of oil require collaboration among powerful actors. The oil business started out as a private business venture, but World War I was a turning point, as militaries depended on gasoline for their ships, tanks, trucks, and cars. The growing political importance of oil—and its geopolitical implications—made it a major factor in colonialism and wars—including the two world wars. Wartime creativity promoted the use of petroleum for a multitude of new products. Improvements in refining technology and chemistry provided the foundation for transforming petroleum into a rapidly increasing number of new petroleum-based products that did not involve burning it, including vinyl flooring, paint, and plastics; fibers such as nylon, acrylic, and polyester; and microplastics like microbeads used in body scrubs and toothpastes. This ability to create new products meant that oil companies could nimbly adapt to changing societal demands.

Petroleum and its multiple spaces influence global politics and economy, but also society and culture. Its impact on space and society depends on national and local efforts and perspectives. This is effectively illustrated through the use of terms such "oil crisis" or the "first oil shock" to describe events in 1973–1974. What was a crisis and a shock to the oil-consuming world, leading to the implementation of car-free Sundays in Germany and the Netherlands, had benefits for some of the producing countries, both within OPEC and beyond. The arrival of oil money in countries like Algeria led to the construction of new oil ports, such as Skikda, and the construction of new buildings and housing districts in capital cities, for example, in Tehran. New global players have also created new geographies of distribution. These shifts were often provoked by military campaigns or technological innovations—at times sponsored and brought about by the petroleum industry and its innovative research.

The presence of extremely powerful corporate and national actors has made the history of petroleum appear more coherent than it is. The petroleumscape's astounding longevity results from the feedback loop of overlaying functions in time. Wars and natural disasters have regularly destroyed or disconnected parts of the petroleumscape, yet industrial structures—refineries, oil tanks, oil ports—have been rapidly rebuilt, even under changing leadership, so that the old or new owners can continue to provide petroleum. Political shifts with territorial consequences, such as colonization, nationalization, or decolonization, have often involved profound disruptions. They have transformed global petroleum chains and flows, but they have not necessarily transformed the petroleumscape itself. Several major oil companies lost refineries and other installations to nationalization, but they rebuilt the missing structures rapidly in new locations to guarantee energy security. For a post-oil future, we need to generate new energy landscapes and new perspectives, yet even catastrophes like oil spills that demonstrate the dangers of the petroleumscape do not necessarily lead to changes: think of the Exxon Valdez catastrophe off Alaska in 1989 or the BP/Deepwater Horizon explosion in the Gulf of Mexico in 2010. At best, some of the petroleum-related disasters have led to localized changes, often increasing the pressure on other petroleum sites around the world.

Petroleum has come to facilitate movement and ease of life, but it has also jeopardized the future of human life on Earth and has advantaged some population groups over others. How we conceptualize petroleum spaces can help bridge the social science and humanities approaches to the study of petroleum's impact on society; it can also bridge the study of the past and the design of the future, given that space is a key agent in path dependence and feedback loops. The concept of the global petroleumscape, which I have introduced and refined in several articles and which will be further explored in a monograph, inspires reflections on the territorialities, spaces, and imaginaries of oil and on the spatial relevance of energy more generally.

The Concept of the Petroleumscape

The concept of the petroleumscape starts with the insight that the diverse spatial emanations of oil—including refineries, storage sites, office buildings, and gas stations—are connected through their relation to this single commodity and its group of industrial players. Connecting the actual places where oil is physically or financially present with the representation of these spaces and of the practices related to petroleum products, this concept aids understanding of the ways in which oil shapes behaviors and secures continuous production and expansion of its spaces. The physical structures and spaces of oil require extensive investment, and once funding has been sunk into the soil or the seafloor, economic and governance systems tend to reinforce earlier investments, reinforcing initial decisions. Oil imaginaries have served the promotion of lifestyles and everyday practices related to the consumption of petroleum products, creating a second feedback loop that reinforces the presence of the petroleum industry through the participation of citizens for whom oil is a constant necessity. The benefits of cheap energy for travel and heating and of easy-to-use building materials have led citizens of different classes, races, cultures, genders, and ages around the world to embrace and develop the new urban and rural, natural and cultural, land- and sea-based petroleum spaces; they have encouraged the collaboration of groups whose goals are otherwise often conflicting. Careful promotion of the benefits and uses of petroleum products have helped reinforce widespread citizen buy-in, creating an energy culture that reinforces the spatial presence of the industry and leads to increased consumption in everyday life.

The concept of the petroleumscape brings architecture and the built environment into an already rich conversation. Conceptualizing the ways in which petroleum has shaped various architectural and urban spaces can be aided by the concept of planetary urbanization developed by the American urban theorist Neil Brenner and the Swiss sociologist Christian Schmid.[3] Planetary urbanization includes the urbanization of the sea, a process that has been driven in part by petroleum interests.[4] Arjun Appadurai suggested that the suffix "-scape" allows for an understanding of the "new global cultural economy as a complex, overlapping, disjunctive order that cannot any longer be understood in terms of existing centre-periphery models."[5] The excellent analysis of Michael Watts (referencing the work of Timothy Mitchell, Andrew Barry, and others) and his term *oil assemblage* connects oil with its territory, identifying a technological zone, while stopping short of discussing the designs and representations of actual buildings, cities, and landscapes.[6] The use of the term "oil," however, is unfortunately imprecise. Research on the term will lead to sources and references that include many different types of oil. The concept of the petroleumscape is more precise as it addresses a specific commodity. The way it is used here encourages discussions of the role of space as an agent as well as consideration of how people have lived in the environments that petroleum has made possible. By studying the impact of oil on built form, we gain insight into the diverse spaces and temporalities of modernity in globally linked spaces.

This reading of the global petroleumscape is in line with Henri Lefebvre's understanding of space as socially produced and then appropriated by the powerful as a tool.[7] In "The Production of Space," Lefebvre argues that space "in addition to being a means of production … is also a means of control, and hence of domination, of power."[8] In the case of petroleum, this dynamic is particularly evident. The *spatial practices* (the life of inhabitants in a space) and *representations of space* (the approaches of built environment professionals) intersect with the *spaces of representation* (the images and associations of the users)[9]; they are multiple (palimpsestic) layers of physical space and professional and public representation. While Lefebvre sets up a complex system, this book focuses on oil-related spatial practices of petroleum production, administration, retail, and consumption, and explores the representations of these petroleum spaces. A fuller analysis of the represented petroleumscape in our everyday culture, including everyday language, is still needed.

The petroleumscape has emerged through the collaboration of diverse actors as shown in the infographic (Figure 1.1) and it has become an actor itself. The presence of the petroleumscape often attracts additional functions—a secondary petroleumscape—which reinforces the presence of the oil infrastructures. Since first proposing the concept of the petroleumscape a decade ago, I have increasingly come to see the petroleumscape as singular (rather than plural), composed of multiple layers and aspects that combine into one overarching landscape. Each of these spatial layers has similar functions and typologies (style, location, or architectural form) interconnecting around the world. Together, these layers

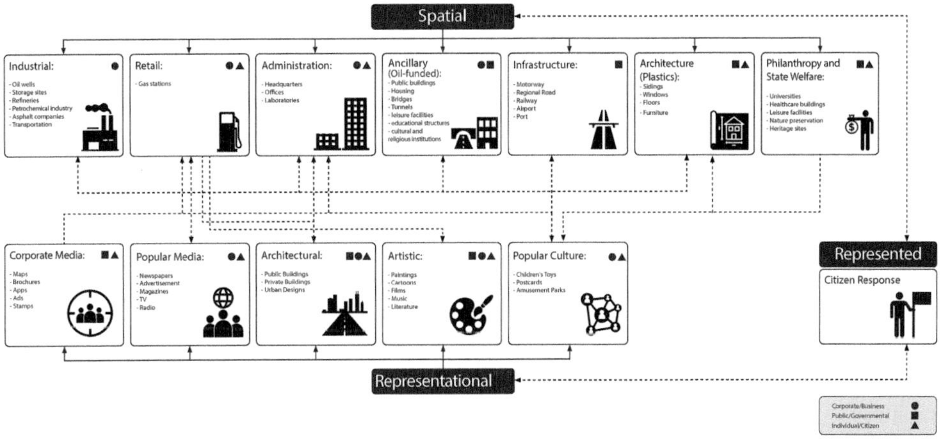

FIGURE 1.1 The hybrid, multiple, shifting, and uneven ways in which many actors collaborate to create the global petroleumscape. *Source*: Carola Hein.

of the industrial, administrative, retail, and infrastructural petroleumscape form a single global landscape. This spatial presence of petroleum structures and the close collaboration of relevant actors has created path dependencies (to use a concept developed in the political sciences) that reinforce the petroleumscape. Over time and depending on the temporal, geographic, and larger political, economic, social, and cultural context, the various petroleum actors, with the support of the general public, have created an energy culture with multiple feedback loops, both spatial and represented, that stabilize the system, make oil a positive and fun factor of everyday life, and effectively prevent companies and countries from making a transition from oil to other energy sources.

Petroleum has transformed spaces, their representation, and the way we live around the world. These spaces are usually experienced, seen, and discussed separately from each other. They are also promoted by diverse petroleum actors in uneven ways. This separation, I argue, has facilitated the spread of petroleum-based practices. Sites of extraction are far away from the locations of consumption. The benefits of mobility, heating, or plastic use are not obviously related to the pollution and environmental disasters or poor living conditions created by the production of oil. The fact that these different spaces are usually studied and promoted separately in terms of national context, typology, and appearance hides the complexity and interconnectedness of the petroleumscape and makes it easier for international corporations to influence national politics and planning processes. The "nationalization" of what is effectively a single international petroleumscape, the focus on select territories and representations of the petroleumscape, establishes separate bubbles and makes it even more difficult to induce change. The spaces and practices, once created, are key to the petroleumscape feedback loop.

Each layer of the spatial petroleumscape has specific characteristics that are distinct in function, location, land use, scale, typology, urban and architectural form, and financial impact, but they are effectively part of one spatial system. Public and private petroleum actors engage with these spaces in often quite different ways, discussed here under the theme of the representational petroleumscape, that is, the way in which these spaces are communicated to the local and global public. The following section explores the spatial and the representational petroleumscape, leaving the represented petroleumscape to later investigation. The interrelation between the spatial and representational petroleumscape facilitates the longevity of the petroleumscape and prevents changes.

The Spatial Petroleumscape

The concept of the spatial petroleumscape aims to overcome the segmented, monodisciplinary, and localized approach to the spaces of petroleum. It identifies the diverse spaces that are inherent parts of petroleum production, transformation, and consumption, and bring them into conversation. Scholars usually study these realms from different and disconnected perspectives and consequently tend to overlook the petroleumscape's systemic character. Understanding the multiplicity of petroleum spaces and their interconnectedness provides insight into the complex interrelationship and mutual constitution of the spatial networks that support petroleum exploration, transformation, and use. The following sections present and discuss the various spatial elements of the spatial petroleumscape through its industrial, retail, administrative, infrastructural, ancillary, architectural, and philanthropic components.

The *industrial petroleumscape* has a large spatial extent, including areas on sea and land. Encompassing ships and ports, drilling platforms, storage places, pipelines, and refineries, it is a truly global space with standardized structures that are identical around the world. Port city regions, where maritime transport, oil refining, and landside transportation intersect, are perhaps the most important nodes in the industrial petroleumscape.[10] Because water is essential both for industrial processes and for shipping, refineries have often been located near rivers. Control over land- and sea-based transportation, including the possession of a shipping fleet, has allowed companies to respond flexibly to their global commodity chains and has served as a geopolitical tool. Industrial structures, such as refineries, require extensive technical knowledge, financing, and expertise. As a result, the industrial petroleumscape is a story of superlatives. Refineries in Pladjoe in Indonesia, Suez in Egypt, and Abadan in Iran were all at some point the largest installations of their kind. The scale of these installations is enormous; the Jamnagar refinery in Gujarat, India, for example, commissioned in 1999, occupies 7,500 acres.[11] Despite their huge scale and territorial impact, these structures are not accessible to ordinary citizens.

Consumers experience petroleum more closely through the *retail petroleumscape*. The early petroleum business provided oil for lamps, but this did not lead to a dedicated retail system. With the emergence of the car, gas stations became the industry's most visible retail structure and one that is clearly part of the petroleumscape. Since their emergence in parallel with the spread of the car as a key mode of transport, they have become more recognizable, highly branded, their network serving as an advertisement for oil companies. From flimsy buildings, often designed in keeping with local interests, they have come to integrate the functions of a supermarket, coffee shop, or even wedding center. They serve both land and sea, and gas stations are even located on floating structures that cater to boats. Other forms of petroleum products have entered the retail chain in various ways, from building materials to clothes, and stores of all sizes around the world serve to promote petroleum products.

Perhaps less recognizable than gas stations, but often even more prominent in particular locales, is the *administrative petroleumscape*, encompassing petroleum company headquarters and research institutes. Prominent local architects have designed many of the buildings in line with the predominant taste in corporate or public architecture. These iconic buildings are often located next to each other and next to government ministries, in capital cities, designed in a style characteristic—and sometimes trendsetting—of their time and often similar to that of key administrative buildings nearby. Oil companies have become architectural landmarks in their respective towns and a leading force in the development of business districts. Historic headquarters buildings in prime locations include the Shell-Mex House in London—the clock tower-crowned building that faces the Thames and backs up to the Strand, linking Buckingham Palace and the City. Exxon, a Standard Oil spinoff, has set trends in design and urban location through its regional headquarters. The spatial prominence is aligned with the power of the institutions. Decisions taken in New York, London, or The Hague, in Tehran, Moscow or Caracas, can change global geographies.

If and when necessary for the functioning of the petroleum industry, oil companies have built an *ancillary petroleumscape*, consisting of diverse structures needed for the functioning of the industry: streets, housing, leisure facilities, or even entire cities, not directly related to the physical or financial flows of petroleum. The city of Abadan in southern Iran is one example. When oil was found in southern Iran, the Anglo-Persian Oil Company (later BP) designed Abadan, a company town between 1910 and 1951. Whereas such investments are necessary in the proximity of extraction sites, they are less needed where housing and office buildings exist and where spaces can be rented, bought, or built to host petroleum clubs (like the Overseas Club), to provide hotel rooms or golf courses to provide spaces for political networking. The case of Caracas with its Avila hotel and golf courses stands as an example: oilmen like Rockefeller and business and political leaders met there with architects and urbanists such as the French Maurice Rotival—who would go on to become a professor at Yale—and Wallace Harrison—Standard Oil's main architect. Both men designed projects for Standard Oil and the Rockefeller family.[12]

The global reach of the petroleum industry requires extensive infrastructures, many of which are not on land owned or controlled by the industry. Road, rail, and pipeline connections as well as bridges, dams, and port infrastructure—the *infrastructural petroleumscape*—require public support, often from multiple countries and through different logistics companies. Private oil companies through their economic weight and technological expertise can help reshape global geographies. Rockefeller's Standard Oil built its empire on petroleum transport and refining, rather than on access to oil sources. The Shell company started as a transport company, building, for example, the SS Murex, the first oil tanker to cross the Suez Canal (completed in 1869) in 1892. By using the Suez Canal, Shell was able to compete with the Rothschilds, who were transporting petroleum by train from Russia, and with Standard Oil, which was sending petroleum in barrels around the world.[13] A close relation between state and oil interests can potentially make it easier to engage with an energy transition—consider Norway—but it can also result in the opposite.

The petroleumscape has constantly evolved. The industry depended on chemical innovation to find new uses for petroleum and to expand its global network. Beginning with the invention of synthetic polymer-based plastics at the beginning of the twentieth century, designers and architects were intrigued by the possibilities for using plastics in building, creating the *architectural petroleumscape*. Initially, the uses they envisioned ranged from the small scale—light switches and furniture—to whole elements—windows and walls. Chemical companies envisioned a profitable use for their products. Experimentation in all-plastic houses peaked in the late 1950s and early 1960s. Probably the most famous of these was the

"House of the Future." Its location in Disneyland in southern California guaranteed that it was seen by millions of visitors. Opened in 1957, the powerful collaboration of research, construction, and design of MIT architects, the Monsanto Chemical Company, and Disneyland inaugurated a visionary plastic house of the future with novel forms and technologies that was marketed as part of a modern lifestyle: clean, functional, and fun. Although all-plastic houses were something of a temporary fad, the rapidly growing plastics industry fueled petroleum extraction, shipping, and refining.

Oil industry actors have long been eager to use architectural and urban space to showcase the way in which they—personally or as a company—are beneficial to society, creating a *petroleumscape of philanthropy.* Investments in health, education, culture, and heritage have become a means of branding and, often, a way to effectively boost the global flows of oil. In parallel with their commercial ventures but consciously and carefully separated from the Standard Oil/Exxon name, the Rockefeller family made architectural gifts to educational, health, and cultural institutions and supported natural and historical sites in key locations, establishing their philanthropic footprint in their headquarters city, New York, but also throughout the US and in many other parts of the world. Their gifts promoted societal transformation and constructed a cultural modernity through physical spaces and carefully built mindscapes in tune with the oil-based transformation of cities and landscapes around the world. Oil corporations similarly invest in diverse structures: on the building scale, these include BP Bridge in Chicago by the world-renowned architect Frank Gehry. They also sponsor educational and cultural spaces, including international schools and museums. This support has been criticized by groups such as Culture Unstained, which aims to end the oil industry's efforts to gain social legitimacy through philanthropy.[14]

The Representational Petroleumscape

While the concept of the spatial petroleumscape engages the physical presence of petroleum flows and funding, the *representational petroleumscape* concept explores the ways in which corporate and national actors advertise the spaces and uses of petroleum, and in which public media, private actors such artists and architects, and agents of popular culture influence perceptions of the spatial petroleumscape. This representation of petroleum has changed extensively in line with private and public needs and interests and with cultural practices and their evolution through time and space.

Just as oil companies adapt the spaces of petroleum to specific local needs, they adapt the display of oil spaces to local contexts. The industrial spaces of oil are the object of company advertisements, government publications, and popular media presentations with the focus and amount of this attention shifting over time. In the petroleumscape's early years, when American media addressed petroleum, the focus was largely on the growth of the industry, its financial impact, and advertising. Questions of safety and working conditions received less attention until 1904, when Ida Tarbell exposed the practices of the Standard Oil Company. Demonstrating the power of "muckraking" journalism, a Supreme Court decision in 1911 ultimately broke up Standard Oil into thirty-four separate companies.[15] But the impact of critical journalism remained limited. Companies and governments quickly claimed the power of the media for themselves. Oil companies made promotional films about their accomplishments, including cities they built and products they created.

As corporations and governments collaborated more intensively in the development of the petroleumscape, many countries started to celebrate their colonial programs, national policies, oil exploration, exploitation, and development. The Netherlands documented its colonial endeavor in Indonesia through multiple films, and British popular media documented the buildings and everyday operations involved in the Persian oil exploration effort and life in the British model city Abadan.[16] Corporate road maps of the 1950s, published by oil companies and freely distributed, showcase the ways in which corporate action weaves together advertising, knowledge, and the public's interest in leisure. As Brian Larkin observes, "Roads and railways are not just technical objects … but also operate on the level of fantasy and desire."[17] As citizens see the apparent benefits of the petroleumscape firsthand, they appear less concerned about negative impacts.

Prior to the mid-twentieth century, artists also depicted the industrial petroleumscape in positive ways. Pride in oil storage and handling seems to have inspired oil paintings of the ports of Rotterdam and Dunkirk that document the emerging petroleumscape.[18] Artistic representations of the benefits of petroleum are prominent in China, Russia, and many countries of the Middle East. Celebrating the scale and the power of oil infrastructure has been key to national narratives in various parts of the world at select moments in time. In countries where the oil company is owned by the state, this becomes even more visible; in China or Iran, for example, images of refineries on money or stamps have been a key part of everyday experience. In line with the contemporary assessment of petroleum use as environmentally damaging, some companies have started to debrand this industrial landscape, for example, by removing the logos from oil storage tanks. This is in line with companies' attempts to promote the idea that their products are environmentally friendly when they are not (which critics refer to as "greenwashing").

Critical artistic approaches to the petroleumscape can help raise awareness of the enormous scale of oil's presence and its representation in order to support new energy values in line with a post-oil society and to create new imaginaries of post-oil life. Engaging with our ongoing research on the petroleumscape in the Netherlands, a young illustrator, Jenna Arts, created two illustrations that capture the Dutchness of oil, translating Vermeer's milkmaid into an oil maid, and adding a gas station into Vermeer's Little Street of Delft (Figure 1.2). These illustrations speak to the extent to which the consumer has become a partner complicit in the feedback loop of the petroleumscape and raise the question: What would the post-oil equivalent of this maid or of this street look like?

FIGURE 1.2 Reinterpreting well-known works of the Dutch artist Vermeer, Jenna Arts aimed to capture the Dutchness of the representational petroleumscape. *Source*: Jenna Arts

Architects, including famous ones, have produced their own commentary on the emerging petroleumscape, often at the invitation of the petroleum or car industry or in support of them. The influential visionary project for a city with skyscrapers and highways, the Cité Voisin plan by Le Corbusier, was sponsored by a car manufacturer. Cars were also a key feature in Frank Lloyd Wright's Broadacre City. The most striking connection between oil, cars, and city planning was made by the City of the Future project for the 1939 World's Fair in New York. It resulted from a collaboration between General Motors, designer Norman Bel Geddes, and Shell Oil, who used the imagery for advertising purposes. Featuring a modern city of highways and skyscrapers, the exhibit went on to inspire planners in shaping real cities and encouraged citizens to imagine how to live in the second part of the twentieth century. At the same time, the petroleum industry, along with other private and public players, was lobbying for oil-friendly public policies like subsidies for highways and against public transit. Together with the heavily promoted desire for the single-family home—and in the US, federal mortgage deals for veterans in the postwar period—these were a recipe for increased driving and suburban sprawl.

Architects have drawn inspiration from the petroleum industry for new designs of non-petroleum buildings since the 1960s. For example, visionary drawings by Archigram and the plug-in city by Peter Cooke in 1964 are reflections on refineries, with an example depicted in Archigram's publication. Actual buildings, such as the Centre Pompidou in Paris designed by Richard Rogers and Renzo Piano and finished in 1967, show the degree to which architecture accepted the aesthetics of refineries. The Norwegian Oil Museum in Bergen, built in 1999, with its oil tank-like features, explicitly recalls oil structures.

Popular culture merits further investigation, forming part of the representational petroleumscape's affective qualities—both positive and negative. Think of children's toys produced without direct collaboration of the oil companies: a puzzle featuring an oil refinery, a toy car with the Esso logo, a Lego car handed out at a Shell gas station, or plastic Barbie houses that might eventually inspire their owners to choose plastic window frames, furniture, or floor coverings. Such objects promote the recognition of oil buildings, logos, and colors from an early age, preparing children for an oil world, albeit without displaying the oil or shipping flows that drive it. Greenpeace understood this relationship when the organization made a film called "LEGO: Everything is NOT Awesome," praising the dissociation of Lego from Shell.[19]

The degree to which the spaces of petroleum have entered the represented petroleumscape merits further investigation. Identifying and analyzing the individual spatial and representational elements of oil constitutes a first step in better understanding the impact of oil and in shaping our future energy spaces and practices.

Book Overview

In three distinct parts, the chapters in this volume provide insights into the territoriality, agency, practices, and imaginaries of petroleum through case studies. Building on unpublished primary sources and considering questions of (energy) landscapes, they bring architecture and urbanism into a sustained dialogue with humanities scholarship. This detailed investigation of the spaces and representations of petroleum gathered in this volume represents a foray in understanding the multiple and complex ways in which petroleum shapes contemporary spaces and practices. The impact of petroleum on global space is too vast to be fully examined in a single volume. A topic as complex as the global petroleumscape requires global and multidisciplinary collaboration based on shared approaches, and the contributions assembled here can only be seen as a preliminary investigation.

Part I on **Oil, Agency, and Territoriality** investigates questions of territories, boundaries, and policy in reshaping global geographies over time. It demonstrates how public and private actors have created the diverse spaces of the petroleumscape, creating spatial patterns and select nodes to facilitate petroleum flows, transformation, and use. Depending on political systems—capitalist, socialist, communist—and the position of the oil company in relation to the state, but also countries' environmental, social, and cultural preferences, the clustering of oil functions has varied through space and time.

This section starts with an exploration of the first petroleumscape in North America. Carola Hein and Alan Lessoff discuss a key example of petroleum-fueled spatial growth and the emergence of new practices focusing on the period from the 1850s to the 1950s. Ben de Vries probes the intricate intersection of national lust for petroleum and the construction, destruction, and reconstruction of petroleum's infrastructure through the case of the Pladjoe petroleum cluster in the turmoil of the 1940s in the colony of the Dutch East Indies, now Indonesia. He shows how petroleum resources tend to encourage military conflict that damages a petroleumscape, but it can be quickly restored and its longevity assured with the collaboration of diverse stakeholders. Stephen Ramos continues this trajectory, exploring the interrelated and dynamic processes of territory, territoriality, and sovereignty in the Persian Gulf region from the pre-oil period through periods of oil exploration, discovery, and industry consolidation to the 1970s. The multiple layering of Gulf territorial signification simultaneously combined British national and imperial objectives with the corporate objectives of interested oil companies and the establishment of constituent pre-nation-state urban, territorial, and legal frameworks. Giulia Scotto analyzes an aspect of the global scale of the petroleumscape as she investigates the *grande disegno africano* implemented by ENI (Ente Nazionale Idrocarburi), Italy's national hydrocarbon agency, showing how the company shaped territories and lived experience in postcolonial Africa, specifically Ghana, Tanzania, and Zambia in the 1960s. The chapter studies typologies, spaces, and scales often neglected by architectural historians. Nancy Couling expands the exploration of territoriality to the sea, where oil installations—drilling platforms and pipelines—form part of the offshore petroleumscape—a fifty-year-old historical format that has urbanized the entire North Sea since the 1960s. She emphasizes the need to expand the petroleumscape concept to include the sea, not only as a place of extraction and transport, but also of urbanization and of the reuse and reimagination of these structures.

Part II on **Oil, Materiality, and Cultural Practices** explores oil's direct and indirect influence on built spaces, commodities, and lived experience through the lens of oil-rich and oil-connected areas in Africa, the Middle East, Brazil, China, and Europe. They explore changes and continuities in colonial and postcolonial frameworks and how struggles for sovereignty have often reproduced colonial notions of progress and modernity. Petroleum companies played an important role in establishing housing districts and promoting consumer goods and new lifestyles as they developed extraction sites and as petroleum funds led to the construction of new cities. The growth of the industry imparted a new direction to and intensified the circulation and exchange of people, commodities, and technical and scientific knowledge, playing a key role in the creation of a new urban space and new kinds of societies, inside and outside oil towns.

Aspects of continuity and similarity in the import of new objects and practices such as cars and consumer items play an important role in Nelida Fuccaro's investigation of oil company towns built by foreign multinational oil companies. These company towns contain the traces of an early petroleum modernity closely associated with Western imperialism, old style colonialism, and industrial urbanism in the post-World War II period. Drielli Peyerl examines the case of Brazil, where the government focused on energy independence, and beginning in the 1930s, "designed" the petroleumscape based on a vision that connected infrastructure construction with drilling, hinterland development, and modernization with promotion of research and education. Laura Hindelang explores the respective approaches and displays of water and oil through infrastructure in Kuwait City, arguing that the visual-spatial absence of oil obscured the two fluids' interdependent conditions of existence, helping water to become the representative liquid of Kuwait's oil-based modernization in the 1950s–1970s. Christoph Strupp complements this investigation with his study of the marketing of Hamburg as a clean—not industrial—port. Oil played a strong role in the port and in local economic politics in the 1950s and 1960s, and effectively it remains an oil port, but highlighting that is no longer seen by the government as in the local interest. Turning to China, Li Hou provides a close study of Daqing's state-sponsored industrialization strategies that bound the lives and life choices of the common people with the fate and choices of the consolidating socialist state, providing a contrast with the predominant association of oil with consumerism, corporate power, and suburbia in the 1960s and 1970s. It shows how a government can use its construction of a petroleumscape for its own promotion.

Part III on **Oil Ecologies and Imaginaries** connects the exploration of oil systems to challenges of the energy transition and projects for overcoming them. A critical part of the transition away from oil is education about its actual history: raising awareness of the extent to which it shapes our everyday lives, the costs and effects of its ubiquity, and the power of the ongoing propaganda pursued by oil companies. We need to design the transition and make the spaces of post-oil just as pervasive, heroic, and appealing as the spaces of oil. In place of petroleum and its narratives, we need to close circles of consumption and production to create circular economies, lessen the use of fossil fuel, reduce petroleum-based plastic waste to stop an energy-intensive lifestyle, guarantee water safety, and reduce hunger in line with the UN Sustainable Development Goals.

Imre Szeman and Caleb Wellum start the section by reflecting on the field of energy humanities and its quest to understand the essential role that energy (not just petroleum) plays in our lives and the significance for the investigation of oil spaces. Pieter Uyttenhove explores carbon (diamonds as well as petroleum) in Antwerp to help us understand the connection between petroleum spaces, their use, preservation, reuse, and clean-up, but also because the imaginary of petroleum, including its smell and dangers, plays an important role in the way the public views the city. He relates the port's carbon history to its history of promoting itself as the biggest and best (as evidenced in statistics). Chiara Geroldi and Gloria Pessina reflect on the power station of Porto Tolle in the Po River delta of northern Italy, one of twenty-three unused or underused Enel power stations, which were fueled by oil to produce electricity and now have been designated for redevelopment. The chapter recognizes the need to see rural power stations as a symbol of petroleum-based modernity connecting remote spaces through petroleum-related infrastructure to consumers. Carola Hein, Christine Stroobandt, and Stephan Hauser consider the important role of port cities in the emergence of the petroleumscape. They aim to show how a careful historical analysis, in this case of the petroleum flows in Dunkirk, can help inform design solutions for the future—both visionary sketches and concrete ideas.

Outlook

The chapters brought together in this volume form a coherent whole while addressing a complex global topic. They cannot be all-encompassing, but aim to create interest. Focused on a shared topic and grounded in local knowledge, the chapters represent a wide variety of perspectives. The book includes case studies from five of the seven top oil-producing countries (the US, Canada, Brazil, Indonesia, and Saudi Arabia) and explores the petroleumscape of several European countries and of China, Kuwait, and Africa. There are many more places that have played important roles in the global petroleumscape that have not been included here. This book is an invitation for further exploration of the spaces of oil in Russia, Iraq, Nigeria, Venezuela, and many other places.

Increased burning of fossil fuels and the creation of carbon dioxide play a big part in contemporary global warming and environmental pollution. Oil leaks can harm the natural environment, regardless of whether they are caused by ship crashes, pipeline breaks, or the spilling of gasoline in an everyday setting. Rising petroleum costs, global warming, the leading industrial countries' dependence on foreign oil, and the impending peaking of oil (if it has not already occurred) have led to an initial rethinking of energy sources, which, in turn, may lead us to once again transform our landscapes and built environment. If we want to change, we need to understand where we stand and how we got to this point. We need to understand how petroleum has been written into spatial practice and representations, both global and local, serial and unique, long and short terms, but always in a way that is interconnected and interdependent.

A close analysis of the spatial, representational, and represented petroleumscape can help create new perspectives on the role of energy spaces and facilitate the development of new systemic approaches. We need to better understand the path dependence that underlies its development over time and the forces that impede us from fully committing to the transition to sustainable energy spaces and practices. We need structural change rather than piecemeal interventions. Such an understanding is important not only for the design of a green energy landscape, but also as a foundation for heritage decisions. Many historic oil sites are now becoming part of our history and our cultural heritage. We need to make informed decisions about the fate of these sites. To do that, we need to tease out a new vocabulary of conceptual understanding: of petroleum spatial systems, tangible and intangible spaces, oil materiality, oil ecologies, the energy humanities. We can design a transition that promotes circularity and sustainability in a socially just way and develop new narratives about eco-friendly materials to promote new creative practices.

Notes

1 Carola Hein, "Between Oil and Water. The Logistical Petroleumscape," in *The Petropolis of Tomorrow*, eds. Neeraj Bhatia and Mary Casper (New York: Actar/Architecture at Rice, 2013), 436–47; "Global Landscapes of Oil," *New Geographies* 2 (2009): 33–42; "Oil Spaces: The Global Petroleumscape in the Rotterdam/the Hague Area," *Journal of Urban History* 44, no. 5 (September 2018): 887–929, https://doi.org/10.1177/0096144217752460; "'Old Refineries Rarely Die':

Port City Refineries as Key Nodes in the Global Petroleumscape," *Canadian Journal of History* 55, no. 3 (2018): 450–79, DOI:10.3138/cjh.ach.53.3.05.

2 Daniel Yergin, *The Prize* (New York: Simon & Schuster, 1991); Brian C. Black, *Crude Reality: Petroleum in World History* (Lanham, MD: Rowman & Littlefield, 2012); Anthony R. Walker, "Oil-Dependent Economies and Port Development," in *Seaport Systems and Spatial Change: Technology, Industry and Development Strategies*, eds. Brian S. Hoyle and David Hilling (Chichester: John Wiley & Son, 1984), 179–97; M. Hubbard, *The Economics of Transporting Oil to and within Europe* (London: Maclaren & Sons Ltd., 1967); Timothy Mitchell, *Carbon Democracy: Political Power in the Age of Oil* (London: Verso, 2013); Matthew T. Huber, *Lifeblood Oil, Freedom, and the Forces of Capital* (Minneapolis: University of Minnesota Press, 2013); Drielli Peyerl, *The Oil of Brazil. Exploration, Technical Capacity and Geosciences Teaching (1864–1968)* (New York: Springer, 2019); Farah Al-Nakib, *Kuwait Transformed: A History of Oil and Urban Life* (Redwood City: Stanford University Press, 2016).

3 Neil Brenner and Christian Schmid, "Towards a New Epistemology of the Urban?," *City* 19, nos. 2–3 (2015): 151–82; "Planetary Urbanization," in *Urban Constellations*, ed. Matthew Gandy (Berlin: Jovis, 2011), 10–13.

4 Nancy Couling and Carola Hein, eds., *The Urbanisation of the Sea: From Concepts and Analysis to Design* (Rotterdam: nai010/BK Books, 2020).

5 Arjun Appadurai, "Disjuncture and Difference in the Global Cultural Economy," *Theory, Culture & Society* 7, nos. 2–3 (1990): 295–310, https://doi.org/10.1177/026327690007002017.

6 Michael J. Watts, "Crude Politics: Life and Death on the Nigerian Oil Fields," *Niger Delta Economies of Violence Working Papers* 25 (2009): 1–27; Janet Stewart, "Making Globalization Visible? The Oil Assemblage, the Work of Sociology and the Work of Art," *Cultural Sociology* 7, no. 3 (2013): 363–84.

7 Henri Lefebvre, *The Production of Space* (Oxford: Blackwell, 1991), 26–27.

8 Ibid.

9 Lukasz Stanek, *Henri Lefebvre on Space: Architecture, Urban Research, and the Production of Theory* (Minneapolis: University of Minnesota Press, 2011); Lefebvre, *Production of Space*; Henri Lefebvre, *A Critique of Everyday Life* (London: Verso, 1991).

10 Hein, "Between Oil and Water."

11 Praveen Duddu, "Top 10 Large Oil Refineries," *Hydrocarbons Technology*, hydrocarbons-technology.com/features/feature-top-ten-largest-oil-refineries-world/, accessed 10 July, 2018.

12 Carola Hein, "Maurice Rotival: French Planning on a World-Scale (Part I)," *Planning Perspectives* 17, no. 3 (2002): 247–65, DOI:10.1080/02665430210129315; "Maurice Rotival: French Planning on a World-Scale (Part II)," *Planning Perspectives* 17, no. 4 (2002): 325–44, DOI:10.1080/02665430210154740.

13 Carola Hein, "Suez et la route du pétrole," in *Suez*, ed. Claudine Piatone (Cairo: Institut Francais d'Archeologie orientale, 2011).

14 Culture Unstained, "Home," https://cultureunstained.org/, accessed 22 December, 2020.

15 Ida M. Tarbell, *The History of the Standard Oil Company* (New York: McClure Phillips, 1904).

16 Willy Mullens 1924, Haghe Film (Den Haag), FLM5699; C.W. A. van Bergen 1930–1932 | Production Company: unknown | Film from the collection of EYE (Amsterdam), https://www.eyefilm.nl/; 1920s Abadan, Iran: Abadan Streets Iran, 1950s – Film 99068; Leisure And Work In Abadan Iran, 1950s – Film 99069, Huntley Film Archives.

17 Brian Larkin, "The Politics and Poetics of Infrastructure," *Annual Review of Anthropology* 42 (2013): 327–43.

18 Carola Hein, "Oil in Oil (and Other Art Media): Painting the Petroleum Port," *PortCityFutures* (20 October, 2020), https://www.portcityfutures.nl/news/oil-in-oil-and-other-art-media-painting-the-petroleum-port.

19 Greenpeace, "Lego: Everything Is Not Awesome," https://www.youtube.com/watch?v=qhbliUq0_r4, accessed 22 December, 2020.

PART I

Oil, Agency, and Territoriality

2

THE ORIGINAL NORTH AMERICAN PETROLEUMSCAPE

Oil-and-Gas Empire, Petrochemical Nation

Carola Hein and Alan Lessoff

The first—and perhaps most all-encompassing—petroleum-created landscape emerged in the US. With its massive refinery and petrochemical complexes; its pipelines crisscrossing the continent; its petroleum-oriented ship channels, offshore rigs, and tanker-loading platforms; with its land use configurations, employment patterns, institutions, political economy, and material culture all tracing their character to oil as a business and as a way of life, the US serves as a case in point for the analysis featured in this book.[1]

By the late 1850s—when petroleum's commercial potential became evident—coal mining and steam technology had already incorporated fossil fuels into transportation, industry, and everyday life. By the mid-century, Philadelphia, Pittsburgh, Chicago, and other eastern and midwestern industrial cities were becoming hubs of coal-railroad-metalmaking complexes comparable to the German Ruhr region or the English Midlands. Even in coal-oriented US cities, considerable flux characterized business relations, labor and class structure, and urban geography throughout the late 1800s. With only loose path dependence on coal, these cities were open to the novel forms of capital investment and energy use that oil and gas represented. In addition, in entire US regions, especially the Southwest and West, urban and commercial networks and energy-use patterns were still in formative stages when the shift of energy regimes from coal to petroleum began to reshape cities and urban culture. The US was thus highly receptive to its landscape becoming the kernel of the global petroleumscape.[2]

Between the 1850s and 1950s, the US became oriented around the production, marketing, and consumption of petroleum. Freeways carved up cities, which stretched outward into a succession of freeways, subdivisions, shopping centers, and office parks, the stereotyped manifestations of North America's petroleumscape. This diffuse, segmented, multicentered urban form appeared with particular clarity in the US Southwest, whose urbanization, largely catalyzed by enormous oil and gas discoveries in the early twentieth century, took shape in tandem with the ubiquitous adoption of the automobile, without a firm legacy from the walking city and railroad eras. The petroleumscape came to pervade rural areas and small towns as well. Farms became part of the petroleumscape, dependent on gasoline and diesel machinery and chemical fertilizers and pesticides. The extraction,

DOI: 10.4324/9780367816049-4

refining, transportation, and consumption of petroleum reshaped rivers, coasts, and seas. The spatial patterns and material culture associated with oil and gas became fundamental to Americans' imagination of themselves as a nation.

While the North American petroleumscape has distinctive aspects, its influence pervades the petroleumscape around the world. Driven by entrepreneurial and corporate capital rather than state enterprise, the US gave birth to vertically and horizontally integrated petroleum companies that came to control the industry in large parts of the world. At a time when European nations were still expanding (or even building) their colonial realms—often with an eye on access to oil—entrepreneurs from the US were constructing a neocolonial system of multinational energy firms, whose networks of extraction, processing, and marketing spanned the globe. This petroleumscape grew to include built infrastructures and urban forms in select locations outside the US, but only to the extent that these served the main purpose of the company: the extraction, refining, and selling of petroleum.

This chapter explores the early North American petroleumscape, from the 1850s into the years after World War II, with an emphasis on the many ways the petroleum industry reshaped cities, urban networks, and relations between regions within the US and across the North American continent. As the chapter moves into the twentieth century, it expands geographically from petroleum's initial base in Pennsylvania and Ohio to what historian Carl Abbott has termed the "Energy West."[3] The petroleum-driven transformation of the territory from Houston to Los Angeles began with spectacular oil booms in Southern California and East Texas in the 1890s and early 1900s. This transformation ended with Houston as the energy industry's global headquarters, and with immense swaths of the continent—from Louisiana and the Gulf of Mexico through Oklahoma and Colorado and onward into Alberta and Alaska—an interconnected system of exploration, drilling, refineries, pipelines, and petrochemicals.

The North American Petroleumscape in the Kerosene Era

The North American oil story began in Western Pennsylvania, near Titusville north of Pittsburgh. There, Edwin Drake (1819–1880) drilled for oil in 1859, setting off the first oil rush. Once oil had been found, the main challenge involved transporting it to refining sites, with horse carts, ships, and pipelines all playing a role. By the early 1860s, railroads reached into the oil fields, allowing the slippery substance to be transported to Cleveland or Pittsburgh to be refined. As consumers adopted new uses of petroleum—in particular, it quickly replaced whale oil for lighting—businesses rapidly expanded storage, refining, and shipping capacities. Production increased from some 4,450 barrels in the first year to 220,000 barrels in 1860 and 2,114,000 barrels in 1861.

The Western Pennsylvania boom thus compelled the rushed creation of the first layer of the spatial petroleumscape, oil's industrial footprint: petroleum's extraction, storage, transformation, and transportation. During these early years, a large number of interests battled for control, among them John D. Rockefeller, a Cleveland merchant who invested heavily in kerosene and naptha refining during the US Civil War. Combining technological innovation with aggressive business methods, Rockefeller and his partners made Cleveland the hub of the early oil industry. By the mid-1870s, their firm, Standard Oil, had acquired a network of refining operations in Atlantic coastal ports. By the 1880s, as Daniel Yergin points out, Standard Oil refineries in Cleveland, Philadelphia, and Bayonne, New Jersey, together produced over a quarter of the world's kerosene. Standard's enormous Bayonne refinery along New York Harbor was a major reason that the New York area came to account for over 40 percent of Standard Oil's production and nearly three-quarters of its exports. Oil refining also had a presence in the East Coast ports of Baltimore, Boston, and Portland, Maine, but their role was negligible next to New York/New Jersey and Philadelphia.[4]

Philadelphia's durable role in both domestic and trans-Atlantic markets was built on the city's history as a port, its status as terminus and headquarters of the Pennsylvania Railroad, and its strategic links to Western Pennsylvania. With extensive unbuilt land on the Schuylkill and Delaware Rivers, the city offered ample rail and water infrastructure as well as access to water for the new industry. Philadelphia also offered a diverse, highly skilled industrial base that produced everything from locomotives and tools to textiles and beer. The boom-and-bust pattern of refineries in Philadelphia and their changing locations illustrate the ways in which the emerging industry explored locational preferences, inserted refineries into cities with favorable, pre-existing industrial bases and transportation networks, but then reshaped the infrastructure of those cities and redirected their development. The pattern underscores the power of consolidation in strategically located cities with strong infrastructural connections through rail, road, shipping, and pipelines.[5]

The location of oil facilities in the Philadelphia region depended on a range of factors, especially access to water, both for industrial processes and for shipping. By 1866, the city listed seven petroleum storage facilities and six refineries. Several more appeared by 1875. Disasters made safety a concern. Environmental considerations became an additional reason to relocate. The Belmont Petroleum Refinery was initially located between the Philadelphia and Reading Railroad and the Schuylkill River, just above the Columbia Bridge where the Pennsylvania Railroad split from the Junction Railroad. This facility produced some two million gallons of petroleum products in 1868, but because it was located upstream from the municipal waterworks, its wastewater endangered the city's drinking water supply. Along with other industrial properties in the vicinity, the Belmont Refinery ultimately closed, and the city bought the land as an extension of Fairmount Park. After the Belmont refinery stopped operations, two new petroleum centers emerged: one would disappear, while the other has lasted until the present era. Both were built on agricultural land within city limits, downstream from the drinking water intake, where it was easier for companies to create industrial complexes and to run new railroads (Figure 2.1).

The Atlantic Refining Company, founded by Pittsburgh and Philadelphia investors, occupied a seventy-acre property on the Schuylkill just below the Point Breeze Gas Works. Like other Philadelphia industrial businesses at the time, the Atlantic refinery preferred to portray itself as modern but also scenic and in harmony with nature, not as a disruptive social and environmental force of Charles Dickens-style foreboding (Figure 2.2). In 1874, Standard Oil purchased the Atlantic Refining Company as part of its attempt to consolidate oil transport and refining into a single enterprise capable of weathering major setbacks, even the destruction of an entire facility. This happened in 1879, when lightning destroyed the Atlantic refinery along with several ships moored on its wharves. Two thousand men lost their jobs, and many of the seamen from the ship lost all their belongings. Standard Oil carried no insurance and had to pay for reconstruction. Yet the company was already big enough to absorb such a loss.[6]

The Pennsylvania Railroad, meanwhile, purchased a large plot of land on the Delaware at Greenwich Point, about 1.75 miles south of the Navy Yard, for leasing to independent oil operators. For a time, these independents and the Pennsylvania Railroad engaged in a shipping rate war against the combined power of Standard Oil, the Baltimore & Ohio Railroad, New York Central Railroad, and Erie Railroad. Ultimately, the Pennsylvania Railroad lost

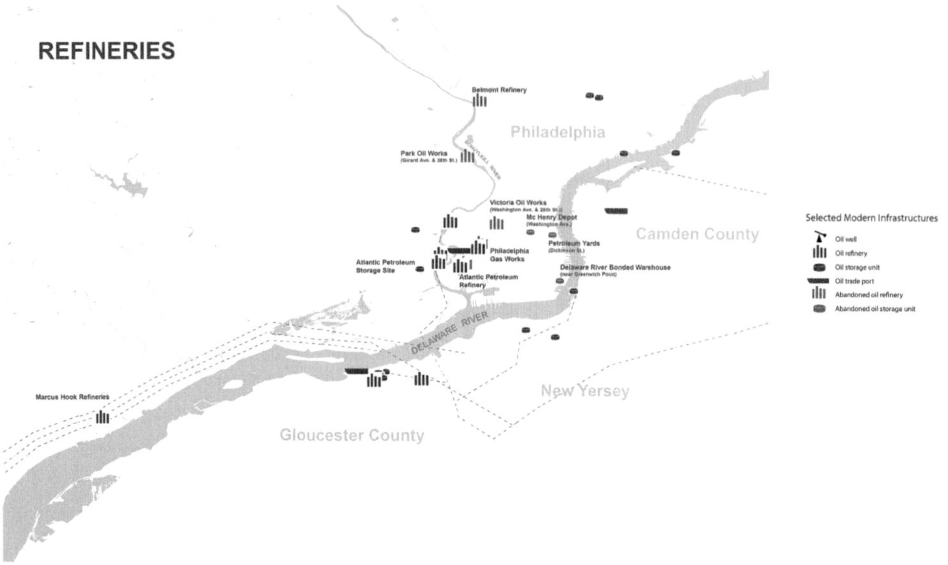

FIGURE 2.1 Locations of historical and modern refineries in Philadelphia and Marcus Hook. Map by Carola Hein and Arnoud de Waijer.

the contest and sold its oil interests to Standard Oil.[7] The Delaware facilities continued to work until 1911, and then disappeared.

The rebuilt Atlantic refinery at Point Breeze benefited from Standard's consolidation and expansion, so much so that for a time, the company's Philadelphia operations produced on a scale comparable to those in New York. By 1891, 35 percent of all US petroleum exports came from the 360-acre Point Breeze plant, which featured a navigable waterfront of 1.7 and 6 miles of private railroad track. The facility burned 350,000 tons of coal each year to refine 40,000 barrels of petroleum daily, which underscores the connection between Philadelphia as a center of coal transport and as a petroleum hub. Continued tensions between Standard Oil and the Pennsylvania Railroad prompted Rockefeller to divert trade back toward New York, which in any case had succeeded Cleveland as the company's headquarters after the formation of the Standard Oil Trust in 1882.[8]

Still, as of 1907, petroleum products exported by the Atlantic Refining Company (owned by Standard Oil) accounted for 22 percent of the city's export trade and were valued at $23,647,194 in foreign gold. Meanwhile, after 1901, the opening of Texas' mammoth oil fields enabled a new competitor to Standard Oil to emerge in Philadelphia. Joseph Pew (1848–1912), cofounder of Sun Oil, built a refinery in Marcus Hook, just outside the city in Delaware County, for refining and resale of Texas crude that his company shipped in coastal tankers to avoid railroads allied with Standard. Another upstart, Union Petroleum, also located at Marcus Hook. In this way, Philadelphia illustrated how even before the 1911 US Supreme Court decision that led to the breakup of Rockefeller's trust, competitors with links to newly discovered Gulf Coast fields were developing their own tanker fleets, pipeline networks, and refineries to cut into Standard's dominance of the East Coast and Atlantic petroleum trade. Although Philadelphia and New York reached the peak of their refinery production as early as 1891, their East Coast location ensured them an enduring

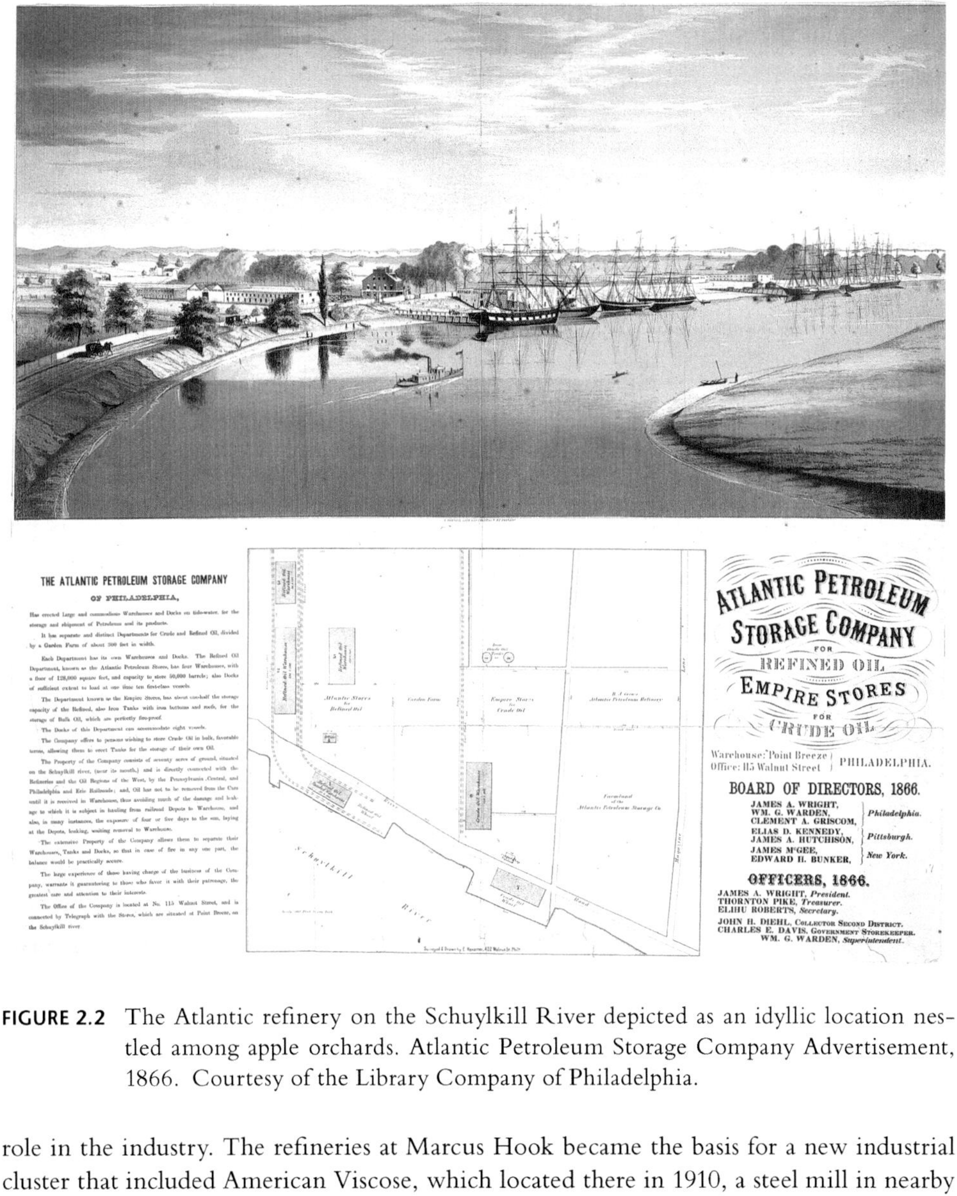

FIGURE 2.2 The Atlantic refinery on the Schuylkill River depicted as an idyllic location nestled among apple orchards. Atlantic Petroleum Storage Company Advertisement, 1866. Courtesy of the Library Company of Philadelphia.

role in the industry. The refineries at Marcus Hook became the basis for a new industrial cluster that included American Viscose, which located there in 1910, a steel mill in nearby Claymont, Delaware, and the Sun Oil Ship Building and Drydock Company (focused on tanker construction) in Chester in 1917. In the 1930s, Sun Oil built a pipeline from Western Pennsylvania to bring refined heating oil to market, further reinforcing the nodal character of Philadelphia and expanding the supporting infrastructure of the petroleumscape.

As Rockefeller and his rivals and critics understood, control of refineries, transportation, and distribution provided more leverage than control of oil fields themselves. Rockefeller's so-called Cleveland massacre of 1872, in which he and his partners acquired nearly all of northeastern Ohio's refining capacity, set in motion the relentless expansion that ended with the Standard Oil Trust in command of over 90 percent of the country's refining capacity at the height of its power. Despite increasingly fierce competition, on the eve of its 1911 antitrust dissolution Standard Oil still processed around two-thirds of petroleum products across the US. Standard Oil leveraged its horizontal dominance of refining to construct a vertically integrated operation of enormous scale and scope, one that brought together all stages of the petroleum business (the so-called upstream, midstream, and downstream functions) over a huge geographic area.[9] Standard Oil's integration of refineries, railways, and pipelines—captured in a widely circulated 1901 caricature of King Rockefeller with a crown of railways—are striking examples of the private control of infrastructure and the role of corporate capital in the creation of urban patterns and built form (Figure 2.3).

While Americans embraced Standard Oil's products, the company became profoundly unpopular, a byword for predatory capitalism. Indeed, the Rockefeller group's abuse of the legal device of the trust to tighten its grip on oil refining in the 1870s is the reason that Americans came to refer to anti-big business movements as *antitrust*. Caricatures of Rockefeller and cartoons about the threat of monopolistic oil helped to make the graphic arts a familiar part of the representational petroleumscape. Throughout the twentieth century, big oil recurred as a target for critical artistic commentary, for example, in the 1970s, when oil embargoes and gas shortages shattered trust in cheap energy, technology, and their supposedly unlimited potential. In the late twentieth century, artists and cartoonists in Europe and the US began to satirize and excoriate the oil industry's arrogance, along with the negative environmental impact of petroleum, even as in the Middle East a regionwide oil boom drew whole new populations into the lifestyles and ways of thinking associated with oil and gas consumption.

Administrative and research facilities emerged as a spatial petroleumscape distinct from oil's landscapes of production, transportation, and marketing. Oil companies supported

FIGURE 2.3 John D. Rockefeller, the "King of the Combination," *Puck*, February 21, 1901. Courtesy of Library of Congress: LC-DIG-ppmsca-25503.

research and development, although at first this tended to be less institutionalized and systematic than the industrial research laboratories created by General Electric, AT&T, and other electrical and telecommunications firms. By the 1910s–1920s, Standard Oil's successor firms, along with automobile companies, were investing in "cracking," anti-knock additives, and other innovations to increase the productivity and usefulness of gasoline for automobiles. And by the 1920s, chemical firms such as DuPont, Dow, and Union Carbide in the US, Imperial Chemical in Britain, and I.G. Farben in Germany were vigorously researching the use of petroleum by-products such as ethylene and propylene in plastics, synthetic materials, pharmaceuticals, dyes, paints, and explosives. By World War II, corporate-sponsored research in both private and academic settings—along with university training programs and professional associations in industrial chemistry and engineering—had become a significant manifestation of the petroleumscape. Although early plastics such as celluloid and bakelite could be made from coal tar and other organic chemicals and materials not necessarily derived from petroleum, DuPont's nylon, introduced in the 1930s, set the stage for the omnipresent postwar expansion of petrochemicals into domestic life and material culture. By the 1960s, the petroleumscape encompassed every household as well as people's bodies and everyday lives.[10]

Corporate offices became an especially visible dimension of the North American administrative petroleumscape. Oil's physical and financial flows intertwined, but their divergent spatial locations gave them a different presence on the landscape and in public consciousness. Oil administration buildings became distinctive urban and architectural spaces, their design in tune with that of other central city office buildings. Companies sited them in prestigious and strategic locations—in New York's financial district, for example, rather than Cleveland or Western Pennsylvania (Figure 2.4). In downtown Houston, the Texas Company (Texaco) headquarters, built in 1915, and Humble Oil's enormous complex, which came to fill an entire city block between 1919 and 1940, became major objects of civic pride, central to Houston's sense of heritage and eventually to its historic preservation efforts. Likewise, for most of the twentieth century, Dallas took as its symbol the enormous, red revolving Pegasus sign—the logo of Mobil Oil—placed in 1932 atop the 400-foot Magnolia Petroleum building, which was the tallest skyscraper west of the Mississippi when it opened in 1922 (Figure 2.5).[11]

The industry also created ancillary spaces, buildings not part of the production and business of oil. These included structures associated with housing, leisure, and the education of oil workers (managerial or working class). While often financed in full or in part by a company, they are rarely identified directly with the oil industry. As needed, corporations built entire neighborhoods or company towns. In West Texas, for example, settlements often began as hastily constructed barracks, boxcars, and trailers before evolving into oil towns, many of which then withered when the wells ran dry.[12]

In keeping with the country's customary emphasis on private civic activity, US petroleum interests have used philanthropy to leave their mark on cities and urban life, even as they exerted unabashed political influence to avoid taxes and regulation. Standard Oil founder John D. Rockefeller and his descendants sponsored educational, cultural, and health institutions, including the University of Chicago, Rockefeller University, and the Rockefeller Foundation. Meanwhile, Rockefeller Center (built 1931–1940), exemplifies the flow of oil wealth into commercial real estate, as do numerous residential subdivisions, hotels, and shopping centers from Houston to Los Angeles (Figure 2.6).

FIGURE 2.4 Architects' drawing from 1923 of the expanded Standard Oil headquarters at 26 Broadway, a well-known New York icon. Courtesy of Prints and Photographs, Library of Congress, Washington, DC, www.loc.gov/item/2014648263/.

Gasoline, Pipelines, and the Energy West

By the late 1800s, Standard Oil found itself beset with foreign competitors, such as the Dutch and British operations whose merger in 1907 created Royal Dutch Shell. European operators sought their own sources in Russia, the Dutch East Indies, and elsewhere. They created their own facilities at European ports, with the goal of limiting Standard Oil's presence in their domestic markets. They even asserted themselves in the US itself. Shell, for example, was an early backer of the Spindletop oil field near Beaumont, Texas, although the Pittsburgh-based Mellon family, whose Texas acquisitions around the same time became the basis for Gulf Oil, managed to push the Europeans out of the US oil fields for the time being (Figure 2.7).

From the start, the transformation of the US Southwest into the heart of the North American petroleumscape was connected to the corporate structures and multinational connections that had already come to characterize the petroleum's industry. The Texas oil business cultivated a self-image as a bastion of gruff, rebellious wildcatters resisting haughty East Coast capitalists. The lore of the Texas oilman as perpetuating the spirit of the Old West in the age of global capitalism and science-based industry—popularized by movies such as *Giant* (1956) and then recast into a garish soap opera by the series *Dallas*

FIGURE 2.5 For over six decades, a giant rotating red neon image of Pegasus, symbol of SOCONY and later Mobil Oil, loomed over the Magnolia Building, Dallas' tallest skyscraper. Courtesy of American Oil and Gas Historical Society, https://www.aoghs.org/petroleum-art/high-flying-trademark/.

FIGURE 2.6 *Cleveland News* cartoonist Bob Satterfield portrayed John D. Rockefeller as Santa Claus delivering $1.85 million to University of Chicago President William Rainey Harper, December 1903. Courtesy of Wikimedia Commons, https://commons.wikimedia.org/wiki/File:Satterfield_cartoon_on_Rockefeller_gift_to_University_of_Chicago_(1903).jpg.

FIGURE 2.7 Heywood #2 Gusher at the Spindletop oil field near Beaumont-Port Arthur, Texas, c. 1901. Courtesy of Prints and Photographs, Library of Congress, www.loc.gov/pictures/item/2010649519/.

(1978–1991)—had an element of reality. Even so, as with every western resource boom since the California Gold Rush of the 1840s, the Texas energy sector found that it needed East Coast capital and expertise to turn discoveries into steady profits. Both the Mellon family behind Gulf and the Pew family behind Sun had connections to the early oil and gas business in Western Pennsylvania. Joseph Cullinan, founder of Texaco, had managed pipelines for Standard Oil and run a Pennsylvania oil equipment company before investing in East Texas at the start of the boom there. And Magnolia Petroleum, originally a Texas-owned firm active in the opening of the oil fields of West Texas and Oklahoma, only remained independent until 1916, when it was jointly acquired by Standard Oil of New York and Standard Oil of New Jersey.[13]

Nevertheless, a striking aspect of the Texas oil boom was the ability of Houston and Dallas-Fort Worth to draw on outside capital and expertise and yet retain ample earnings. Capital generated by oil became the basis for a homegrown energy industry, for diversified regional economies, and for universities, museums, research institutes, and other cultural components of dynamic, autonomous urban regions. The Texas experience thus provides a sharp counterexample to the so-called curse of oil, the reputation of this form of resource extraction for consigning its sources to neocolonial status. A factor in this ability to turn an oil boom into durable wealth was the unusual level of solidarity among Houston and Dallas business and civic elites. The business and professional classes in these and other Texas cities put together oligarchic, pro-growth governing coalitions that maintained control of the major Texas and southwestern cities during their formative periods. The cohesiveness of commercial-civic rule in Dallas or Houston was to some degree a product of the desire among urban business classes in the Southwest to avoid the sort of internal colonial dependence on northeastern capital, expertise, and control that seemed to weigh upon the post-Reconstruction South.[14]

Oil production spread to Texas, Oklahoma, and California while fossil fuels and petroleum products became more pervasive aspects of modern industry and life. The initial discoveries in East Texas at the start of the twentieth century produced vast quantities of crude oil and natural gas. Not much of the East Texas oil was of the quality needed for kerosene, still petroleum's main use. But it was good enough to power the oil-fired engines increasingly adopted by railroads and steamship companies. The adoption of oil power by industries ranging from brickyards to breweries accelerated the emergence of industrial sectors in cities such as Houston and Los Angeles. Likewise, oil furnaces began to replace coal in household heating. This was apart from the massive new demand for gasoline. Initially an unwanted by-product of refining for kerosene, gasoline became valuable by the early 1900s, when the internal combustion engine outpaced steam and electricity as motive power for automobiles. Over the next two decades, innovations in refining and in processing made gasoline an increasingly important part of the oil industry, even as the spread of electric lighting reduced demand for kerosene.[15]

The Pennsylvanians brought to Texas an understanding of the potential of natural gas, a pervasive presence in oil fields, but one often regarded by drillers as a dangerous nuisance to be burned off. The Pews, for example, had first become well-known through an effort in the 1880s to supply Pittsburgh with natural gas as a substitute for the manufactured coal gas that fueled early gas lighting systems. By the mid-1880s, Standard Oil, which initially began constructing pipelines from oil fields as an alternative to dependence on railroads, had made its own investments in pipelines for natural gas, first as an industrial fuel and later for domestic uses. As soon as western oil fields began to open, cities as far-flung as Kansas City and Denver welcomed natural gas pipelines as a cleaner alternative to the coal gas monopolies resented by many urban residents. Natural gas overtook manufactured gas in the 1920s–1930s, by which time the Southwest outproduced Appalachia, the first source for the fuel, by a factor of 4–1. The strategic character of gas and oil pipelines as well as a series of political scandals prompted the federal government, along with producing states such as California, Oklahoma, and Texas, to extend regulatory authority over them. The Texas Railroad Commission, which gained oversight of petroleum fields and oil and gas pipelines in 1917–1919, became a lever that Texas interests used to ensure that they and not "foreigners" profited from the state's energy resources. The New Deal era and then World War II prompted the federal government to expand its regulation of the gas and oil pipelines, which had come to span the continent. Federal and state regulators gained sway over the routes of pipelines, the quantity and quality of gas, its allocation, and its price (Figure 2.8).[16]

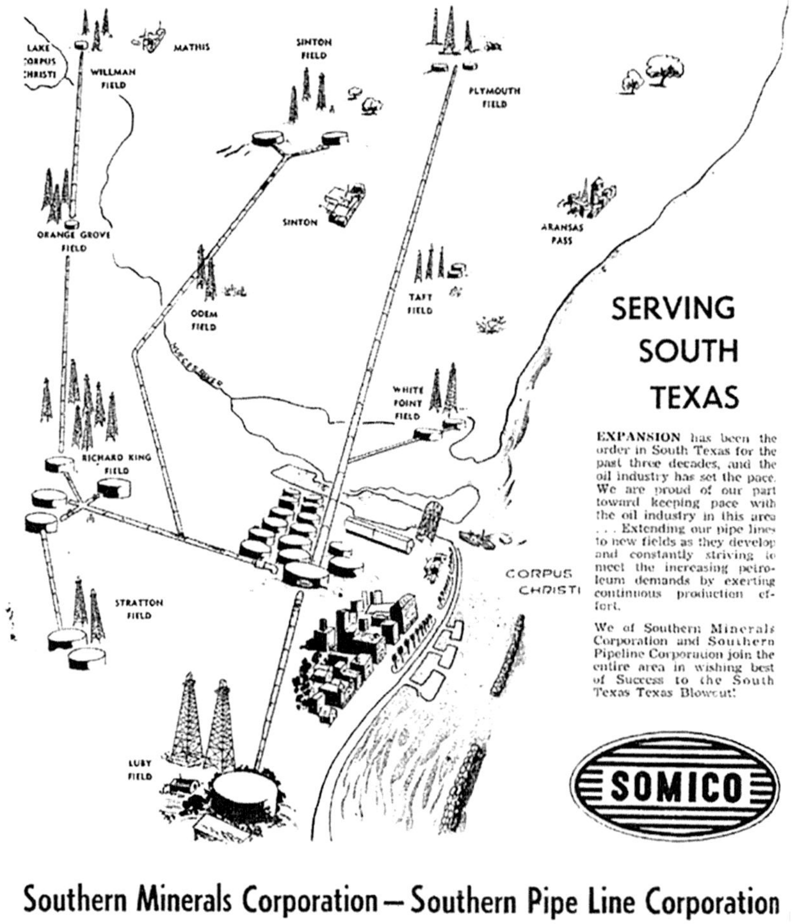

FIGURE 2.8 Advertisement for the Southern Minerals Corporation (SOMICO) in an oil industry convention brochure, 1951. Founded in the 1930s to supply natural gas to high-energy industries then locating around the Corpus Christi Ship Channel, SOMICO illustrates how Texas' petroleum sector promoted regional industrial diversification. Courtesy of Kilgore Collection, Special Collections, Texas A&M University—Corpus Christi.

The Texas Coast: From Waterscape to Petroleumscape

Starting in the early twentieth century, the Gulf Coast witnessed a massive reengineering of land and water in the service of oil and gas. In 1896, the US Congress authorized the dredging of Buffalo Bayou, the slow-moving river that runs through Houston into Galveston Bay. The goal was to transform the shallow bayou into a deep channel navigable by ocean-going ships. At that time, the project was understood as a recognition that inland ports—with their ample land for rail connections, freight yards, and industrial sites—were becoming more viable for industrial-era commerce than traditional ocean-side ports such as Galveston. Besides being vulnerable to catastrophes such as the horrific Hurricane of 1900 (to this day, the deadliest natural disaster in US history), Galveston, a long, thin barrier island two miles off the mainland, had limited space for railroads, freight, and factories, along with limited supplies of fresh water. As with the Manchester Ship Canal and Chicago's Sanitary and Ship Canal, the Houston Ship Channel reflected a new understanding among urban commercial interests that industry and railroads had overturned the traditional logic of harbor location. Counterintuitively, it now made more sense to bring ships inland to industrial complexes, railroads, and freight yards than to bring all that to the sea.

By the time the fifty-two-mile Houston Ship Channel opened in 1914, pipelines, storage tanks, and refineries had become more relevant than the cotton gins, grain elevators, and other agricultural operations that the Port of Houston was initially envisioned as specializing in. Seventeen railroads met at Houston, the city where Gulf, Sun, Texaco, and other post-Spindletop producers concentrated their operations. Nine refineries lined the ship channel by 1930, while gas and oil pipelines began to attract high-energy-using industries: cement and fertilizer works, metals manufacturers, and both organic and inorganic chemical plants. Already by 1950, twenty-seven chemical plants lined the Houston Ship Channel, a trend that continued over the next decades, until the ship channel emerged as the country's largest petrochemical complex (Figure 2.9). The larger area stretching from Houston to Beaumont contained over half the country's petrochemical capacity by the 1980s. Houston also became the center for oil equipment manufacturers such as Hughes Tools, founded by Howard Hughes Sr., father of the billionaire adventurer. Houston-based construction firms such as Brown and Root innovated methods for constructing pipelines, petrochemical plants, and offshore oil platforms. Beyond its unparalleled transportation and storage facilities, therefore, Houston offered a comprehensive aggregation of support industries and technical and business services to the energy sector. The city became the country's—and eventually the world's—energy capital because the sector's knowledge economy concentrated there, in exploration, drilling, logistics, and refining.[17]

By the mid-twentieth century, the Port of Houston had become what environmental historian Christopher Sellers calls a "petropolis." An autonomous jurisdiction under the authority of the Port of Houston Authority, the port stretched twenty-five miles, about half the total length of the Houston Ship Channel. The enormous zone covered by refineries, storage tanks, and chemical factories provided unforgettable urban vistas, comparable to the skylines of New York or Chicago. Refinery towers and columns, writes the architectural historian Barrie Scardino Bradley:

> gleam in the sunlight, just as they shine with lights at night, offering a picture of a contained, surreal city the way medieval pilgrims must have found the seventy-two towers of San Gimigano in Italy rising out of the land.[18]

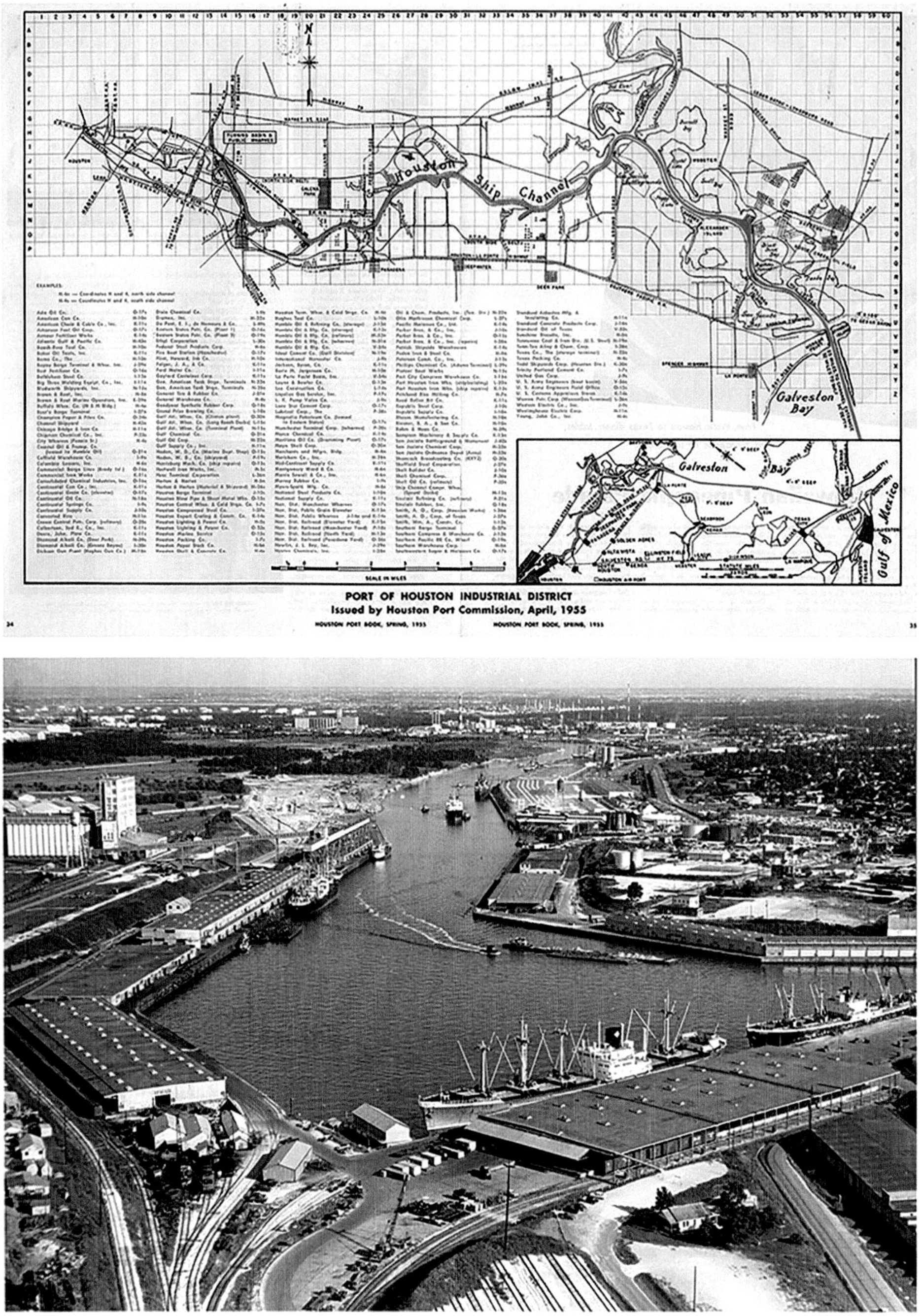

FIGURE 2.9 Map of the fifty-two-mile Houston Ship Channel (a), along with bird's-eye view of the ship channel's turning basin, mid-1950s (b), by the mid-twentieth century, the focal point of the North American petroleum and petrochemical industries. Courtesy of University of Houston Digital Library.

Almost no one lived within the Port Authority's city of petroleum and petrochemicals. In Houston and other burgeoning Texas metropolises, energy companies could leave the ancillary city of houses, shopping, and entertainment to real estate developers. White migrants from Texas and the southern US flowed into ramshackle neighborhoods on either side of the ship channel, eager for steady pay and skilled work offered at the port and its industries. The renowned songwriter Rodney Crowell recalls that in the last months of World War II, his maternal grandparents quit sharecropping in West Tennessee and made their way with his mother to Houston, where his grandfather was thrilled to find a "twenty-seven-dollar-a-week night watchman job" at Hughes Tool. Shortly thereafter, they were joined by Crowell's father, a troubled seventh grade dropout and disappointed country musician who had grown up in Arkansas and Kentucky. In 1955, Crowell's parents moved from a rented house on Avenue P, three blocks south of the ship channel, to Norvic Street in the working-class suburb of Jacinto City, about five miles to the ship channel's north (Figure 2.10). The neighborhood, Crowell recalls, was filled with "cookie-cutter bungalows whose poor workmanship, lack of imagination, and cheap materials destined them for an early demise," but were nonetheless "barely affordable to the workforce employed by the refineries and chemical plants lining the ship channel like so many poison gas spewing space stations."[19]

African Americans and Mexican Americans also sought opportunities in the oil-and-gas ports of the Gulf Coast, even though labor market segregation confined these groups to the least skilled and most dangerous jobs until the 1960s–1970s. In Houston, diverse and dynamic Black and Latino neighborhoods emerged around the sprawling city, including in the Second and Fifth Wards on either side of the ship channel. In Corpus Christi, the Tejano

FIGURE 2.10 Aerial photo of Houston's Hughes Tool plant, 1941, a famed example of the numerous oil industry equipment firms located along the Houston Ship Channel. Such firms attracted migrants from small towns and rural areas across Texas and the South. Courtesy of Bob Bailey Studios Photograph Collection, e_bb_3158, Dolph Briscoe Center for American History, University of Texas at Austin.

barrio stretched across the Westside to the edge of that city's ship channel, while the African American Northside expanded into formerly Anglo American neighborhoods such as Hillcrest, whose proximity to refineries accelerated white flight from the area. The proximity of Black and Hispanic neighborhoods to one another and to the port and its industries helped these groups to some degree overcome mutual suspicions and job market competition to cooperate in labor union organizing and in anti-segregation and civil rights activism.[20] (Fuccaro, this volume, discusses diverse streams of labor that flowed to oil towns in Kuwait and other parts of the Middle East and the political complexity that arose, despite corporate power that pushed a more uniformly celebratory view.)

Two hundred miles to Houston's south, Corpus Christi exemplifies how the opening of more and more oil fields in the decades after Spindletop catalyzed the reconstruction of much of coastal Texas into petroleumscape. In 1919, a terrifying hurricane and storm surge hit this portion of the South Texas coast. Similar to how the 1900 Galveston storm affected the Houston Ship Channel debate two decades earlier, this catastrophe convinced Texas and federal officials to dredge a ship channel twenty miles from the Texas barrier islands across shallow Corpus Christi Bay to a protected turning basin at Corpus Christi, then a coastal town of around 10,000 people, where four railroads already met. When it opened with much fanfare in 1926, the Port of Corpus Christi was—again similar to Houston—initially envisioned as an outlet for South Texas cotton, grain, and cattle. But oil and gas discoveries throughout the city's hinterland soon shifted the port toward oil and petrochemicals. By the 1960s, five refineries were located along the ship channel as well as organic and inorganic chemical producers, metals manufacturers, and food processors drawn by natural gas pipelines and the promise of ample freshwater supplies. Corpus Christi's autonomous Port Authority repeatedly extended and deepened the channel, until it reached thirty-four nautical miles and a depth of forty-two feet. Corpus Christi developed its own equipment construction and other support businesses, although control of these over time concentrated

FIGURE 2.11 Aerial postcard of the Corpus Christi Ship Channel in the 1940s. Envisioned as an outlet for South Texas agriculture, it became dominated by oil and gas and by high-energy industries starting in the 1930s. Courtesy of Corpus Christi Public Library.

in Houston, as did specialized services such as petroleum geology. This movement of professional and technical services as well as management from Corpus Christi to Houston especially, but also to San Antonio and even Wichita, Kansas (home of Koch Industries, which came to own much of Corpus Christi's energy sector), reduced the Texas coast's second most important port to the status of a satellite of the US West's energy capitals (Figure 2.11).[21]

The Representational Petroleumscape of Extractive Empire

As in Texas, southern California's emergence as an energy metropole appears in literature and lore as an updating of the region's frontier and pioneer self-image, with oil a successor in spirit to the gold and land rushes of the mid-1800s. Upton Sinclair's *Oil!* (1926) and the film loosely based on it, *There Will Be Blood* (2007), capture much of this atmosphere. As in Houston, the reengineering of Los Angeles's port at San Pedro began not with oil but with California agricultural products. The southern California oil boom that began in the 1890s led to rigs on beaches, in parks, and even in residential neighborhoods. It also became the foundation for Los Angeles as the archetype of distended, multicentered metropolis, with its subdivisions of bungalows and ranch houses that stretch through valleys and into former orchards. The residential Los Angeles of the Anglo middle class hinged on a network of petroleum-based industrial suburbs, which were more diverse in class and ethnicity than Southern California's archetypal residential suburbs, but they barely registered in Los Angeles's self-promotion or public image.[22]

Beyond serving as an archetype for the petroleumscape's urban form, Los Angeles became a launching point for another group of operators who extended the US petroleumscape outward through the Americas. Edward Doheny, the Irish American surveyor and itinerant prospector on whom Sinclair's protagonist was loosely based, expanded the US presence in oil fields near Tampico, Mexico, starting in 1901. Doheny held on to his Mexican interests through the twists and turns of the Mexican Revolution that began with the overthrow of Porfirio Díaz's regime in 1911 (Figure 2.12). By the time Mexico expropriated and nationalized foreign operations in the country in 1938, about 65 percent of its production was controlled by Royal Dutch/Shell, with US companies controlling most of the rest. Foreign companies hurriedly withdrew families from their complexes in the Tampico region. Well-maintained towns for expatriate managers and engineers surrounded by poorly built and poorly serviced settlements for local workers had already become a recognizable urban manifestation of the oil industry's

FIGURE 2.12 Oil tank cars at loading docks in Tampico, Mexico, 1910s. Courtesy of Hazard-Dyson Collection Eberstadt Collection, e_enr_050, the Dolph Briscoe Center for American History, the University of Texas at Austin.

transnational dimensions. Although incensed by Mexican nationalization of their holdings, European and American oil operators grudgingly accepted their ouster from the country, in part because they had already developed an alternative set of fields in Venezuela.[23]

In sum, the North American petroleumscape—including that in Mexico and Canada—came to incorporate and perpetuate an older reality about the American West, its tendency to grow as offshoots or hinterlands of either a Greater California or a Greater Texas. Houston, Dallas, and Oklahoma oil millionaires, for example, underwrote and oversaw the formation of Colorado's energy sector in the 1950s, to the extent of putting up the funds for Denver bank buildings, corporate offices, and even that city's Petroleum Club. As geographer D.W. Meinig remarked in 1969, oil brought Texas a new "cultural self-consciousness and commercial imperialism." Houston-based producers and processors, equipment manufacturers, and suppliers had made "oil districts far beyond the bounds of Texas" into extensions of Houston's energy sector. Even the oil fields of Alberta grew initially as manifestations of US-based enterprise. Until the late twentieth century—when Canadian and provincial authorities succeeded in shifting much of the city's energy sector toward locally controlled independents—Calgary had replicated Denver's role as a "junior headquarters energy capital" for "international oil majors," while Canadian-owned companies gravitated toward the provincial capital, Edmonton.[24]

Even more than Alberta's, Louisiana's energy sector took shape as a dependency of New York or Texas enterprise, an extension of the original administrative petroleumscape. The extractive internal colonialism that Louisiana experienced prefigured and mirrored the extractive neocolonial petroleumscape that the petroleum business spread to much of the world (Figure 2.13). The story can be told through the life and legend of Huey Long, recounted in song, art, and film as well as literature in the case of Robert Penn Warren's great novel *All the King's Men* (1947). Exploitation by Standard Oil and other corporate interests—backed by a pliable New Orleans and Baton Rouge political establishment—made the state susceptible to Long's style of demagogic resentment, political brutality, and garish corruption. By the 1920s, when Long became his state's dominant political figure,

FIGURE 2.13 Well debris in DeSoto Parish, Louisiana, 1925. Environmental devastation quickly became a source of resentment and political tension in Louisiana. Courtesy of the Photograph Collection, State Library of Louisiana.

the sense was widespread that Louisiana had been abandoned to the environmental degradation and class exploitation associated with the oil industry. Outrageously favorable leasing and tax arrangements and other favors and concessions meant that oil brought most residents few benefits in terms of improved public services, education, or opportunities.

Drillers found oil in the vicinity of Lafayette and Shreveport, Louisiana, in the early 1900s, not long after the strikes over the Sabine River in East Texas. By the 1920s, supplies had been discovered across southern Louisiana. In 1909, meanwhile, Standard Oil began constructing a large refinery above the Mississippi near Baton Rouge; a pipeline from Oklahoma fed this plant even before it began to draw from nearby fields. The Baton Rouge location gave the company "no little satisfaction," as a detailed history of the company remarks, since Standard thereby sidestepped the "ill will" and state regulation that hampered it in Texas. Two more refineries were constructed closer to New Orleans before the Great Depression. Then, in the postwar era, the federal government, needing aviation fuel, synthetic rubber, and other petroleum products, subsidized refining and petrochemical facilities in stretches along the lower Mississippi.[25]

As in Texas, chemical plants and metals manufacturers followed, so that by 1947, Louisiana contained twenty-five refineries and 147 chemical operations (Figure 2.14). Most were located in struggling agricultural areas, disproportionately inhabited by African Americans. The so-called Chemical Corridor between Baton Rouge and New Orleans occupied former sugar plantation regions, resulting in the jarring spectacle of rundown plantation houses—relic of an earlier extractive economy—scattered among toxic dumps and factories spewing smoke. Portions of the lower Mississippi degraded into what geographer Craig Colten labeled a *sacrifice zone*, where "negligent behavior by industry and government authorities" came to seem tolerable. The area became notorious for chemical plant explosions, toxic leaks, groundwater contamination, and toxic dumping, though the threat to fisheries and urban water supplies did prompt efforts to limit discharges into rivers.[26]

Starting in the 1940s, the spread of the oil industry to offshore platforms created tensions involving Louisiana's shrimping and fishing industry. Local officials attempted to reconcile these tensions, at least symbolically, for example, by recasting Morgan City's Shrimp Festival as the Shrimp and Petroleum Festival, starting in 1967. The environmental scholar Barbara Allen, a native of Morgan City, observes that Louisiana's Chemical Corridor produced 12.5 percent of the country's hazardous waste by the late twentieth century, "sixteen thousand pounds of hazardous waste for every citizen in the state." Even before the 2010 BP/Deepwater Horizon spill, which brought worldwide attention to the petroleum-induced devastation of the Mississippi Delta and the Gulf, Louisiana was a center for environmental justice activism "against multinational corporations and their close allies, Louisiana's regulatory agencies," Allen remarks. Still, what another environmental scholar Ashley Dawson labels "petro-governance" remained difficult to counter. Even after Hurricane Katrina in 2005 and then the BP disaster in 2010, oil and petrochemical companies, determined to protect their networks of pipelines, storage tanks, and production facilities, resisted coastal and wetlands restoration measures that might enable New Orleans and lower Louisiana even minimally to adjust to global warming—which is, of course, induced by the very activities that are also degrading the Mississippi Delta and the Gulf.[27]

As the business of producing, processing, transporting, and marketing petroleum spread across the North American continent, it brought about epic geographic and environmental changes. Parallel changes took place in people's homes, routines, and material culture. Home appliances, advertising and public relations, roadside attractions, gas stations, freeways, traffic lights, and parking lots extended the petroleumscape into every aspect of society and made

FIGURE 2.14 1940s postcards of the Cit-Con Oil Refinery (a) and the Mathiesen Alkali Works, Lake Charles, Louisiana (b). The spread of petroleum refining, petrochemicals, and high-energy-using industries complicated efforts at waterway protection, coastal restoration, and storm and flood control. Reference courtesy of Historic Photographs of Southwest Louisiana, Frazer Memorial Library, McNeese State University.

fossil fuels seem inexorable and inescapable, a future from which there could be no turning back. Americans interact with the constructed landscape of fossil fuels as though there is no alternative. Usually, they treat it as though it is hardly there, in a manner akin to the fish in David Foster Wallace's well-known speech "This is Water." When confronted with the fact that they swim in water, they respond mystified, "What the hell is water?" The story's lesson—that "obvious, important realities" can be "hardest to see and talk about"—is "just a banal platitude," Wallace concedes. But, he adds, "banal platitudes can have life or death importance."[28] Having remade itself over the century covered by this chapter into an empire of oil and gas, a nation organized around and defined by fossil fuels, the US may be the most daunting socio-political challenge of the worldwide environmental crisis. Its petroleumscape is a massive obstacle in the way of the pressing necessity of switching energy paths.

Notes

1 Carola Hein, "Oil Spaces: The Global Petroleumscape in the Rotterdam/the Hague Area," *Journal of Urban History* 44 (September 2018): 887–929; Carola Hein, "Between Oil and Water: The Logistical Petroleumscape," in *The Petropolis of Tomorrow*, eds. Neeraj Bhatia and Mary Casper (Houston: Actar/Architecture at Rice, 2013), 436–47; Carola Hein, "Global Landscapes of Oil," *New Geographies 02: Landscapes of Energy*, ed. Rania Ghosn (Cambridge: Harvard University Press, 2009), 33–42.

2 Joel Tarr and Karen Clay, "Pittsburgh as an Energy Capital," in *Energy Capitals: Local Influence, Global Impact*, eds. Joseph A. Pratt, Martin V. Melosi, and Kathleen A. Brosnan (Pittsburgh, PA: University of Pittsburgh Press, 2014), 5–29; Christopher Jones, *Routes of Power: Energy and Modern America* (Cambridge: Harvard University Press, 2014), ch. 1–2.

3 Carl Abbott, *How Cities Won the West: Four Centuries of Urban Change in Western North America* (Albuquerque: University of New Mexico Press, 2008), 248.

4 Ida M. Tarbell, *The History of the Standard Oil Company* (New York: Peter Smith, 1950[1904]), 32ff.; Daniel Yergin, *The Prize: The Epic Quest for Oil, Money, and Power* (New York: Free Press 2009), 30; Ralph W. Hidy and Muriel Hidy, *History of Standard Oil Company (New Jersey): Pioneering in Big Business* (New York: Harper & Brothers, 1955), 101, 123–24; Jones, *Routes of Power*, ch. 3–4.

5 Carola Hein, "Refineries (Oil)," *Encyclopedia of Greater Philadelphia*, philadelphiaencyclopedia.org/archive/refineries-oil/#22535, accessed June 20, 2020; Carola Hein and Dirk Schubert, "Resilience and Path Dependence: A Comparative Study of the Port Cities of London, Hamburg, and Philadelphia," *Journal of Urban History* 47, no. 2: 389–419. Article first published online: June 11, 2020; Issue published: March 1, 2021, https://doi.org/10.1177/0096144220925098; Nathaniel Burt and Wallace C. Davies, "The Iron Age, 1876–1905," in *Philadelphia: A 300-Year History*, ed. Russell F. Weigley (New York: Norton, 1982), 471–523; Philip Scranton and Walter K. Licht, *Work Sights: Industrial Philadelphia, 1890–1950* (Philadelphia, PA: Temple University Press, 1986).

6 "Our Oil Trade, Petroleum the Rival of King Cotton," *Philadelphia Inquirer*, October 16, 1867, 5; "Struck Oil, A Great Fire at Point Breeze," *Philadelphia North American*, June 12, 1879, 1.

7 Ron Chernow, *Titan: The Life of John D. Rockefeller* (New York: Random House, 1998), 171; Grant Segall, *John D. Rockefeller: Anointed with Oil* (Oxford: Oxford University Press, 2001), 57–58; Albert J. Churella, *The Pennsylvania Railroad*, Vol. 1: *Building an Empire, 1846–1917* (Philadelphia: University of Pennsylvania Press, 2012), 359ff.

8 Burt and Davies, "Iron Age," 482–83; Hidy and Hidy, *Pioneering in Big Business*, 12.

9 Hidy and Hidy, *Pioneering in Big Business*, 417; Yergin, *The Prize*, 79. Alfred D. Chandler Jr., *The Visible Hand: The Managerial Revolution in American Business* (Cambridge: Harvard University Press, 1977), 321ff., 350–3, 418ff.

10 Keith Chapman, *The International Petrochemical Industry: Evolution and Location* (Oxford: Blackwell, 1991), esp. ch. 2–3; Jeffrey Meikle, *American Plastic: A Cultural History* (New Brunswick: Rutgers University Press, 1995), ch. 1–5.

11 Barrie Scardino Bradley, *Improbable Metropolis: Houston's Architectural and Urban History* (Austin: University of Texas Press, 2020), 89–90; Harvey J. Graff, *The Dallas Myth: The Making and Unmaking of an American City* (Minneapolis: University of Minnesota Press, 2008), 36–40; "Petroleum Art: Mobil's High-Flying Trademark," https://aoghs.org/petroleum-art/high-flying-trademark/, accessed August 23, 2020.

12 Roger M. Olien and Diana Davis Olien, *Oil Booms: Social Change in Five Texas Towns* (Lincoln: University of Nebraska Press, 1982), ch. 2; David G. McComb, *The City in Texas: A History* (Austin: University of Texas Press, 2015), ch. 21.

13 These two paragraphs draw on Yergin, *The Prize*, 70–79, 215–17; Diana Davids Olien and Roger M. Olien, *Oil in Texas: The Gusher Age, 1895–1945* (Austin: University of Texas Press, 2002), 29–74; Keith L. Bryant and Henry C. Dethloff, *A History of American Business*, 2nd ed. (Englewood Cliffs, NJ: Prentice-Hall, 1990), 164–65.

14 Robert Fairbanks, *For the City as a Whole: Planning, Politics, and the Public Interest in Dallas, 1900–1965* (Columbus: Ohio State University Press, 1998); Amy Bridges, *Morning Glories: Municipal*

Reform in the Southwest (Princeton, NJ: Princeton University Press, 1998). On the so-called "oil curse" and the divergent experiences of urban regions with it, Joseph Pratt et al., conclusion to Pratt, Melosi, and Brosnan, *Energy Capitals*, 181–96.

15 Olien and Olien, *Oil in Texas*, 59; Clay McShane, *Down the Asphalt Path: The Automobile and the American City* (New York: Columbia University Press, 1994), ch. 5–6.

16 Yergin, *The Prize*, 76. Jones, *Routes of Power*, ch. 4, Hidy and Hidy, *Pioneering in Big Business*, 172ff. Mark H. Rose, *Cities of Heat and Light: Domesticating Gas and Electricity in Urban America* (University Park: Penn State University Press, 1995), ch. 1–2; Christopher James Castaneda, *Regulated Enterprise: Natural Gas Pipelines and Northeastern Markets* (Columbus: Ohio University Press, 1993); William R. Childs, *The Texas Railroad Commission: Understanding Regulation in America to the Mid-Twentieth Century* (College Station: Texas A&M University Press, 2005), esp. ch. 5.

17 These two paragraphs: Joseph A. Pratt, "A Mixed Blessing: Energy, Economic Growth, and Houston's Environment," and William C. Barnett, "A Tale of Two Cities: Houston, the Industrial Metropolis, and Galveston, the Island Getaway," in *Energy Metropolis: An Environmental History of Houston and the Gulf Coast*, eds. Martin V. Melosi and Joseph A. Pratt (Pittsburgh, PA: University of Pittsburgh Press, 2007), 21–33, 185–204; Martin V. Melosi and Joseph A. Pratt, "The Energy Capital of the World? Oil-Led Development in Twentieth-Century Houston," in Pratt, Melosi, and Brosnan, *Energy Capitals*, 30–57; Joseph J. Pratt and Christopher J. Castaneda, *Builders: Herman and George R. Brown* (College Station: Texas A&M University Press, 1999); Joseph A. Pratt, *The Growth of a Refining Region* (Greenwich, CT: JAI Press, 1980).

18 Bradley, *Improbable Metropolis*, 151–52; Christopher Sellers, "Petropolis and Environmental Protest in Cross-National Perspective: Beaumont-Port Arthur, Texas, versus Minatitlan-Coatzacoalcos, Veracruz," *Journal of American History* 99 (June 2012): 111–23.

19 Rodney Crowell, *Chinaberry Sidewalks: A Memoir* (New York: Random House), 2011, 10–12.

20 Tylna Steptoe, *Houston Bound: Culture and Color in a Jim Crow City* (Berkeley: University of California Press, 2016), esp. ch. 2; Emilio Zamora, *Claiming Rights and Righting Wrongs: Mexican Workers and Job Politics during World War II* (College Station: Texas A&M University Press, 2009); Alan Lessoff, *Where Texas Meets the Sea: Corpus Christi and Its History* (Austin: University of Texas Press, 2015), esp. 45–48, 131–39.

21 Lessoff, *Where Texas Meets the Sea*, 41–45, 65–69, 267–71. Mary Jo O'Rear, *Storm over the Bay: The People of Corpus Christi and Their Port* (College Station: Texas A&M University Press, 2009).

22 Sarah Elkind, "Los Angeles: The Energy Capital of Southern California," in Pratt, Melosi, and Brosnan, *Energy Capitals*, 77–89; Jason Cooke, "Energy Landscape: Los Angeles Harbor and the Establishment of Oil-Based Capitalism in Southern California, 1871–1930," *Planning Perspectives* 32 (January 2017): 67–86; Fred W. Viehe, "Black Gold Surburbs: The Influence of Extractive Industry on the Suburbanization of Los Angeles, 1890–1930," *Journal of Urban History* 8 (November 1981): 3–26.

23 Myrna Santiago, "Tampico, Mexico: The Rise and Decline of an Energy Metropolis," in Pratt, Melosi, and Brosnan, *Energy Capitals*, 147–58; Yergin, *The Prize*, 194–220, 254–60.

24 D. W. Meinig, *Imperial Texas: An Interpretive Essay in Cultural Geography* (Austin: University of Texas Press, 1969), 80; Matthew N. Eisler, "At Arm's Length: Energy and the Construction of a Peripheral Prairie Petrometropolis," in Pratt, Melosi, and Brosnan, *Energy Capitals*, 114. Carl Abbott, *The Metropolitan Frontier: Cities in the American West* (Tucson: University of Arizona Press, 1993), 159–61; Abbott, *How Cities Won the West*, 248–50.

25 Hidy and Hidy, *Pioneering in Big Business*, 401–02, 420. Craig E. Colten, "An Incomplete Solution: Oil and Water in Louisiana," *Journal of American History* 99 (June 2012): 91–99; Craig E. Colton, "Making a Lemon out of Lemonade: Louisiana's Petrochemical Corridor," in Pratt, Melosi, and Brosnan, *Energy Capitals*, 58–76.

26 Colten, "An Incomplete Solution," 91, 94; also Colten, *An Unnatural Metropolis: Wrestling New Orleans from Nature* (repr. Baton Rouge: Louisiana State University Press, 2005), 125–31.

27 Barbara L. Allen, *Uneasy Alchemy: Citizens and Experts in Louisiana's Chemical Corridor Disputes* (Cambridge: MIT Press, 2003), 1–2; Ashley Dawson, *Extreme Cities: The Peril and Promise of Urban Life in the Age of Climate Change* (New York: Verso, 2017), 115; Tyler Priest, "Shrimp and Petroleum: The Social Ecology of Louisiana's Offshore Industries," *Environmental History* 21 (July 2016): 488–515.

28 David Foster Wallace, "This Is Water," repr. at https://fs.blog/2012/04/david-foster-wallace-this-is-water/, accessed July 20, 2020.

3

PETROLEUMSCAPE AS BATTLEGROUND

Pladjoe, Pearl in the Crown of the Bataafsche Petroleum Maatschappij (BPM/Shell) in the Dutch East Indies

Ben de Vries

Oil fields can easily turn into battlefields. This happened more than once in the former colony of the Dutch East Indies (Nederlands-Indië, now Indonesia) in the turmoil of the 1940s, when Japanese, Dutch, Allied, and Indonesian forces fought fierce battles over the control of local oil facilities—with good reason, because in the 1940s the Dutch East Indies was one of the world's biggest oil exporters. Centered on the oil industry city of Plajoe, this portion of the *global petroleumscape*,[1] a multilayered network of spaces made possible by petroleum and facilitating its flow, was designed to support a country on the other side of the world. It was then reshaped by war and revolution, in the context of changing global actors. This chapter tells a largely unknown story about the construction, destruction, and reconstruction of the petroleumscape in what is now the province of South Sumatra in Indonesia. It is a story of how oil fields can become battlefields and of how even in peacetime a petroleumscape is shaped by attempts to protect and defend sources of wealth and power.

In the mid-1880s, oil was discovered at Telaga Said I, at a shallow depth of 400 feet in a thick jungle in the northern part of the huge island of Sumatra. In the mid-1890s, better quality oil was found in a swampy area in southern Sumatra near Peraboemoelih (Figure 3.1). As a result, nearby Palembang, an ancient city with harbor facilities, quickly mushroomed into a vibrant oil industry city. At the same time, the small kampoeng of Pladjoe (Plaju), about 8 km further along the Moesi (Musi) River, became a spider in an enormous web of petroleum-related infrastructures. In 1907, the Royal Dutch Petroleum Company or Koninklijke Nederlandsche Maatschappij tot Exploitatie van Petroleumbronnen in Nederlands-Indië (1890) formed a subsidiary called the Batavian Oil Company or Bataafsche Petroleum Maatschappij (BPM/Shell) and built at Pladjoe the largest, most productive, and modern refinery in Southeast Asia at the time. The scale of operations grew and BPM planned a comprehensive oil company town with administration buildings, refineries, pipelines, roads, railroad tracks, and jetties for mooring tankers. This chapter examines how BPM and the municipality of Palembang, the main oil actors in this region, spatially and economically planned its huge industrial oil footprint at Pladjoe and safeguarded its facilities against brutal intrusions and destruction during the Japanese occupation of the Dutch East Indies during the Pacific War (1942–1945) and Indonesia's struggle for independence (1945–1949).

DOI: 10.4324/9780367816049-5

FIGURE 3.1 Oil drilling tower in Soeban Djerigi, near Peraboemoelih, South Sumatra, 1915. *Source*: Collection Spaarnestad Photo, Haarlem.

Pladjoe: The Pearl in BPM's Crown

Pladjoe provides a case of how oil fields can change hands and suffer destruction, leading to shifts in the petroleum networks in which they are located, but then these sites can be rapidly rebuilt and continue to function. To understand the development of the petroleumscape in Pladjoe, it will be helpful to consider a brief history of Dutch oil exploitation in Indonesia before examining not only architecture and the built environment, but also oil's representations and social meanings. The second part of this chapter addresses how the existence of the petroleumscape encouraged both military conflict and efforts at protection.

Oil production in the Dutch East Indies began with a rich Dutch farmer's son, Aeilko Jans Zijlker, who went to the Indies to find his luck after an unhappy love affair. Zijlker, administrator of the East-Sumatra Tobacco Company, discovered oil first in North Sumatra. What he found was a kind of oil seep, of rather poor quality, good only for lamp oil. After negotiating royalties with the owner, the Sultan of Langkat, in 1883 Zijlker obtained the concession of the area and permission to drill.[2] He was incredibly lucky as the results proved to be much better than expected. After quickly assessing its potential, the Dutch King William III named the company the Royal Dutch Petroleum Company or Koninklijke Nederlandsche Maatschappij tot Exploitatie van Petroleumbronnen in Nederlandsch-Indië. This freestanding company was officially established on June 16, 1890, with a large financial guarantee of about US$360,000 (value at the time).[3] As a result of this amazing success, Zijlker was sent to do more exploration and drilling around his *oil kampoeng*, but his luck did

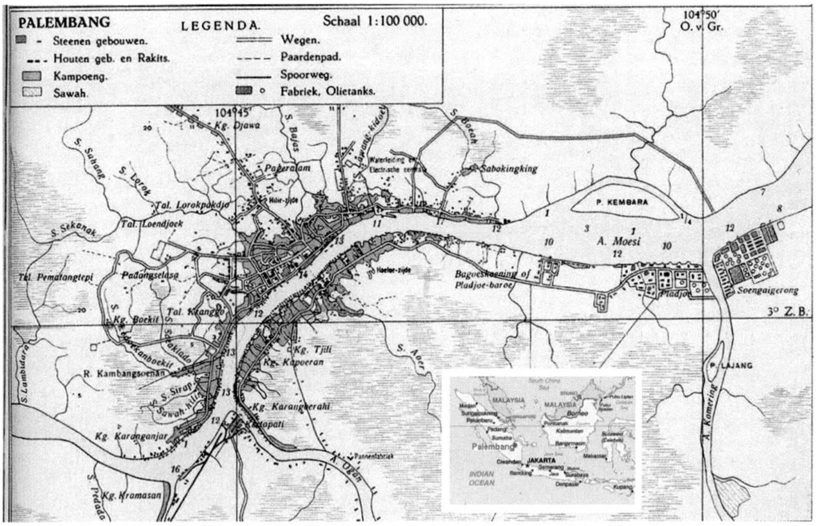

FIGURE 3.2 General map oriented to the northwest, of Palembang and environs, in 1938, with Kertapati (left-lower), Pladjoe and Soengei Gerong (right-middle), and Bagoes Koening (middle), all on the south bank of the Moesi River. *Source*: J. R. Van Diessen, Comprehensive Atlas of the Netherlands East Indies.

not last and he died in December of the same year. Nevertheless, this site would become the origin of one of the biggest multinationals of the world today: Royal Dutch/Shell Group, the name reflecting a 60:40 Dutch-British ownership structure.[4]

When oil was found in southern Sumatra, it was experienced Jan Willem IJzerman, head of the French-owned Moeara Enim Petroleum Company, who stepped in first and invested in a site near the ancient port city of Palembang, capital of the province of South Sumatra.[5] Palembang, founded in 683, saw a period of general prosperity in the later seventeenth century as an important exporter of pepper and tin, and it was a major river transport hub, strategically located along the Malacca Straits, near Singapore and Batavia (Jakarta). All over the world at that time, oil production and transport relied heavily on water.[6] From Palembang harbor to the open waters of the Bangka Straits was about fifty nautical miles. The waterfront city, known as "Venice of the Far East," stretched along both sides of the 750-km-long muddy Moesi River (Figure 3.2). The main economic and political hubs, like the Kraton—the former sultanates' fortified palace and its surroundings—could be found on the eastern bank, called the Hilir-side (Ilir) ("downstream"). On the western, "upstream" riverbank or Hoeloe-side (Ulu), about 8 km from the center of Palembang, IJzerman established an oil refinery in 1897. At that time, the place was nothing more than a small kampoeng carrying the local name Pladjoe, although at this strategic river location in the eighteenth and nineteenth centuries there had been local wooden palisades forming a defensive barrier. There was no overland transport between Pladjoe and the city center of Palembang, as there were no bridges yet. People had to take a ferry. Hence, Pladjoe was located across the river, geographically proximate but socially distant from the city, and it transformed steadily into a well-organized Western enclave—a symbol of modernity and European might.[7]

FIGURE 3.3 Private house of the Engelberts family. Mr. Engelberts was a technical manager of the cracking installation—a unit used to break hydrocarbons—at Pladjoe, 1947. *Source*: Collection Mieke Engelberts.

The refinery on the west bank of the mouth of the tributary produced marketable secondary products which could be efficiently transported by oil tankers from the refinery's jetty to Java, Singapore, Europe (through the Suez Canal), the US, or Japan (especially to Yokohama). The refinery received an abundance of crude oil from the rich oil fields of Talang Djimar, Peraboemoelih, and Moearaenim. Half of the Indies' oil production was produced at this refinery, especially aviation crude.

On the west side of the sprawling production refinery—which took up about 1,500 × 2,000 m—BPM created an ancillary petroleumscape in the form of a compound with houses for executives, employees, and workers (Figure 3.3). Most of these stand-alone villas were grouped in blocks of eight or ten and were almost identical in colonial design, varying only slightly depending on the rank of the employee.[8] Most of the houses had a driveway, a garden, a front veranda and a large overhang that provided protection from the tropical sun. The houses were well ventilated and each one had an annex at the back with a separate bathroom, a kitchen, and rather simple rooms for indigenous people employed as servants. The European employees' houses were situated along palm-bordered avenues and featured impeccably manicured lawns, easily accessible by private cars. This neat and orderly area was continually being extended to the south and southwest. The streets had no names, only numbers. An address, for example, was "4th Street" and "House 122."

The compound, with its almost rectangular grid plan of about 500 × 1,000 m, was fully equipped with all kinds of facilities. BPM had its own administration building, hospital, post office, hotel (pasanggrahan), bakery, shop, and pasar (Figures 3.4–3.6). There was a church, mosque, European primary schools, a Chinese primary school, library, a clubhouse or "soos" with cinema and theater, sports fields for hockey and tennis, and of course, there was a swimming pool.[9] To this day, one can see this Dutch legacy and enjoy the green and spacious compound and the striking tropical modernity of its so-called Indies

FIGURE 3.4 BPM hospital at Pladjoe. *Source*: Expatriate Archive Centre, The Hague.

FIGURE 3.5 Postage stamp of Pladjoe, May 20, 1934. *Source*: Author.

Architecture—a hybrid mix of a Dutch modern style adapted to the tropical climate—despite the fact that some of the colonial buildings are dilapidated. Most of the original buildings in this exclusive settlement still serve the same function as before, but are owned and used by high-ranking staff of Pertamina, the Indonesian state-owned oil company.

In 1924, BPM launched a three-hour-long "Bataafsche Petroleum Film" made by the most well-known Dutch filmmaker of the 1920s, Willy Mullens, in cooperation with C. W. A. Van Bergen, who worked for the Bataafsche marketing organization. This carefully planned propaganda film, one of the earliest moving images of BPM employees in the Dutch East Indies, depicted the modern colonial society in which BPM fully participated.[10] During the official launch of the documentary film on 18 June at the majestic concert hall

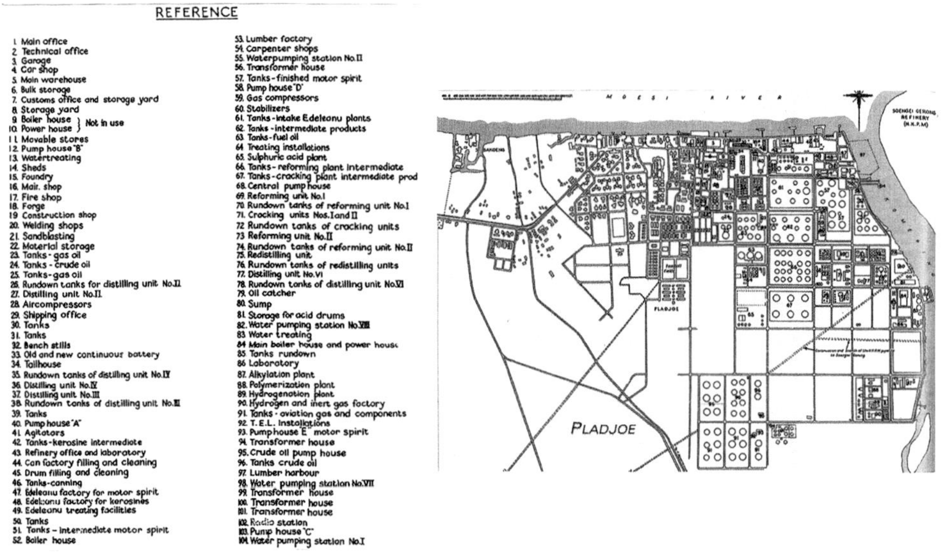

FIGURE 3.6 The Pladjoe compound, July 1945. This map is based on BPM maps dating from 1936 and 1940. The reference shows only the technical installations. *Source*: J. R. Van Diessen, Comprehensive Atlas of the Netherlands East Indies.

of the Kurhaus in Scheveningen, near The Hague, BPM Managing Director Dr. August Philips provided a welcoming speech to the crowd of 1,600 invitees, among them many important representatives from the business community, the government, the diplomatic corps, and the press. Philips was very proud of the company and stressed the positive impact of oil on the lives of people of the Indies. The film documented company activities with the aim of gaining public support. After the premiere, the corporate film toured the Netherlands and was shown at various societies and associations, and also abroad.

Outside white-collar neighborhoods of gloriously shaded streets and lots, the urban space of the colonial society was more differentiated and divided by ethnicity, race, and class. The socio-ethnic categorization was reflected in the spatial layout of the refinery town. In Pladjoe, at the beginning of the 1920s, there were about 250 (475 in 1929) European employees and about 4,500 local Indonesian and Chinese or Asian workers (Figure 3.7).[11] Among the contract laborers, 80 percent came from South Sumatra. European residential quarters were situated in the western part of the compound, at quite a distance from industrial noise and pollution. Europeans were also accommodated in another salubrious BPM housing complex called Bagoes Koening (or Pladjoe-baroe: New Pladjoe), a little further west of Pladjoe. Groups of indentured laborers (or "free coolies," a term that at the time was not recognized by the Dutch as a racial slur) stayed in barracks (tangsi) with vernacular-style roofs and a common kitchen and bathroom, and in so-called koelie bangsals and in kranie houses, located in the south part of the compound.

In order to facilitate the steadily growing petroleum industry and to improve the rather low standard of living, the municipality of Palembang started to construct new ancillary structures and assets, like asphalted public roads, railways, ports, and two airfields. One of the airfields was in Talangbetoetoe, 15 km northwest of Palembang, and the other one was a secret military airfield, 75 km west of Palembang, near Peraboemoelih, called P2, which was still under construction in 1941. It was perfectly

FIGURE 3.7 Part of the residential area for Asian laborers of BPM at Pladjoe. *Source*: Expatriate Archive Centre, The Hague.

camouflaged by the dense tropical forest and not known to Japanese troops at the time of the invasion.

An important stage of the expansion can be seen in the municipal traffic commission's implementation of a new town plan in 1937, designed by the Dutch architect Thomas Karsten, with help of urban planner Hans Lüning. Land was reclaimed from rivers and used to build new public housing, inland transport systems, a ring road, several bridges over the Moesi River, and the Wilhelmina Bridge (1939) over the Ogan River, replacing the river ferry. This important arch bridge connected both oil refineries from the eastern to the western bank, where the new Kertapati railway station (1939) and the coal harbor were located. On a regular basis in the late 1920s, ocean steamers with a maximum loaded draft of 7 m navigated the river. Meanwhile, the new and continuously expanding oil industry led to the rapid influx of newcomers and a subsequent population increase in Palembang.[12] Also, a new European residential quarter was begun on the higher grounds of Talang Semoet, 1.5 km west of the Kraton. The population rose from 50,703 people in 1905 to 73,726 in 1920, and 108,000 in 1930, of which 2,000 were Europeans and 16,000 Chinese. The population would eventually rise to about 140,000 in 1940 and to 208,379 in 1951.

The booming oil industry not only had an enormous influence on the socioeconomic structure of the city, but on the whole region as well, and in a way that was opportune for the oil industry. More than thirty-five oil-mining concessions in the area needed a complete infrastructural network of iron pipelines to connect hundreds of drill towers, oil pumps, and storage tanks, which subsequently transformed the existing semi-urban spatial landscape into an industrial one. They all stand as material witnesses to the invasiveness of petroleum. Nothing remained like it was. Extraction, refining, transformation, and the consumption of petroleum all required new spaces, and the expensive and extensive petroleum infrastructure made a huge impact on the landscape, rivers, nearby sea, city, and buildings. Eventually, these spaces were all connected to this single commodity and its group of industrial players. In addition, BPM built a road network, a vivid symbol of modernization, and forty-two schools throughout the southern province before the Pacific War.[13] Such an extended petroleumscape with various oil actors required multiple headquarters for administration and supervision, like the one in the political heart of Batavia, designed in a modernist colonial style in 1937 by Dutch architect Thomas Nix. It did not reference the global head office of BPM/Royal Dutch Shell in The Hague, which was both hypermodern and traditionalistic, inspired by the Renaissance and by seventeenth-century Dutch Classicism, built in 1917 by the well-known architects the Van Nieukerken brothers.

Most important, a network of BPM petrol stations extended across Sumatra. According to a 1929 Shell road map, there were in total 185 fuel stations strategically positioned on the island, including seventy-one in the South Sumatra region, two in Palembang, one in Djambi, and one in Pladjoe itself (Figures 3.8 and 3.9). Although most of the stations were rather basic shed-type gas stations, they had a recognizable corporate standardized design: a combination of a BPM sign with typical lettering and a Shell pecten logo, a humble pump island, and prefabricated steel panels.

Just like in the Netherlands, the gas stations handed out free road guides, which served as public relations material produced by the "Handelszaken" (Commercial Affairs) unit of BPM. The 1929 guide was covered with two slogans: on the left, "Shell motor oil worth every drop" (*Elke Druppel is van Waarde*), and on the right, "Shell autoline for balance and energy" (*voor Evenwichtigheid en Arbeidsvermogen*). Of course, these representations promoted petroleum usage and driving. BPM/Shell used the same advertising strategies throughout the world. In this way, petroleum imaginaries shaped citizen behavior and slowly oil became a heroic partner in creating contemporary society and identity.[14] Oil became central to modern life and was taken for granted in a way that made it almost invisible despite its ubiquity.

Monopoly of Mighty BPM

Further north, around the residency of Djambi (Jambi) in Central Sumatra, were situated the oil fields and small distillation companies of Tempino, Kenali Asam, Badjoebang, Betoeng, and Mangoendjadja. The fields were owned by the Nederlands Indische Aardolie Maatschappij (NIAM), a public-private partnership between BPM and the Dutch East Indies government, set up in 1921. This joint venture was an economic novelty. It was the first time the government became an actor through shares in the Sumatran oil industry. As was usual in these mixed enterprises, the private partner, BPM, provided the operational management and disposed of the production.

The actual refining process did not take place in Djambi, but in Pladjoe.[15] Starting in 1935, and with help of Persian pipe drillers, oil was transported from Tempino to Pladjoe through a pipeline for 270 km through hilly country covered by dense forests, along rice fields, intersecting many watercourses. In 1933, partly with NIAM's support, roads and an airport were built and the city was linked to a railroad. In 1936, a road was opened between Palembang and Padang by way of Djambi. And in 1938, Djambi accounted for about 14 percent of the total East Indies oil production.[16]

BPM feared and had to cope with increasing competition from the American Standard Vacuum Oil Co. (henceforth Stanvac, today ExxonMobil), working through an affiliate: the Nederlandsche Koninklijke Petroleum Maatschappij (NKPM). Although the Dutch Indies government tried to keep foreign oil companies outside the colony through the Mining

FIGURE 3.8 Shell road guide (Wegengids, 1929) with a map of the residencies of Palembang, Benkoelen, Djambi, and the Lampongsche districts. *Source*: Bijzondere Collecties, University of Amsterdam.

Act, Stanvac struggled to compete with BPM. Hence, they started in 1912 to operate their own refinery at Soengei Gerong, directly opposite Pladjoe, on the east side of the Komerine River.[17] It developed into the second-largest oil facility in the Asian region, receiving 3,500 barrels of crude per day by pipelines from the fields of Talang Akar.[18] The plant at

FIGURE 3.9 Private luxury car being fueled at BPM station. Over time, oil influenced lifestyles and became an everyday necessity. *Source*: Collection Nationaal Museum van Wereldculturen. Coll. no. 60037451.

Soengei Gerong challenged BPM's control over global flows of oil. In 1930, another serious rival emerged on stage: Caltex (today Chevron). They also worked through a subsidiary, the Netherlandsch Pacific Petroleum Maatschappij (NPPM), and secured extensive exploration concessions in Central Sumatra and on the Riau Islands.[19]

BPM, operating exclusively in the Dutch East Indies, had quite a strong position, managing large export refineries and controlling an enormous distribution network across the archipelago.[20] Apart from the ones on Sumatra, there were refineries on the island of Java in Tjepoe and Wonokromo. The island of Borneo (now Kalimantan) had a huge refinery in Balikpapan, a few oil fields in Tandjoeng, and a strategic oil enclave on the island of Tarakan. Further to the east, there were concessions on the Moluccan island of Ceram. Finally, BPM held 40 percent of the shares in the Nieuw Guinee oil fields of Klamono and Mogoi-Wasian. The other 60 percent were managed together by the N.V. Nederlandsche Nieuw Guinee Petroleum Maatschappij (NNGPM), Stanvac, and Caltex. Around 1930, BPM owned 85 percent of the oil production; but by the end of the 1930s, this percentage had declined to 55 percent due to heavy competition with Stanvac, which by then owned 30 percent of the promising oil market. In 1938, oil and related products determined an output value of 23.6 percent of the total export, a value of about US$190 million. The six million tons of oil produced yearly in the Dutch East Indies accounted for only 2.7 percent of the world's total.[21] Nonetheless, the Dutch economy depended heavily on the revenues of the oil resources coming from its crown colony.

An Epic Battle for Oil

Once constructed, the Dutch Indies' petroleum landscape attracted the interest of the expanding Japanese Empire, which was eager to seize the area from its European colonizer and use it for its own purposes. In the event of war, the Netherlands, as one of the smaller

European powers, would not be able to defend its territory in Europe or its colossal colonial empire overseas. Implicitly, they put their faith in British military protection.

After the German victory over the neutral Netherlands in May 1940, Japan stepped up pressure by sending two missions at ministerial level to Batavia demanding a bigger share of the East Indies' oil export.[22] During these delicate oil talks, the Dutch government was advised by Jan Carel Baron van Eck (member of the executive board), on behalf of Koninklijke/Shell and Fred Kay of Stanvac. The Dutch did not give way, ignored the Japanese demands, and sent the Japanese delegation home empty-handed. But time was running out. Without doubt, oil was the main quest of the armies of the Japanese Emperor and the reason for the invasion of the Dutch East Indies. Oil was vital for the Japanese war machine, although they had almost no experience and expertise in the industry. The weakness of her war potential lay chiefly in the fact that Japan's home production of natural and synthetic oil amounted to only some 10 percent of the country's annual requirements. Of Japan's oil imports in 1939, about 53 percent came from the US, 38 percent from the Dutch East Indies, and 9 percent from other countries.[23]

Indonesia's oil would quickly become a focus of international conflicts and destruction. Japan had to ensure a means of replenishing her storage tanks.[24] Tokyo needed yearly at least 7.9 million tons of oil to win their Greater East Asia War.[25] Sumatra could deliver 5.3 million; so for a long-drawn-out conflict, the Sumatran oil was crucial (Figure 3.10). Therefore, Japanese troops were instructed to put the most important oil centers in their hands quickly without any loss or destruction (Figure 3.11). At the same time, Dutch troops were instructed to destroy the oil facilities before Japanese troops could seize them. Authorities in Batavia had boasted that they had 500,000 tons of oil in storage in Palembang and that if the Japanese forces started to move upriver; they would release 10,000 tons of oil per day into the Moesi River and burn up the Japanese convoy.[26] The tension was tangible everywhere. The Dutch colony was definitely in peril.[27]

FIGURE 3.10 Aerial photograph of Pladjoe's invaluable and large export refinery, taken around 1930. The two refineries of BPM and Stanvac, being international, were largely independent of local Indonesian conditions. *Source*: Collection Nationaal Museum van Wereldculturen. Coll. no. 10006848.

FIGURE 3.11 Japan's need for Palembang's oil. *Source*: Shell Historical Heritage and Archive, Johan Fabricius, Brandende Aarde.

Meanwhile, after the fall of France, Japan occupied air and naval bases in French Indochina. Immediately, in August 1941, the American President, Franklin D. Roosevelt, announced a total embargo on the export of oil, including a freeze on all bank transfers. Great Britain and the Dutch government-in-exile in London followed with the same retaliatory measures instantly.[28] On December 8, a day after the attack on the American navy fleet in Pearl Harbor, Batavia declared war on Japan. One month later, on January 11, 1942, the Japanese Navy conquered the oil island of Tarakan and the oil center of Balikpapan on Borneo. Another month later, on February 14, at the time that the "invincible" British fortress in Singapore was about to fall, a Japanese invasion fleet of some 10,000 men led by Vice Admiral Jisaburõ Ozawa was sailing from Cam Ranh Bay in French Indochina toward South Sumatra. Their operational plan was to seize and secure the oil refineries at Pladjoe and Soengei Gerong and the superior aerodrome Talangbetoetoe, where American B-17 Flying Fortress bombers were temporarily stationed.[29] In order to secure these sites and halt the frustrating scorched-earth policy, Japan flew in—by complete surprise—over 500 men of the first Paratroop Raiding Group, under Colonel Seiichi Kume.[30] Prior to the airborne insertion, Japanese aircrafts dropped countless propaganda leaflets over Pladjoe issuing a warning that anyone who set oil fields or refineries on fire would be executed (Figure 3.12). The leaflet had the size of a postcard, and was printed in the Royal Dutch orange color, and written in Dutch, ordering: "Dutchman: protect the oil facilities" (Hollander: *Bescherm het petroleum-terrein*).

To defend Palembang, there were about 2,000 Dutch troops, plus reinforcements of Australian units and British anti-aircraft sections. The Territorial Command of the Royal Netherlands Indies Army (KNIL), under Lieutenant Colonel Laurens N. W. Vogelesang,

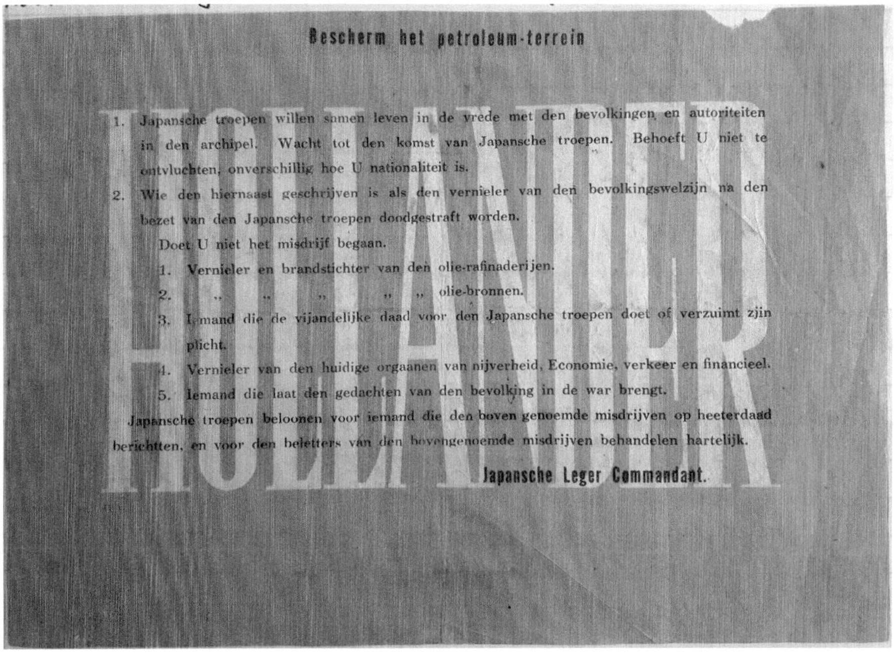

Bescherm het petroleum-terrein

HOLLANDER

1. Japansche troepen willen samen leven in de vrede met den bevolkingen en autoriteiten in den archipel. Wacht tot den komst van Japansche troepen. Behoeft U niet te ontvluchten, onverschillig hoe U nationaliteit is.
2. Wie den hiernaast geschrijven is als den vernieler van den bevolkingswelzijn na den bezet van den Japansche troepen doodgestraft worden.

Doet U niet het misdrijf begaan.

1. Vernieler en brandstichter van den olie-rafinaderijen.
2. ,, ,, ,, ,, ,, olie-bronnen.
3. Iemand die de vijandelijke daad voor den Japansche troepen doet of verzuimt zjin plicht.
4. Vernieler van den huidige orgaanen van nijverheid, Economie, verkeer en financieel.
5. Iemand die laat den gedachten van den bevolking in de war brengt.

Japansche troepen beloonen voor iemand die den boven genoemde misdrijven op heeterdaad berichtten, en voor den beletters van den bovengenoemde misdrijven behandelen hartelijk.

Japansche Leger Commandant.

FIGURE 3.12 Airborne leaflet propaganda tried to discourage the Dutch destroying their own oil installations. *Source*: Collection Museon, The Hague.

wanted to wait as long as possible before destroying the refineries. Influenced by the oil companies, only a limited destruction was planned, keeping the main installation intact.[31] Suddenly, in the middle of the night of February 15, Vogelesang received orders from the KNIL headquarters in Bandoeng that priorities had shifted and that his military units must demolish the refinery at once. In the rush, huge tanks of oil products were not damaged at all and neither was the refinery. After this half-hearted action, the Dutch army hastily retreated and quietly slipped away in the darkness.[32] Pladjoe was soon completely in Japanese hands and they managed to put out the fire in the boilers, shut valves, turn cranks, and disarm most of the demolition charges placed in the complex by the Dutch before they were driven out (Figure 3.13).[33] From this point on, Pladjoe was called "Refinery No. 1" and managed by Nihon Sekiyu (Nippon Oil), and BPM was renamed Dai Ici Seiyuzo. But there was more damage inflicted on the region. In Palembang, the harbor facilities, rubber factories, and warehouses were set ablaze; in Peraboemoelih, drilling sites were destroyed; in Talang Djimar, the wells were nailed shut; and in Djambi, the oil wells were demolished.

The refinery of Soengei Gerong, on the contrary, was successfully defended by Dutch troops. After the order of Vogelesang, military units deliberately destroyed 80 percent of the refinery, including the oil tanks, using a time-delayed demolition charge.[34] Nevertheless, the damage done to the refineries and machineries as a result of the scorched-earth tactics was easily fixed by a group of Japanese drilling crewmen and oil engineers flown in to complete the work. After six months, production resumed.[35]

FIGURE 3.13 Japanese soldier watching the burning oil refineries at Palembang, February 15, 1942. *Source*: Image bank WWII, NIOD, Amsterdam.

Japanese Occupation

At the start of the Japanese occupation, the oil fields in Palembang fell directly under control of the Japanese military, the 25th Army. In the beginning, they made use of 150 oil company employees and evacuees who were forced to work for them on the refineries for about one year as so-called "Nippon workers." But starting in May 1942, all BPM personnel were sent to prison camps. Thus the Japanese could only rely on local oil personnel who were hardly capable of maintaining the machinery.[36] As reinforcement, in April 1942 the Japanese brought about 300 oil laborers from Java to work at the refineries, where they were also trained at a newly established oil engineering school. In total, about 1,200 workers went to this school during the Japanese occupation,[37] stimulating a long-term process of "Indonesianization": the number of foreigners who could be employed by foreign oil companies was reduced and positions once held by foreigners were filled with high potential Indonesian nationalists.[38] Despite these efforts, the considerable lack of investment and the unprofessional handling of oil obtained from the oil fields led to a dramatic drop in oil production during the war, as the annual reports of the Koninklijke pointed out. Wartime figures of the total crude oil production in the Dutch East Indies still showed a rise in 1942 from 3,250,000 to 6,500,000 metric tons in 1943. But inevitably, a steep decline started in 1944. Production fell to 3,750,000, in 1945 to 850,000, and then to 302,000 metric tons in 1946, a historic low.[39]

At the same time, all Dutch symbols, including those of BPM/Shell, were banned and replaced by concepts that conformed to the policy of Tokyo's Greater East Asia Co-Prosperity Sphere, which went beyond the political, and also embraced cultural signs and urban symbols like flags, names of houses, monuments, colonial buildings, railway stations, and streets. In March 1943, the Japanese even allowed a huge rally in Batavia to celebrate the one year anniversary of their victory by pulling a statue of "national hero" Jan Pieterszoon Coen in public space from its plinth. This iconoclastic action was filmed by the Japanese and used as propaganda material afterwards. Moreover, Batavia was renamed Djakarta. For safety

FIGURE 3.14 Pladjoe burning after the "Meridian I" attack by British Royal Navy aircrafts on January 24, 1945. The more compact refinery of Soengei Gerong, the target of the follow-up "Meridian II" strikes on January 29, sits in the bottom-left of the picture with the Komerine River separating the two. *Source*: www.armouredcarriers.com.

reasons, Shell moved its headquarters to Curaçao in the Caribbean. Immediately, the Japanese Propaganda Corps took over the empty BPM head office in the city center.

In November 1943, about 2,000 Dutch, Dutch Indies, and British prisoners of war and conscripted laborers from Java or *rōmosha* (economic soldiers) were forced to load ships with oil and work relentlessly on the construction of two new airfields on remote areas covered in dense vegetation around Palembang. The first was in Betoeng/Ketiau, about 75 km southwest of Palembang, and was intended to replace the war-damaged Dutch airfield P1 (Talangbetoetoe/Kenten). Construction on the second airfield began in May 1944 in Pangkalan Balai, about 40 km northwest of Palembang, and it replaced P2, the strategic Dutch airfield near Peraboemoelih.[40] At least 20 percent of the men died during construction work.[41]

At the end of the Pacific War, massive Allied bombardments started on Palembang and its oil refineries. First, the attack by American B-29s in the night from August 10 to 11, 1944, was launched all the way from Ceylon (Sri Lanka), but with negligible effect. Only a single building was destroyed, but mines dropped in the river sank three Japanese ships, preventing oil from being transported for a month. Plans were adapted, and in January 1945, a more precise series of British air strikes were undertaken on the Japanese-held twin refineries that were supplying at that time half of Japan's oil and three quarters of their aviation fuel (Figure 3.14). As a result, columns of thick black smoke rose 10,000 feet for days. Along with these "Palembang Raids" that diminished oil production, Japanese oil tankers were constantly attacked, which had a tremendous effect on the Japanese ability to transport oil to Japan. These collective Allied efforts eventually brought Japan's war machine to a halt.[42]

Recapturing the Oil Facilities

During the brief interlude of Japanese control, the Indonesian nationalist movement had gained in power. Nationalist leaders, pressed by zealous Indonesian youth groups (pemoeda), seized the opportunity created by the unexpected Japanese surrender signed on August 15, 1945. On August 17, 1945, two days after Japan's unconditional surrender, 300,000 Japanese troops in Indonesia took note as two prominent nationalist leaders Achmad Soekarno and Dr. Mohammed Hatta proclaimed the independent Republic of Indonesia unilaterally. This revolutionary action prevented the oil installations from automatically falling back into Dutch hands. The British/Allied supreme command, led by General Sir Philip Christison, realized that the Indonesian people were fighting for their own cause and issued a statement that implied a de facto recognition of the republic. As a consequence, no Dutch troops were allowed to land and reconquer the former colony.

Meanwhile, the situation at the oil fields was complicated. At the request of the Allied forces, the refineries at Pladjoe and Soengei Gerong were temporarily put under the supervision of Japanese soldiers in September 1945. They had to occupy these oil installations, including the nearby oil fields, in order to prevent the freedom fighters from destroying the oil wells. All the other oil fields in South Sumatra were taken over by the Indonesian forces. On the spot, Indonesian (oil) freedom fighters (lasjkar minyak) quickly founded their own semi-militarized oil workers association: Persatoean Pegawai Minyak (PPM). This initiative came from Dr. Adnan Kapau Gani and Dr. Mohammad Isa, who both held prominent seats in the regional government of the republic in Palembang.[43]

Far away in London, the CEO of the Koninklijke, Dr. Barthold T. W. van Hasselt, watched the nationalist developments with alarm. He was eager to start the reconstruction of the refineries and make money again, like in the old days.[44] He tried to convince the British and Dutch military command of the enormous interests at stake in the oil business.[45] In his opinion, the (former) Japanese troops should be replaced by British troops. He needed at least 4,000 men to protect the oil installations, but they were not available, because the spearhead of the military operation was on Java. Apparently, the British were not inclined to help the Dutch.[46]

Then something remarkable happened. Gani, the Republican Governor of South Sumatra, proposed that BPM itself should take the oil centers back from Japanese control as quickly as possible.[47] He wanted two things in return. First, BPM had to pay the republic interest on its profits.[48] Second, all the Indonesian workers that were occupying the refineries at that moment should be hired by BPM and paid with food, textiles, and household items[49] (Figure 3.15). This was a profitable oil deal, because through this peaceful consultation and smart oil diplomacy, at the end of September 1946, the republic gained the international recognition they urgently needed, and a confident BPM succeeded in regaining control of their most valuable oil installations without the assistance of either the Dutch government or the army.[50] As a bonus, "trade-soldier" Gani hoped to consolidate his political base and improve his own economic position.[51] Soon, 3,600 laborers were back at work at the refineries.

Nevertheless, truce and peace were elusive. The situation escalated in the end of October after an inflammatory speech by Republican General Raden Soedirman. The Republican revolt was reverted and led on New Year's Day 1947 to heavy bombing of Palembang by combined Dutch navy and air force units. This punitive strike was undertaken without any

FIGURE 3.15 Laborers did their shopping in the so-called Land Civo on the Pladjoe compound, 1948. *Source*: Collection Ruud Spangenberg.

notice or warning, causing many civilian losses and leaving the city center in ruins.[52] After this "Battle of Five Days and Five Nights" in the city heart of Palembang, Republican troops were forced to withdraw to a radius of 20 km around the petroleum city. Subsequently, their political leaders fled and joined the radical camp in the Djambi area. Economic occupation and military warfare followed in rapid succession.

Operation Product

After the landing of Dutch troops in March 1946, progressive Dutch Lieutenant Governor General Dr. Hubertus van Mook in Batavia and several politicians from the then-cabinet Beel in The Hague became more amenable to the oil interest in South Sumatra and they were increasingly inclined to intervene. The First Dutch Offensive (Agresi Militer Pertama or Eerste Politionele Actie) soon followed in July–August 1947 and was appropriately named "Operation Product." The main goal was to occupy the vital economic areas as soon as possible and restart the main companies. Moreover, there was plenty of work for the tens of thousands of people in the overcrowded areas. In this way, an important recruitment basis for the Indonesian freedom fighters would disappear. In fact, the mission would lead to better working conditions, would diminish the chance of industrial strikes, and would increase oil production (Figure 3.16).

The Dutch Y-Brigade led by the hotheaded Colonel Frits Mollinger had to occupy the key oil centers around Palembang that were occupied by the Republicans. From a military and economic point of view, this concerted campaign was a success. By mid-August a substantial recovery of the oil production was realized—the refineries were once again receiving oil from the oil fields and would work their way up to 2.5 million barrels per day.[53] During military operations, small units of BPM technicians were operating on the heels of the Dutch forces in order to reboot oil extraction in the recaptured areas as soon as possible.[54] The Second Dutch Offensive in December 1948 had hardly any effect on oil production at Pladjoe. BPM's crown jewel was functioning at full speed. Postwar figures of

FIGURE 3.16 Indonesian workers on the Pladjoe plant go home after work, 1948. *Source*: Collection Ruud Spangenberg.

FIGURE 3.17 The Choice. Uncle Sam, the US, had to choose between the political picture drawn on the left side, where continuity, colonialism, prosperity, order, and full oil tankers were omnipresent; and that on the right side, where revolution, chaos, burning oil installations, and waving Republican and communist flags prevailed. *Source*: Elseviers Weekblad, 16 August 1947.

the total crude oil production in the Dutch East Indies showed a sharp rise: although the production was in 1946 still 302,000 metric tons, in 1947 it was already 1,115,000, in 1948 4,326,000, in 1949 5,930,000, and in 1950 it had reached the prewar level of 1938.[55]

Ironically, the Dutch military successes were a pyrrhic victory, because they caused US support to shift to the republic and forced the Dutch to negotiate the transfer of sovereignty to Indonesia. The US tried on an anticolonial attitude and pressured the Dutch government to recognize the Indonesian cry for freedom. The US played their joker card. At the time, they were funding the costly reconstruction of postwar Europe, including the Netherlands. So, if the Netherlands refused to cooperate and kept on fighting the republic, they would risk the promised annual revenues of the US-sponsored Marshall Plan (Figure 3.17).

At the same time, there was a housing shortage and a large demand for adequate housing for military units in Palembang. BPM and Stanvac, as corporate enterprises, wanted to help out the colonial state, for once. They gave up between fifty and sixty European houses to the Dutch army in order to protect Palembang against Republican forces. Moreover, the BPM built fourteen permanent and sixteen nonpermanent European-style houses, two apartment blocks for forty bachelors, and twenty-seven extra hotel rooms. In return, the central government invested half a million dollars in building fifty houses in the elite neighborhood of Talang Semoet.[56]

Oil Diplomacy

In the beginning of 1948, the management of BPM, including Johan Frederik van Diermen, the headstrong administrator of Pladjoe, was exploring the possibilities for regaining control of the oil fields near Djambi. In a secret report addressed to Prime Minister Dr. Louis Beel, BPM did not demonstrate support for a military action against Djambi for fear of damage to and sabotage of the oil installations. Instead, BPM paid US$5,000 to Republican security forces of the oil fields and installations near Djambi in order to prevent destruction.[57] BPM pragmatically went one step further, and in May 1948 began direct negotiations with the moderate leaders and "oil barons" Isa, Sutowo, Gani, and Pattiasina, who had fled to Djambi after the earlier battle around Palembang. As a result of these "peace talks," the vital oil concessions fell back into Dutch control. BPM staff was relieved, but Van Diermen's conduct led to outrage at the headquarters of the Koninklijke in The Hague and among the staff of Stanvac, as they were not informed at all.[58] This was not the first time BPM had engaged in such covert dealings; it had happened before with the Gani deal at Pladjoe. The Minister on Overseas Territories Lubbertus Götzen informed the prime minister right away and told him that "Van Diermen could not care less who has the supreme authority, if only he gets oil."[59]

At times during the Cold War, BPM became convinced that communism might get a foothold in the Indies and they might lose their monopoly position once again. Van Diermen wrote several letters to the Dutch prime minister to share his concern. One synopsis concluded:

> the red peril is now prevailing all over the world, and the power of international communism is expanding to the South East Seas. If the third world war should break out, Soviet Russia will occupy South East Asia. To prevent that peril, now is the time.[60]

Directly after the landing of Dutch paratroopers on December 29, 1948 and January 5, 1949—performing a heavy-handed action—rebels caused severe damage to the NIAM oil fields near Rengat and Air Molek. This sabotage action was executed despite the Van Diermen deal with the Republicans and was carried out by separatist groups striving for independence; the Indonesian revolution was full of regional dynamics. BPM staff had predicted the destruction in many conversations with Dutch politicians.[61] However, both BPM and Dutch politicians were in favor of recapturing this important oil center.[62] Not least, because the US oil company A. S. Cowie & Co., Inc. located in California was eager to buy local concessions that were now in the hands of the republic for US$1 million.[63] During the military action, small technical BPM units operated again as "advance parties" in the slipstream of the Dutch troops and were able to minimize the damage and start repairing works.[64] Back in Dutch hands by mid-1949, crude oil was being piped again at prewar production levels.

Remarkably, in their effort to safeguard the oil business and restore the prewar situation of peace and prosperity, BPM's captains of industry, Dutch army commanders, and politicians in both Batavia/Jakarta and The Hague worked closely together, constantly using oil as a military weapon, since it was needed directly to fuel military operations in Asia. It was obvious that European managers of the oil refinery, the central Republican government, and the local government of Palembang all had an interest in this oil company town. As a result, they were constantly in conflict with each other.

With the advent of Indonesian independence in December 1949, the balance was drawn. It was clear that BPM could not do business the way it had before the war. Pladjoe and its infrastructure was saved from total destruction, but Djambi was regained only in 1949 and the refinery in Pangkalan Brandan in North Sumatra, the origin of Royal Dutch/Shell Group, was lost forever due to fierce local resistance in the northern province of Atjeh. Despite this outcome, postwar oil production would eventually triple, compared to the prewar average.[65]

In the beginning of the 1950s, the need for road renovation and construction in the war-damaged city of Palembang was more urgent than ever, especially in the Hoeloe area, which had been heavily damaged during the war and the revolution. In 1950, the municipality implemented a reconstruction plan, designed by Dutch architect Hans Lüning, which was an adaptation of the town plan of Thomas Karsten dating from 1937.

In the mid-1950s, when anti-Dutch sentiments prevailed in postcolonial Indonesia, political leaders in Jakarta decided *not* to nationalize BPM, unlike more than 700 other Dutch-owned enterprises and industries, a decision made because the company was "indispensable" to the Indonesian economy and because of BPM's combined Anglo-Dutch ownership.[66] Subsequently, Europeans who worked in the oil refinery continued to play a leading role in the postindependence period. Finally, in December 1965, when the political situation in Indonesia was rather tense, Koninklijke/Shell pulled out by selling all rights of exploration and production for US$110 million to Pertamina.[67] Once again, this was a clever oil deal, and besides, Shell would return soon.

Conclusion

This concise history illuminates how, from the 1920s on, the South Sumatran city of Palembang with its harbor and perfect river infrastructure gained new importance from the booming oil industry and the growth of a petroleumscape. In relative isolation, Pladjoe could expand and modernize undisturbed, in part because of the continued presence and rule of Western managers. After the corporate battle with competitors, like Stanvac and Caltex in the 1920s and 1930s, the battle for "black gold" with diplomatic means followed, and finally a military battle for oil was fought in the 1940s.

While oil attracted military conflict, the goal of the invader was to keep production going. It aimed to inflict little destruction and rebuild quickly. The presence of a global or multinational company interested in the space, with funding from other parts of the world, facilitated this process. While these sites were fought for halfway around the world, the armies also made sure that production continued. They depended on the sites and made sure that they were rapidly rebuilt when damaged. While foreign governments had to abandon their oil structures and territories, the major companies continued to work across borders. The presence of the oil infrastructure effectively guaranteed the continuation of the carefully built palimpsestic petroleumscape.

Notes

1 Carola Hein, "Oil Spaces: The Global Petroleumscape in the Rotterdam/The Hague Area," *Journal of Urban History* 44, no. 5 (September 2018): 887–929.
2 H. Gabriëls, *Koninklijke Olie: de eerste honderd jaar 1890–1990* (The Hague: Shell, Den Haag, 1990), 9–18; Frederik Carel Gerretson, *History of the Royal Dutch*, Vol. 1 (Leiden: Brill, 1973), 58–59; J. Ph. Poley, *Eroïca, the Quest for Oil in Indonesia (1850–1898)* (Netherlands: Springer, 2000), 82.
3 Joost Jonker and Jan Luiten van Zanden, *A History of Royal Dutch Shell: Vol. 1* (New York: Oxford University Press, 2007), 23; Keetie E. Sluyterman, *Kerende kansen. Het Nederlandse bedrijfsleven in de twintigste eeuw* (Netherlands: Boom, 2003), 53–54.
4 Woonkyung Yeo, "Palembang in the 1950s: The Making and Unmaking of a Region" (PhD diss., University of Washington, 2012), 79–80; Anderson G. Bartlett a.o., *Pertamina: Indonesian National Oil* (Jakarta: Amerasian Ltd, 1972), 48.
5 Frederik Carel Gerretson, *History of the Royal Dutch*, Vol. 2 (Leiden: Brill Archive, 1958), 39–64.
6 Carola Hein, "Between Oil and Water. The Logistical Petroleumscape," *The Petropolis of Tomorrow*, eds. Neeraj Bhatia and Mary Casper (New York: Actar/Architecture at Rice, 2013), 436–47.
7 Ida Liana Tanjung, "The Indonesianization of the Symbols of Modernity in Plaju (Palembang), 1930s–1960s," in *Cars, Conduits, and Kampongs: The Modernization of the Indonesian City, 1920–1960*, eds. Freek Colombijn and Joost Coté (Leiden: Brill, 2015), 302.
8 Freek Colombijn, *Under Construction, The Politics of Urban Space and Housing During the Decolonization of Indonesia, 1930–1960* (Leiden: KITLV Press, 2010), 80, 126; Idaliana Tanjung, "Plaju, Kota Minyak di Ulu Palembang simbol pribumi vs modernitas," 39, Paper Seminar on "Street Images: Decolonization and Changing Symbolism of Indonesian Urban Culture between 1930s and early 1960s," Yogyakarta, August 8–9, 2005.
9 Peter J. M. Nas, "Palembang: The Venice of the East," *Issues in Urban Development: Case Studies from Indonesia*, ed. Peter J. M. Nas (Leiden: CNWS, 1995), 133–34, 140; Ben de Vries, "Epische strijd om Indische olie," *Elsevier*, 26 March 2016, 80–83; Ben de Vries, "Demi Minyak Hindia, Riwayat De Bataafsche Petroleum Maatschappij, pendahulu Shell, yang sukses mengambil minyak di hindia Belanda sampai akhir era 1940-an," *Historia* 32, tahun III (October 2016): 34–37.
10 *Nieuwe Rotterdamsche Courant*, 19 June 1924, and *Het Vaderland*, 29 May 1924, 19 June 1924, and 13 July 1924; V. Hediger and P. Vonderau, eds. *Films that Work. Industrial Film and the Productivity of Media* (Amsterdam: Amsterdam University Press, 2009), 243–53; G. A. Jansen Hendriks, "Een voorbeeldige kolonie: Nederlands-Indië in 50 jaar overheidsfilms, 1912–1962" (PhD diss., University of Amsterdam, 2014), 62–63.
11 Johan Fabricius, *Brandende Aarde* or *East Indies Episode: An Account of the Demolitions Carried Out and of some Experiences of the Staff in the East Indies Oil Areas of the Royal Dutch-Shell Group During 1941–1942* (London: Shell Petroleum Co., 1949), 98–99; Mestika Zed, *Kepialangan, Politik, Dan Revolusi: Palembang 1900–1950* (Jakarta: Pustaka LP3ES Indonesia, 2003), 77–79.
12 Pauline Dublin Milone, *Urban Areas in Indonesia: Administrative and Census Concepts* (Berkeley: University of California Press, 1966), 108–39; W. J. W. Wellan, "De Stad Palembang in 1935: Tweehonderdvijfenzeventig Jaar Geleden, als een Phoenix uit haar Asch Herrezen," *Koloniaal Tijdschrift*, 24, no. 3 (1935), 232.
13 Ida Liana Tanjung, "Indonesianization," 306.
14 Hein, "Oil Spaces," 888, 897–901.
15 Elsbeth Locher-Scholten, *Sumatran Sultanate and Colonial State Jambi and the Rise of Dutch Imperialism, 1830–1907* (Ithaca, NY: Southeast Asia Program, Cornell University, 2004), 280–81, 309.
16 L. J. Touwen, "Voordeel van veelzijdigheid. De economische ontwikkeling van Palembang en Djambi tussen 1900 en 1938," *Economisch en Sociaal-Historisch Jaarboek* 54 (1991): 154–55, 177.

17 Agus Setiawan, "The Political and Economic Relationship of American-Dutch Colonial Administration in Southeast Asia: A Case Study of the Rivalry between Royal Dutch/Shell and Standard Oil in Netherlands Indies (1907–1928)" (PhD diss., SHSS Bremen, 2014), 3–4, 40–41.
18 Bartlett et al., *Pertamina*, 48–49.
19 Hiroyoshi Kanō, *Indonesian Exports, Peasant Agriculture and the World Economy, 1850–2000* (NUS Press, 2007), 226–27, 230.
20 Alex Hunter, "The Indonesian Oil Industry," *The Economy of Indonesia: Selected Writings*, ed. Bruce Glassburner (Ithaca, NY: Cornell University Press, 1971), 256–57.
21 H. Baudet and M. Fennema, *Het Nederlandse belang bij Indië* (Utrecht-Antwerpen: Het Spectrum, 1983), 158.
22 Fabricius, *East Indies Episode*,14; Loe de Jong, *Het Koninkrijk der Nederlanden in de Tweede Wereldoorlog*, Vol. 11a (RIOD, 's-Gravenhage/Staatsuitgeverij 1985), 692; Stephen Howarth and Joost Jonker, *A History of Royal Dutch Shell: Vol. 2* (New York: Oxford University Press, 2007), 17, 63; Willem Remmelink, *The Operations of the Navy in the Dutch East Indies and the Bay of Bengal*, Compiled by the War History Office of the National Defence College of Japan. Senshi Sōsho (War History Series), Vol. 26 (2018) (Leiden: Leiden University Press), 4–14; Gene Eric Salecker, *Blossoming Silk Against the Rising Sun: U.S. and Japanese Paratroopers in the Pacific in World War II* (Mechanicsburg, PA: Stackpole, 2011).
23 De Jong, *Het Koninkrijk*, Vol. 11a first half, 507.
24 C. van den Hoogenband and L. Schotborgh, *Nederlands-Indië Contra Japan, Deel VI: De Strijd Op Ambon, Timor En Sumatra* (Departement Van Defensie, Hoofdkwartier Van De Generale Staf, Krijgsgeschiedkundige Afdeling, 1959); Lionel Wigmore, *The Japanese Thrust. Australia in The War of 1939–1945*, Vol. IV (Canberra: Australian War Memorial, 1957). The Japanese dilemma, part 1, 11, *The Road to War, from Second World War Official Histories.*
25 De Jong, *Het Koninkrijk*, 11a first half, 507; De Jong, *Het Koninkrijk*, 11a second half, 694–95.
26 Willem Remmelink, *The Invasion of the Dutch East Indies.* Compiled by the War History Office of the National Defense College of Japan. Senshi Sōsho (War History Series), Vol. 3 (2015), 270.
27 Herman Theodore Bussemaker, "Paradise in Peril. Western Colonial Power and Japanese Expansian in South-East Asia, 1905–1941" (PhD diss., University of Amsterdam, 2001), 754.
28 Online article on dutcheastindies.webs.com/DEI_oil.html, "Japan's Need for Oil and the Embargo (1940–1941)," *Fabricius, East Indies Episode*, 16–18; De Jong, *Het Koninkrijk*, 11a second half, 695–96, 702.
29 De Jong, *Het Koninkrijk*, 11a second half, 655; J. J. Nortier, "De gevechten bij Palembang in februari 1942," *Militaire Spectator* (1985): 312; Loe de Jong, *The Collapse of a Colonial Society: The Dutch in Indonesia during the Second World War* (Leiden: KITLV Press, 2002), 286, 388–89.
30 Erwin Langewis, "Hoe de Japanners de Indische archipel veroverden, 75 jaar geleden: Java zit in de Japanse val," *Historiek.net*, 23 February 2017; Johannes Jan Nortier, P. Kuijt, and Petra M. H. Groen, *De Japanse Aanval op Java: Maart 1942* (1994); Remmelink, *Invasion*, 288–89, 316, 329; De Jong, *Collapse*, 35.
31 Howarth and Jonker, *Royal Dutch Shell*, 65–66.
32 Gene Eric Salecker, "Deadly Dash Forward, Japanese Parachute Forces Mounted an Assault on the Island of Sumatra soon after Pearl Harbor," *WWII History* (October 2016): 49.
33 Remmelink, *Invasion*, 329–30; David van Reybrouck, *Revolusi* (De Bezige Bij. Amsterdam (2020), 167–68.
34 Fabricius, *East Indies Episode*, 105–6, 120–25; Peter Keppy, *Sporen van Vernieling: Oorlogsschade, Roof en Rechtsherstel in Indonesië 1940–1957* (Amsterdam: Boom, 2006), 42–43; J. J. Nortier, "De aanval op Palembang in februari 1942", *Militaire Spectator*, part 2 (1985), 361; Remmelink, *Invasion*, 344.
35 De Jong, *Het Koninkrijk*, Vol. 11a, first half, 677; J. J. Nortier, "Japanese parachutisten, Samoerai van de Tweede Wereldoorlog," *Militaire Spectator* (1983): 152, 519–23; Remmelink, *Invasion*, 284.
36 Jean Aden, "Oil and Politics in Indonesia 1945 to 1980" (PhD diss., Cornell University, 1988), 40.
37 Bartlett et al., *Pertamina*, 56–57; De Jong, *Collapse*, 50.
38 J. Thomas Lindblad, "Van Javasche Bank naar Bank Indonesia. Voorbeeld uit de praktijk van *indonesianisasi*," *Tijdschrift voor Sociale en Economische Geschiedenis* 1 (2004): 28–46; "The Importance of *Indonesianisasi* during the Transition from the 1930s to the 1960s," *Itinerario* 26, no. 3/4 (2002): 51–71.
39 Hunter, *Indonesian Oil Industry*, 259, 266.

40 J. van Dulm a.o., *Geïllustreerde Atlas van de Japanse Kampen in Nederlands-Indië 1942–1945* (Torino, Italy: Asia Maior, Purmerend, 2001), 74–79.
41 Loe de Jong, *Het Koninkrijk*, Vol. 11b, second part ('s-Gravenhage/Staatsuitgeverij), 694–95.
42 Howarth and Jonker, *Royal Dutch Shell*, 69.
43 Bartlett et al., *Pertamina*, 71–73.
44 Howarth and Jonker, *Royal Dutch Shell*, 226–28.
45 Archief Indonesië in overgangstijd. Overgekomen archieven uit Batavia. Algemene Secretarie (AS), 2e Afdeling, no. 3503, 12 April 1946. National Archives (NA), The Hague.
46 Collectie Nederlands-Indië 1945–1950, Sumatra, 207-I, 23 October 1946. Archive Netherlands Institute Military History (NIMH).
47 Collectie Nederlands-Indië 1945–1950, Sumatra, 207/A, nota situatie Palembang, ondertekend 27 August 1946, NIMH.
48 Aden, "Oil and Politics," 70–71.
49 Secretaris-generaal Overzeese Gebiedsdelen aan Lt. G-G, 24 August 1946. AS, number 3503. NA.
50 Bartlett et al., *Pertamina*, 73–74; Ida Liana Tanjung, 312–13; Yeo, "Palembang," 84–86.
51 Bambang Purwanto, "Economic Decolonization and the Rise of Indonesian Military Business," *Indonesian Economic Decolonization in Regional and International Perspective*, eds. J. Thomas Lindblad and Peter Post (Leiden: KITLV Press, 2009), 44–46.
52 Anne-Lot Hoek, "De verzwegen moorden van Palembang. Geschonden oorlogsrecht," *Vrij Nederland*, September 2017, 42–49; Zed, *Palembang 1900–1950*, 110.
53 Annual Report (Jaarverslag) of the *Koninklijke*, 1947.
54 Collectie Nederlands-Indië 1945–1950, Sumatra, 207/4, December 1948, NIMH.
55 Hunter, *Indonesian Oil Industry*, 259.
56 Colombijn, *Under Construction*, 259–60, 341–43.
57 Directeur van het kabinet G-G aan Lt. G-G, 2 August 1947, AS, no. 3505, NA; Locher-Scholten, *Sumatran Sultanate*, 287.
58 Götzen aan minister-president Beel, 26 May 1948. Rijks Geschiedkundige Publicatiën (RGP) XIII, no. 319.
59 Minister zonder portefeuille Götzen aan Beel, 7 June 1948. RGP XIV, no. 12.
60 Archief Indonesië in overgangstijd. Overgekomen archieven uit Batavia. AS, no. 353.83.1.329.15. Synopsis, part 1, 19 June 1948, "Het communisme in de Indonesische Republiek". Part 2: "Internationale aspecten en Kominform," NA. Indonesian leaders surpressed a communist revolt in Madioen (East Java) on 18 September 1948.
61 Ibid.; Locher-Scholten, *Sumatran Sultanate*, 287–92.
62 Notulen van vergadering gehouden ten huize van de legercommandant op 20 September 1948. RGP XV, 138.
63 Chef directie Verre Oosten te Batavia (Elink Schuurman) aan ambassadeur te Washington Van Kleffens en minister van Buitenlandse Zaken Van Boetzelaer van Oosterhout, 1 May 1948, RGP XIII, no. 230.
64 Collectie Nederlands-Indië 1945–1950, Sumatra, 207/4, December 1948, ondertekend door algemeen commandant Operatie Ekster, NIMH; Petra M.H. Groen, *Marsroutes en dwaalsporen. Het Nederlands militair-strategisch beleid in Indonesië 1945–1950* (1991), 143.
65 W. Brand, "Heroriëntatie van vroeger in Indonesië werkende Nederlandse bedrijven," in *Handelswereld en wereldhandel*, ed. H. Baudet (Rotterdam: Honderd jaren Internatio, 1963), 170.
66 Henk Biersteker, "Hoe Shell uit Indonesië verdween," *Hervormd Nederland*, 29 July 1995, 51; Gabriëls, *Koninklijke Olie*, 171; Aden, "Oil and Politics," 177.
67 Howarth and Jonker, *Royal Dutch Shell*, 240; J. Thomas Lindblad, *Bridges to New Business. The Economic Decolonization of Indonesia* (Leiden: KITLV Press, 2008), 159; Khong Cho Oon, *The Politics of Oil in Indonesia: Foreign Company-Host Government Relations* (Cambridge: Cambridge University Press, 1986), 138–39; Sluyterman, *Kerende kansen*, 219.

4

MAPPING THE PERSIAN GULF PETROLEUMSCAPE

The Production of Territory, Territoriality, and Sovereignty

Stephen J. Ramos

On the evening of March 15, 1954, Sir Rupert Hay, who served as British Political Resident in Bahrain from 1946 through 1954, addressed the Royal Geographical Society in London on the topic of "the Persian Gulf States and their Boundary problems."[1] Society President J. M. Wordie introduced Hay as one with "intimate knowledge of the Persian Gulf States and their problems," which would be presented "as delicately as may be—because it is delicate ground where sea and land are concerned and where oil is underneath."[2] In the tradition of colonial knowledge production throughout the British Empire, Hay went on to describe in detail the natural resources and geographic conditions of what were then the ten Trucial States and Muscat.[3] Hay concluded that oil had brought new wealth to Gulf "shaikhdoms," particularly after the collapse of the pearl trade, but that it had also brought new problems, the most important being the "fixation of boundaries by land and sea" (Figure 4.1.).[4]

The Trucial States were part of a larger British protectorate system: the British would back particular ruling families through "truces" or pacts within their "sphere of influence," with the understanding that British interests would be supported in the region.[5] Throughout the nineteenth and early twentieth centuries, these interests were mainly focused on peaceful passage along the maritime trade routes that connected Britain with its colonies in India and East Asia. In 1908, oil discovery in Persia brought British attention ashore. Within the British colonial political geography, the founding of the Anglo-Persian Oil Company moved the region from a peripheral western outlier of the Anglo-Indian orbit to a strategic center for British oil exploration and extraction. Upon entering the global petroleumscape, the Trucial States' colonial spatial order shifted, and the British required a new territorial regime that could properly domesticate, discipline, and organize the new extraction economy.[6] Gulf territorial production simultaneously combined British national and imperial objectives with the corporate objectives of interested oil companies, each overlaid on indigenous territorial codes. The British established a composite cartographic order of constituent pre-nation-state urban, territorial, and legal frameworks. These processes have their own palimpsestic qualities and draw attention to more complex issues of causality for spatial and territorial configurations in the global petroleumscape.

DOI: 10.4324/9780367816049-6

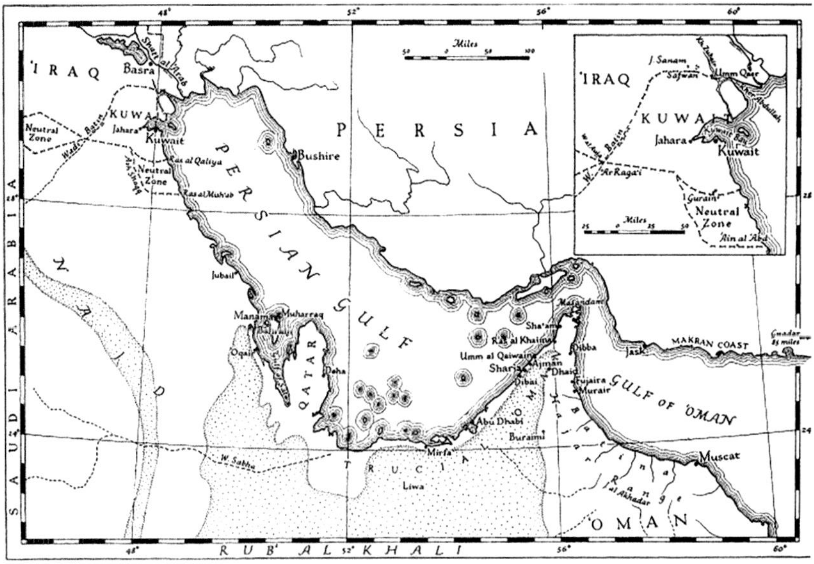

FIGURE 4.1 Map of the Persian Gulf in 1954, produced by Sir Rupert Hay.[7] The Buraimi Oasis is located just southeast of Abu Dhabi and Dubai is labeled "Dibai."

By 1922, Kuwait, Bahrain, and the Trucial States had all signed agreements with the British, committing to granting oil exploration concessions only to companies appointed by the British government. Negotiations for these commitments included representatives from the British government, oil companies, and Gulf rulers.[8] The trilateral, multifaceted interests of each of these parties foretell the complexity of the unique political and territorial configurations that such negotiations would produce. The situation was further complicated when corporate oil interests from allied countries such as the US and France entered the region soon after.[9] The regional urban forms that emerge from this period are a product of the intertwined legacies of imperial rule and extractive capitalism.[10] Large-scale trade infrastructure, such as ports and airports funded by new oil wealth, was also a generative element in Gulf urban and metropolitan formation.[11] Subsequent territories, and the administrative territorialities that governed them, emerged as Trucial States moved through various subject relationships with international powers into their own respective, postcolonial, nation-state projects. Corporate interests catalyzed these regional transformations in territory and territoriality. As historian Rosemarie Said Zahlan states, "Oil firms forced rulers to define their frontiers, and thus to think of political power in [new] territorial terms, for the simple reason that they needed to know exactly what territories their concessions contained."[12]

The history of oil discovery, extraction, and distribution in the Gulf, and the wealth and sociocultural change that it brought to the region are well documented.[13] Recent research estimates that Gulf monarchies, combined, produce over 17 million barrels of crude oil per day, nearly 20 percent of the global total, and they control approximately 37 percent of the world's proven crude reserves.[14] Global production network (GPN) analysis literature maps the oil industry's strategic integration and geographic dispersion, and how the Gulf region serves as an important producer node within this geography.[15] The global petroleumscape framework offers a comparative methodology to explore these broad historic spatial practices and representative modes of oil—its *territoriality* and *materiality*—to demonstrate the extent to which oil has constituted the essence of global modernity in the twentieth and early twenty-first centuries.[16] As Hein writes, "Only in appreciating the power and extent of oil can we engage with the complex emerging challenges of sustainable design, policy making, heritage, and future built environments beyond oil."[17]

The study of a world beyond oil is situated in a growing, interdisciplinary subfield of the post-oil imaginary, a futurism where transitions and inflection points lead the world to new energy distribution systems, presumably with their own spatialities and materialities.[18] But if we follow Hein's metaphor, surely this new aspirational "post" phase will simply be another layer atop the accumulated, fragmented energy palimpsest. In her assessment of the Middle East, Janet Abu-Lughod describes how its cities "contain accretions of successive types of settlement," which in aggregate are all the more complex and challenging to comprehend.[19] So, as far into the future as modes of extractive oil capitalism may continue, its legacy will continue beyond, shaping the world as another layer in the energy palimpsest. This is particularly true for the global petroleumscape's political influence on the negotiation of nation-state projects for both producer and consumer countries. The "power interdependencies" and the continued "centrality of the *nation-state* itself in the calculation oil accessibility and security"[20] are issues of energy power and historic geopolitics that will not simply vanish in the New Age. Path dependencies tether past to future,[21] and in many parts of the world, the global petroleumscape has translated colonial dependencies into corporate ones.[22] Over time, this fusion has led to the nation-state projects of the Gulf region. Political geographer Josh Barkan writes, "Put in ontological terms, it is through time that the potentiality of the world is actualized in new political forms"[23]—and the regional political and spatial transitions in the Gulf throughout the twentieth century are rich histories that help to illustrate this point. By fully appreciating the process of territorial production in the service of the global petroleumscape, we can better understand the degree to which Gulf state (quasi-colonial and postcolonial) and corporate interests were forged together at the petroleumscape's inception, and how territory and its organization (territoriality) were essential processes in the production of this new international petroleum order.

The global petroleumscape proposition presents a transversal historic material ethnography of oil and a reflexive political ethnography of its territoriality in the Gulf region.[24] Just as Gulf region oil spatialities were produced by the geopolitical collaboration of late British colonialism and corporate oil, new spatial and political configurations arise from the multiparty collaborations that form the present nation-state and "global" conditions that continue to enable contemporary extractive capital flows.[25]

The chapter begins with a discussion of the concepts of territory, territoriality, and sovereignty as essential components in the negotiation and struggle for mapping. While the concept of mapping has grown recently to signify exploratory methodologies in spatial analysis and in the social sciences,[26] here the process of mapping also considers its historic agency as both purveyor and arbiter of colonial and corporate power in a way that was essential to the production of the Gulf petroleumscape. British political philosopher Mark Neocleous describes the mapping of the modern state as an act of "cartographic violence in the construction of territory."[27] This description is useful within the broader British imperial project[28]and also in the fixing of oil territories through state and corporate oil concession negotiations. By reviewing the construction and demarcation of Gulf territory and its affiliate territoriality administration of the regional transition during the ascendancy of the oil age, we can then begin to think about how these processes might continue to layer and metamorphosize "after oil." Put otherwise, by analyzing the complexity of the Gulf's historic layers of territory, territoriality, and sovereignty, we may begin to imagine its future through its past.

Mapping Layered Territory, Territoriality, and Sovereignty

In political geography, the concept of territoriality is used to explore the relationship between political systems and territory.[29] Territory is both a political and geographical concept. Territory is not merely a natural phenomenon, but rather an organized space partitioned through political processes.[30] US geographer Robert Sack defines territoriality as the "attempt by an individual or group to affect, influence, or control people, phenomena, and relationships, by delimiting and asserting control over a geographic area," namely, a territory.[31] Alternatively, Swiss geographer Claude Raffestin defines territoriality as a system of relations translated physically and spatially, which constructs more than territory, and is constructed by factors that include and exceed territory[32]; territories, for Raffestin, are *produced* in the Lefebvrian sense.[33] Although often framed within problematic cultural evolutionist claims, there is a distinct historic move from sociopolitical organizations of space, involving kinship, inheritance, or both, toward a more strictly defined political and spatial definition, which finds its maximum expression in the emergence of the nation-state.[34] Taxonomies of premodern functional spatial configurations of territory include networks, nodes, hierarchies, surfaces, and movement.[35] For these societies, nodal points might include wells, shrines, and ancestors' graves, which embed legibility and meaning in spatial configurations.[36]

With the emergence of the nation-state, an essential core attribute of territorial definition is the concept of property and exclusive, precise demarcation of borders with internal coherence and continuity. Rooted in Aristotelian logic and Greek geometry, Western property rights unite in a functional matrix, which in aggregate match the territorial demarcation claims of the state.[37] Contracts gain authority over previous kinship understandings, and *grosso modo*, spatial definition and hierarchy, along with its administration, primed territories for industrial capitalism, for which the nation-state has proved to be the most effective guarantor. European colonial and imperial projects have extended this territorial definition of power relationships unevenly throughout much of the world, but as suggested by the palimpsest metaphor, these territorial configurations do not wholly erase the indigenous territorialities they sought to replace: "It is perhaps more accurate to say that it [European territoriality] was superimposed upon these existing forms, becoming tightly integrated with them in some areas while resting much less comfortably and stable in others."[38] The transition from a politically organized unit to a nation-state is a process in constant flux, as is the perennial challenge to represent and document this process.[39] Because territorial states are in a constant state of restructuring at the "intersection of local and global material conditions," bringing together domestic and foreign considerations and their different historic circumstances becomes the "overriding task."[40] Nation-state territorialities preside over land (surface and subsurface), sea, air, ice, and rivers, which are all important for the extraction and distribution of oil.[41]

If territoriality is the presence of territorial definition and control, sovereignty is the authority to exert such control, often made explicit in law.[42] But the relationship between sovereignty and law is complex. Critical literature on corporate sovereignty demonstrates the constitution of corporate and state sovereignty as conjoined and expressed over territories in specific spatial arrangements, such as imperial trading companies and free trade zones. These biopolitical processes are exercised over the border between the inside and outside of constituted political spheres[43]: "In each case, it is not just the territorialization itself (the trade route, the free trade zone) but larger assemblages of power that enable these territorializations to come into being."[44] For the global petroleumscape, often designed from a distance through colonial and corporate power, oil territory and imposed sovereignty was precisely the authority to fix and define territorial border demarcation, and immediately transgress those borders should concession interests and oil discovery require it for oil companies' externally-oriented resource extraction. This is the territorial contract of petroleumscape corporate sovereignty. Further, the new oil territoriality would essentially build on the nodal assemblage of the region's past, with oil wells, refineries, and circulatory infrastructure constituting new territorial "points and lines." Here the palimpsest's layers become difficult to discern in terms of which layer is at the fore at any given historic moment, or if they are simultaneously performing in concert.

In the Gulf region, British trade maritime routes began their territorial designation along the Trucial Coast (the southeastern Persian Gulf) in the early nineteenth century, and they proceeded inland with greater interest at the dawn of oil discovery and exploration.[45] British intervention in regional boundary disputes, even for those states that were officially deemed protectorates, was never guaranteed. None of the three types of regional sovereignty laws—tribal, Islamic, and international—supported British territorial boundary demarcation authority.[46] For the Trucial States, the right of the British to demarcate boundaries was "questionable, even if barely questioned."[47] In the 1920s, the British signed oil exploration agreements with oil companies and Gulf leaders to establish contractual legality as the definitional frame for territorial extraction rights, inscribing abstracted, linear geometric boundaries onto the shifting, sinuous region—land and sea—so as to avoid potential future conflict.[48] Oil discovery creates a new territorial mandate. Again, Zahlan offers helpful insight: "Now that every inch of land had become potentially valuable, it was inevitable that conflicts over territorial definition would arise."[49] In the oil exploration agreements, British imperial and corporate legality would serve as the arbiter of territorial definition.

The relationship between state formation and oil extraction in the UAE is further complicated by the fact that formation follows discovery in Dubai and Abu Dhabi (the latter has significantly more oil reserves), and the remaining Emirates are subordinated to these two with either significantly fewer or no oil reserves.[50] In this context, to imagine a post-oil future is to completely upend the nation-state's foundational territorial construction, and to a large degree, its very essence.[51] A review of the historic context of Britain's relationship to the region and the antecedent of a cartographic culture helps to inform the process of territorial construction in the Gulf's pre- and post-oil periods.

British Cartographic Culture: Trade Territory

Trade networks are inherently political territorial constructions in the service of imbricated public and private interests, and thus, "territorial enterprises."[52] British interests in the Gulf region begin in 1600, when Elizabeth I gave London merchants a monopoly charter to compete with the Portuguese in the spice trade, and the merchants soon acquired the name of the "East India Company."[53] The Dutch challenged the British for trade opportunities, but in 1765, after various conflicts among the European powers, the East India Company established control over the Mughal provinces. As imperial interests expanded, it was clear that the Company could earn great revenue as territorial administrator and land tax collector. The great achievement of the British Empire was to construct a single geographic state of India from its multitude of component polities and cultures through cartographic "colonial imaginings."[54] British "cartographic culture" was not merely a sum of material map-making practices, but rather forms of territorial representation as the scientific legitimation of empire through the "rational construction of space."[55] British cartographic practices in India were constitutive of Enlightenment spatial ordering to legitimate imperial aspirations of wealth extraction. These same British practices also inform future territorial organization in the Gulf as the bountiful pearl industry of the late nineteenth century gave way to twentieth-century oil discovery.[56] British cartographic practices also illuminate the interpenetration of the company's private entrepreneurial trade interests with British imperial strategy, and the defense and administration of each.

Piracy in the Gulf practiced by local Omani and Qawasim groups complicated European eastern trade routes throughout the eighteenth century, and created significant problems for the East India Company toward the end of the century.[57] For the General Treaty of Peace in 1820, which included leaders from Bahrain and northern Oman (including areas of present-day UAE), all agreed to end piracy attacks. This began a series of truces in 1835, 1843, and 1853, whereby the British, through indirect rule, would "watch over and protect" the southern Gulf region.[58] This policy of informal or indirect rule was based on the principle of "informal control if possible, formal control when necessary."[59] The "collaborators" in the lower Gulf were those tribal leaders in power at the time of trucial signing, and although British commitments to the region would eventually become more direct, before World War I, if trade routes went uninterrupted, Britain would only commit to maritime protection of the coastal pearling settlements of their sheikh cosignatories.[60] If modern India is the creation of British cartography, the very origin and naming of the "Trucial Shaikhdoms" as a geographic entity is created from British imperial interests and their imperiled material trade concerns due to Gulf piracy.[61]

With the opening of the Suez Canal in 1869, French, Russian, German, and Ottoman expeditions began entering the Gulf region. In response, the British signed exclusive agreements with the Trucial States in March 1892, establishing that no territorial sovereignty could be ceded to any other international power without British consent.[62] By 1916, similar treaties were signed by Bahrain, Kuwait, and Qatar, establishing a clear mare nostrum British administration of the Gulf region. At the beginning of the twentieth century, oil was discovered in Persia and Saudi Arabia, and shortly thereafter in Kuwait and Bahrain.[63] Oil exploration was then carried out in all the British Gulf protectorates, which would require a more fixed territorial survey and demarcation rationale along with a new system of extraction infrastructures. But rather than replace an existing indigenous spatial order, oil territory and territoriality were superimposed on the desert and the sea.[64] What follows is an overview of that pre-oil territory upon which the Gulf global petroleumscape was overlaid.

Pre-Oil Gulf Territory

Within the tribal territorial dynamism that characterized the region before oil interests, coastal urban settlements were embedded in larger territorial relationships with the hinterlands and their societies. The region was generally organized around nomadic Bedu tribes in the interior who would charge a protection tax ("zakat") for safe passage through their territory ("dar"), smaller villages where tribes engaged in complementary fishing/agricultural activities, and the coastal towns where pearling and trade were centered.[65] These coastal settlements mark the origin of subsequent urbanization processes of the region.[66] It is this dynamic territorial configuration that better describes the "people-centered" nature of the territory, where violence and flux were ever-present.[67] While authority emanated from coastal settlements, the measure of a coastal ruler's strength was the degree to which he could command Bedu support.[68] The tribal territorial field involved multiple factions and constantly shifting allegiances, wherein dependence on tribal support meant that the Bedu were the "preponderant force" of the Trucial hinterland region, "able to hold Rulers, merchants, and common people to ransom," up to the early 1950s.[69] Territorial sovereignty and demarcation were fluid and negotiated, and open movement across territory was an essential quality of the region's pre-oil territoriality.[70]

Restricted or exclusive territorial rights were reserved for areas that had been improved by labor and investment, such as irrigation land systems, extensive tree farming, wells, buildings, fish traps, and boats.[71] "Usufructuary rights" could be negotiated with tribal leaders for their use, but this was markedly different from those unimproved areas of the territory, in that ownership was established by means other than common land use.[72] These "improvements" were essentially responsive forms of infrastructure intended to harness the productive power of the territory according to varying contexts inland and on the coast. These investments required economic power, and this further entrenched the town as the center of territorial rule: "That no sheikh could rule his people without a command of economic power explains why all rulerships were town-based, at the heart of economic activity in the Gulf."[73] Economic power allowed coastal rulers to buy Bedouin alliances and to invest in the infrastructure required for various forms of production and subsistence, where usufructuary rights served as an additional form of negotiated allegiance. Eventually, the power dynamic shifted to coastal settlements. British Trucial protégé rulers were able to claim territory in their hinterlands, thanks to British backing through oil concession stipends that helped them buy Bedu protection; implicit British Navy backing also helped ensure their protection.[74] The British would enforce stricter corporate territory demarcation in the post-Ottoman negotiations and oil discovery at the early twentieth century. But infrastructure, such as wells and irrigation systems, coded the pre-oil indigenous territory in patterns of nodal functional composition, similar to those of the premodern societies mentioned above, and sovereignty, though located at coastal settlements, was continually negotiated with the Bedu periphery.[75]

The palimpsest metaphor for the global petroleumscape should suggest neither stasis nor harmony. The clashing of these territorial codes and understandings are constant and ever-present. The Buraimi Oasis, described below, helps to illustrate how the production and reproduction of global petroleumscape territory and territoriality was a violent, protracted international geopolitical struggle, occurring at multiple scales and time frames.

Contested British Boundaries: The Buraimi Oasis

If the exclusive agreements of the region the British signed between 1892 and 1916 were useful for keeping other interested international powers away from the region, in the first part of the twentieth century, it was the British relationship with Ibn Saud—considered the founder of modern Saudi Arabia—that would prove most complicated.[76] Ibn Saud led the country's territorial reunification from 1902 through 1932, and he ruled as Saudi monarch from 1932 through his death in 1953. Under his regency, oil was discovered in Saudi Arabia in 1938 by a US geologist working for the Standard Oil Company of New York, in co-operation with the Saudi government.[77] Following World War II, a group of nine villages known collectively as the Buraimi Oasis, located at the southeastern corner of the Arabian Peninsula, was the site of regional conflict between Saudi expansionist interest and the territorial claims of Abu Dhabi, under British trucial protection.[78] Oil prospecting was at the heart of the territorial conflict.[79]

In July 1913, the British had signed the first Anglo-Ottoman Convention to establish sovereignty for Kuwait, Qatar, and Bahrain, establishing the "Red Line" and the "Green Line" around Kuwait as stepped geometric demarcations of British protective spheres.[80] The following year, a second set of lines—the blue and violet lines—had distinguished the British sphere of influence from the Ottoman in the south for Aden along rather arbitrary geometric meridian lines. In 1915, Ibn Saud signed an agreement with the British to contain his territorial interest in the Trucial States and officially recognize them in return for British recognition of Saudi rights to the Najd and Al-Hasa (the eastern region of what is now Saudi Arabia).[81] In the 1930s, oil exploration and concession rights brought back territorial disagreement between Ibn Saud and the British, and in 1935, after several negotiation attempts that included more nuanced understandings of tribal territoriality that were more sympathetic with Saud's position, the British again reverted to an abstract geometrical line termed the "Riyadh Line."[82] Saud rejected the line out of hand. The protracted negotiations were interrupted by World War II, and then resumed and intensified afterward. Saudi land claims moved eastward, reaching, according to the British, "grotesque proportions" in 1949.[83] The British believed that their historic presence and maritime protection in the region had earned them influence and rights to its oil, a belief contested by Saud and the Persians (Figure 4.2).

A series of local, regional, and international meetings attempted to resolve the British-Saudi territorial dispute, but they were unsuccessful. The conflict reached its apex in August 1952 when an armed Saudi group led by Turki bin Otaishan occupied the eastern Buraimi Oasis village of Hamasa, within the territory that the British considered to belong to the Abu Dhabi protectorate.[84] The Buraimi villages marked the essential infrastructural feature for the grazing, nomadic economy, but there were also oil reserves identified in the area Saud's troops occupied, which would become the mineral and infrastructural land determinant in the next phase of the region.[85] Saudi border disputes challenged British territorial interests for oil exploration and British political control of the region. The international scope of the Buraimi occupation was further complicated by the US oil interests' alignment with Saud in their joint oil enterprise, the Arabian-American Oil Company (ARAMCO).[86] The British could not openly declare opposition to the US, so the proxy conflict among US and British oil interests played out in the territorial conflict in Buraimi, between Saudi Arabia and Abu Dhabi.

In 1954, the Buraimi incident went to the International Arbitration Tribunal in Geneva, which also considered Oman's claims and Muscat's partial claims to the oasis lands.[87] Negotiations broke down, and Britain reverted back to their geometrical land demarcation of a modified "Riyadh Line."[88] By the end of 1955, the British-backed security force known

FIGURE 4.2 Cover image from the Standard Oil Bulletin published in July 1933. Standard Oil of California (Socal) established the Bahrain Petroleum Company (BAPCO) in 1929, and oil was discovered there in 1931. The company went on to sign an exploration agreement with Saudi Arabia, where oil was discovered in 1938 (Chevron Corporation).

as the Trucial Levies (established in 1951 when local men were recruited to protect British nationals and their interests and to support their purported opposition to protectorate slave trade) entered the Buraimi village to evict the small Saudi army and unilaterally reestablish the "Riyadh Line." Saudi troops withdrew from Buraimi in 1955, but Saudi Arabia and Abu Dhabi would not resolve their territorial dispute until the 1974 regionally-brokered Treaty of Jeddah,[89] after British withdrawal from the region in 1971.

The Buraimi incident drew the British into direct territorial conflict with Saudi Arabia, and by association, US oil interests.[90] This was the first incident after British protectorate status was established for the Trucial States in 1949, and it explicitly conflated larger British strategic interests with the British Petroleum Development Trucial Coast (PDTC) oil exploration interests, much in the way that British interests and those of the East India Company had been conflated earlier. The modes of territorial negotiation moved from engagement with embedded Bedu knowledge of tribal tax and grazing land territorial outlines to the more abstract geometric lines drawn through British imperial fiat, one not ratified by official international diplomatic protocols. The historic allegiance between tribes and Ibn Saud made the British wary of that embedded knowledge, and they felt that any deference to tribal territorial understandings would undermine the authority they felt they were due, through over a century of serial, bilateral trucial agreements with select coastal leaders.[91] Some historians claim that after India's 1947 independence, Britain began to develop a distaste for what they understood as the costs and responsibilities of empire.[92] The Buraimi incident suggests that they were not entirely through with imperial pursuits.

Against the background of regional territorial boundary disputes, in 1955 the British sent land surveyor Julian Walker to the Trucial States to negotiate, organize, and delineate boundaries among sheikhs, so as to establish more precise, clear territorial boundaries and essentially insist on what it had enjoyed tacitly up to that point. Walker's boundary-marking expeditions through the desert harnessed British diplomatic networks and resources in the service of corporate extraction insurance. If rumblings of independence movements in Iran and Egypt were afoot in the region and ideologically spreading, then the British sought the security of fixed, internationally binding, and recognized borders in the service of the corporate petroleumscape.

"Delicate Ground"

When Sir Rupert Hay addressed the Royal Geographical Society in London, he described border disputes within the Trucial States and Muscat, including one that flared in 1948 when a Dubai raiding group killed fifty-two Manasir allies of Abu Dhabi in a border battle at Khor Ghanada.[93] Dubai and Sharjah had fought over similar issues in 1940 and required leaders from Ras al-Khaima to help broker a truce.[94] Hay also mentioned the Buraimi dispute in his address. Throughout 1954, internal territorial disputes were registered among Trucial States such that at the beginning of the following year, in January 1955, Julian Walker, then British Assistant Political Agent in Bahrain, set out on the "thankless job" of boundary settlement among Trucial rulers so that oil exploration could proceed.[95]

Before he set out, Walker claims to have found only scarce survey information produced by PDTC in the 1930s. In reviewing the British Agency communication from 1937 throughout the Trucial States, a British land agent found that rulers had only approximate knowledge of exact territorial limits.[96] In the autobiographical literature of retired British diplomats that participated, these technical cartographical acts are described as having been reluctantly executed under the auspices of beneficence and regional development.[97] What comes through clearly is the discomfort and concern of the protégé rulers in having to fix, within a limited time, land boundaries and their substrata resource claims, which were, in no uncertain terms, the symbolic and material sources of power.

For the first three months of 1955, Walker traveled on foot and via camel and Land Rover over Dubai's rugged terrain, interviewing tribesmen and trying to decode local and tribal geographic knowledge and translate it into map form. He used territorial signifiers, those features that demarcated ownership through improvement, to help identify and establish territorial boundaries and ownership. Irrigation channels and wells were the infrastructure that marked the territory and helped Walker identify ownership and political relationship based on its use as a clientelistic medium.[98] Walker acted as both anthropologist and cartographer to settle boundaries as closely as possible to the satisfaction of those he interviewed, while knowing that the map itself would have to be completed with or without their approval. In his later accounts of his work, although recalling that he tried to represent as faithfully as possible the "local way of life" in his imposition of territorial "straight lines" in his surveying, Walker repeatedly underscores the fixity, dependability, and order in his work as giving transcendence and legacy to an otherwise (to his eye) chaotic territoriality.[99]

Upon completing his work, Walker wrote up the report and submitted it to the British Foreign Office for *their* approval, again, in keeping with the standard practices of colonial knowledge production.[100] To Walker's surprise, the British government decided to put his report findings and boundary settlements on hold, believing his methodology and evidence in defining boundaries was "dangerously similar to that which the Saudis had put forward to bolster their claims against Abu Dhabi and the [Muscat] Sultanate."[101] At the conclusion of the Buraimi settlement, once Saud's troops were removed, the British Foreign Office took up Walker's border proposal once more and asked that he return to negotiate the border proposal for final approval of Trucial rulers. There was a renewed urgency as PDTC wanted to begin drilling in Sharjah in 1957, and the British Foreign Office wanted boundary definitions approved and finalized by the rulers beforehand.[102]

An Oil Urban Infrastructure

While the Buraimi incident would not be settled until 1955, by late 1954 the Trucial Levies were able to build an airstrip at the Oasis to secure British access to the area.[103] The airstrip provided the British Royal Air Force a base in case the conflict was to escalate. It would also establish a new kind of territorial claim through a new mode of infrastructural improvement as counterpoint to the traditional Buraimi wells. The airstrip served (along with its primary function) as one more territorial strategy deployed in hopes of defending British interests without military intervention. If engagement with tribal land knowledge systems was a "no-win" option for the British in securing their territorial interests vis-à-vis Saud, then the linear form and fixity of road corridors, ports, and expanded buildings would serve as the urban infrastructure to accommodate oil extraction, development, and modernity. Timothy Mitchell refers to these infrastructures as "objects of development," and for the Gulf sheikhs they had the same appeal in their promise of ruling legitimacy.[104] If they did not directly signify complete territorial sovereignty and control, then they at least pointed that way in domesticating what had, up until recently, been a competitive territory of ever-present violence. Coastal cities were the command centers for sheikh leadership with British backing, and regional urbanization infrastructure organized the territory to strengthen that central function for cities.[105] Thus, the political and spatial organization of territory is the appropriate scale to understand Gulf urbanization, and the role of an urbanization infrastructure, built mainly by British engineering companies, codified that political and spatial co-organization (Figure 4.3).[106]

Urban infrastructure, principally oil-related, in the form of fixed circulatory infrastructure such as roads, ports, creek dredging, and airstrips (or creek airplane landings, as was the case for Dubai) spread across the Gulf petroleumscape, beginning in Kuwait once oil

FIGURE 4.3 A 1952 California Texas Oil Company (Caltex) advertisement. Caltex was a joint venture between Socal and the Texas company in 1936. The declarative "From the Middle East...Serving Europe, Africa, Asia, Australasia" announces a new centrality for the region in the global petroleumscape (Chevron Corporation).

was discovered there in 1938.[107] The British were able to harness their influence with rulers of the region through recommended British economic advisors. The advisors helped rulers modernize their states, suggesting that they spend oil exploration stipends and later oil revenue on the kinds of infrastructure projects that British engineering companies in the region were offering.[108] British companies such as Sir William Halcrow's engineering firm moved much of their work to the Gulf region beginning in the mid-1950s, and they would remain in the region well after the British government officially left, having secured corporate interests to which they still felt historically entitled.[109]

The British withdrew from the region by 1971, but their legacy of oil infrastructure construction as a strategy for territorial claims continued into the nation-state period thereafter. A territory disagreement between Iraq and Kuwait was temporarily settled when the British ended their military protection of Kuwait in 1968, only to return in 1973.[110] Iraq deployed military troops to the Kuwaiti northern territories and tried to force a treaty that would give them the right to build "refineries, depots and tanks for the storing of oil and water, bridges, harbors, airports, and railway lines" in the territory without having to pay.[111] The incident was eventually settled through the Arab League, but it clearly demonstrates the use of oil infrastructure as a strategy to secure territorial claims, as the British did and as tribes did in their own "improvement" investment codes previously. The palimpsest of territory, territoriality, and sovereignty in the Gulf petroleumscape persists.

Conclusion

Claude Raffestin, discussing territoriality and the work of Henri Lefebvre, observed that "everyday life most often occurs in territorial morphologies that are not contemporaneous to the relations of which territoriality is woven."[112] For the Gulf petroleumscape, this both illuminates and reiterates the point that regional territorial production and reproduction, along with the respective management structures of territoriality, will likely continue beyond that future when oil dependency recedes. As a global petroleumscape case study, the Gulf illustrates the historical layering of spatial, material, and mapping practices. Within Gulf urban history, there are ongoing debates concerning how to properly gauge the causality of oil discovery in the region: Is oil the essential determinant of all socioeconomic, political, and cultural phenomena?[113] Is there a definitive before and after oil—or a cumulative, catalytic carryover of previous relationships amassed more slowly over time? In many ways, these regional stories are most convincingly told through oil discovery as a central narrative, as an alternative, material history woven into traditional nation-state formation narratives.[114] In the Gulf, the breadth and depth of regional oil centrality is proportional to the challenges it will face in transitioning to an after-oil period.

Notes

1 Rupert Hay, "The Persian Gulf States and Their Boundary Problems," *Geographical Journal* 120, no. 4 (1954): 433–43.
2 Ibid., 443.
3 Matthew H. Edney, *Mapping the Empire: The Geographical Construction of British India 1765–1843* (Chicago, IL: University of Chicago Press, 1997); Nelida Fuccaro, "Knowledge at the Service of the British Empire: The Gazetteer of the Persian Gulf, Oman and Central Arabia," in *Borders and the Changing Boundaries of Knowledge: Transactions 22*, eds. Inga Brandell, Marie Carlson, and Önver A. Çetrez (Istanbul: Swedish Research Institute in Istanbul, 2015), 17–34; Timothy

Mitchell, *Rule of Experts: Egypt, Techno-Politics, and Modernity* (Berkeley: University of California Press, 2002).

4 Hay, "Persian Gulf States," 443.

5 On December 2, 1971, seven of the Gulf States—Abu Dhabi, Ajman, Dubai, Fujairah, Ras al-Khaimah, Sharjah, and Umm al-Quwain—formed the United Arab Emirates (UAE) federation. Ras al-Khaimah officially joined two months later on February 11, 1972. See Shinhab M. A. Ghanem, *Industrialization in the United Arab Emirates* (Aldershot: Avebury, 1992).

6 Claude Raffestin, "Space, Territory, and Territoriality," *Environment and Planning D: Society & Space* 30 (2012): 121–41.

7 Map included in Hay, "Persian Gulf States."

8 Hassan H. Al-Alkim, "The United Arab Emirates and the Subregional Powers," in *A Century in Thirty Years: Shaykh Zayed and the United Arab Emirates*, ed. Joseph A. Kechichian (Washington, DC: Middle East Policy Council, 2000); Rosemarie Said Zahlan, "The Impact of the Early Oil Concessions in the Gulf States," in *The Gulf in the Early Twentieth Century: Foreign Institutions and Local Responses*, ed. R. I. Lawless (Durham: Centre for Middle Eastern & Islamic Studies, 1986).

9 Morton, Michael Quentin, *The Petroleum Gulf: The Discovery and Development of Offshore Oil in the United Arab Emirates* (Abu Dhabi: UAE National Archive, 2016).

10 Alex Boodrookas and Arang Keshavarzian, "The Forever Frontier of Urbanism: Historicizing Persian Gulf Cities," *International Journal of Urban and Regional Research* 43, no 1 (2018): 14–29.

11 Stephen J. Ramos, *Dubai Amplified: The Engineering of a Port Geography* (Burlington: Ashgate, 2010); Stephen J. Ramos, "An Historical Examination of Territory and Infrastructure in the Trucial States," in *Gateways to the World: Port Cities in the Persian Gulf*, ed. Mehran Kamrava (Oxford: Oxford University Press, 2016).

12 Zahlan, "Early Oil Concessions"; William Hale, "Introduction," in *The Gulf in the Early Twentieth Century: Foreign Institutions and Local Responses*, ed. R. I. Lawless (Durham: Centre for Middle Eastern & Islamic Studies, 1986), 3.

13 Toby Craig Jones, *Desert Kingdom: How Oil and Water Forged Modern Saudi Arabia* (Cambridge, MA: Harvard University Press, 2010); Timothy Mitchell, *Carbon Democracy: Political Power in the Age of Oil* (New York: Verso, 2011); Daniel Yergin, *The Prize: The Epic Quest for Oil, Money, and Power* (New York: Simon & Schuster, 1991).

14 Christopher M. Davidson, *After the Sheikhs: The Coming Collapse of the Gulf Monarchies* (Oxford: Oxford University Press, 2013), 42.

15 Gavin Bridge, "Global Production Networks and the Extractive Sector: Governing Resource-Based Development," *Journal of Economic Geography* 8 (2008): 389–419.

16 Ibid.; Carola Hein, "Oil Spaces: The Global Petroleumscape in the Rotterdam/The Hague Area," *Journal of Urban History* 44, no. 5 (September 2018): 887–929, https://doi.org/10.1177/0096144217752460.

17 Hein, "Oil Spaces," 887.

18 Imre Szeman, "System Failure: Oil, Futurity, and the Anticipation of Disaster," in *Energy Humanities: An Anthology*, eds. Imre Szeman and Dominic Boyer (Baltimore, MD: Johns Hopkins University Press, 2017).

19 Janet Abu-Lughod, *New York, Chicago, Los Angeles: America's Global Cities* (Minneapolis: University of Minnesota Press, 1999), 3; Boodrookas and Keshavarzian, "Forever Frontier," 6.

20 Szeman, "System Failure," 59.

21 Hein, "Oil Spaces"; Andre Sorensen, "Taking Path Dependence Seriously: An Historic Institutionalist Research Agenda in Planning History," *Planning Perspectives* 30, no. 1 (2015): 17–38.

22 Hein, "Oil Spaces," 888.

23 Joshua Barkan, *Corporate Sovereignty: Law and Government Under Capitalism* (Minneapolis: University of Minnesota Press, 2013), 12–13.

24 Roger Owen, "Globalization at the Beginning of the Twentieth and Twenty-First Centuries," in *Globalization and the Gulf*, eds. John Fox, Nada Mourtada Sabbah, and Mohammed Al-Mutawa (Oxford: Oxford University Press, 2006), 79.

25 Barkan's use of the term "global" is appropriate: "'Global' names an essential element in the way the potentiality of the past has been materialized in the present, along with the implicit spatiality structuring that process," *Corporate Sovereignty*, 14.

26 James Corner, "The Agency of Mapping: Speculation, Critique, and Invention," in *Mappings*, ed. Denis Cosgrove (London: Reaktion Books, 1999); Denis Cosgrove, ed. *Mappings* (London: Reaktion Books, 1999).

27 Mark Neocleous, "Off the Map: On Violence and Cartography," *European Journal of Social Theory* 6, no. 4 (2003): 409–25.
28 Edney, *Mapping the Empire*.
29 Robert D. Sack, *Human Territoriality: Its Theory and History* (Cambridge: Cambridge University Press, 1986).
30 Steward Elden, *The Birth of Territory* (Chicago: University of Chicago Press, 2013); Jean Gottmann, *The Significance of Territory* (Charlotte: University Press of Virginia, 1973); Henri Lefebvre, *The Production of Space*, trans. Donald Nicholson-Smith (Oxford: Blackwell, 1992 [1974]); Raffestin, "Space, Territory, and Territoriality."
31 Sack, *Human Territoriality*, 19.
32 Raffestin, "Space, Territory, and Territoriality," 126.
33 See also Hein, "Oil Spaces"; Lukasz Stanek, *Henri Lefebvre on Space: Architecture, Urban Research, and the Production of Theory* (Minneapolis: University of Minnesota Press, 2011).
34 Elden, *Birth of Territory*; Sack, *Human Territoriality*; E. W. Soja, "The Political Organization of Space," in *Commission on College Geography Resource Paper 8* (Washington, DC: Association of American Geographers, 1971), 1–54.
35 Peter Haggett, *Locational Analysis in Human Geography* (New York: St. Martin Press, 1966), 6.
36 Soja, "Political Organization of Space," 13.
37 Denis Cosgrove, *Geography and Vision: Seeing, Imagining and Representing the World* (London: I. B. Tauris, 2008); Derek Gregory, *Geographical Imaginations* (Oxford: Blackwell, 1994).
38 Soja, "Political Organization of Space," 10.
39 Pirouz Mojtahed-Zadeh, *Security and Territoriality in the Persian Gulf: A Maritime Political Geography* (Richmond, Surrey: Curzon Press, 1999).
40 John Agnew, "The Territorial Trap: The Geographical Assumptions of International Relations Theory," in *Mastering Space: Hegemony, Territory, and International Political Economy*, eds. John Agnew and Stuart Corbridge (London: Routledge, 1995), 67; Robert W. Cox, "Social Forces, States, and World Orders: Beyond International Relations Theory," *Millennium* 10, no. 2 (1981): 126–55.
41 Nelida Fuccaro, "Between Imara, Empire and Oil: Saudis in the Frontier Society of the Persian Gulf," in *Kingdom Without Borders: Saudi Arabia's Political, Religious, and Media Frontiers*, ed. Madawi Al-Rasheed (Oxford: Oxford University Press, 2008); Soja, "Political Organization of Space," 33.
42 Giorgio Agamben, *Homo Sacer: Sovereign Power and Bare Life* (Stanford: Stanford University Press, 1998).
43 Barkan elaborates, "Sovereignty is the paradoxical structure of power that emerges out of a complex of concrete practices that attempt to govern life by establishing and transgressing the boundaries of law...corporate sovereignty emerges from and is exercised over the border between the inside and outside of constituted political spheres," *Corporate Sovereignty*, 7–8.
44 Ibid., 13.
45 Boodrookas and Keshavarzian, "Forever Frontier"; Nelida Fuccaro, *Histories of City and State in the Persian Gulf: Manama Since 1800* (Cambridge: Cambridge University Press, 2009).
46 J. C. Wilkinson, "Nomadic Territory as a Factor in Defining Arabian Boundaries," in *The Transformation of Nomadic Society in the Arab East*, eds. Martha Mundy and Basim Musallam (Cambridge: Cambridge University Press, 2000).
47 G. Balfour Paul, *The End of Empire in the Middle East: Britain's Relinquishment of Power in Her Last Three Arab Dependencies* (Cambridge: Cambridge University Press, 1991), 113.
48 Dubai's oil discoveries were located on its offshore Fateh Fields, located in the Gulf, 60 miles off the Dubai coast. See Fuccaro, "Between Imara, Empire and Oil"; Morton, *The Petroleum Gulf*; Ramos, *Dubai Amplified*.
49 Zahlan, "Early Oil Concessions," 69.
50 Christopher M. Davidson, *The United Arab Emirates: A Study in Survival* (Boulder, CO: Lynne Rienner, 2005); Davidson, *After the Sheikhs*.
51 See discussion of state formation, urbanization, and citizenship in relation to oil in Nelida Fuccaro, "Pearl Towns and Early Oil Cities: Migration and Integration in the Arab Coast of the Persian Gulf," in *Migration and the Making of Urban Modernity in the Ottoman Empire and Beyond*, eds. Ulrike Freitag, Malte Furhmann, Nora Lafi, and Florian Riedler (London: Routledge, 2011), 99–116.

52 Antoine Picon, "'What Has Happened to Territory?' Special Issue *Territory: Architecture Beyond Environment*," *Architectural Design* 80, no. 3 (2010): 94–99.
53 Edney, *Mapping the Empire*.
54 Benedict Anderson, *Imagined Communities: Reflections on the Origin and Spread of Nationalism* (New York: Verso, 1983); Edney, *Mapping the Empire*.
55 Edney, *Mapping the Empire*, 36.
56 Ibid.; Christopher A. Bayly, *Empire and Information: Intelligence Gathering and Social Communication in India, 1780–1870* (Cambridge: Cambridge University Press, 2000); Felix Driver, *Geography Militant: Cultures of Exploration and Empire* (Oxford: Blackwell, 2001).
57 Richard F. Nyrop, *Area Handbook for the Persian Gulf States* (Washington, DC: American University, 1977).
58 Fauke Heard-Bey, *From Trucial States to United Arab Emirates: A Society in Transition*(London, New York: Longman, 1982).
59 James Onley, *The Arabian Frontier of the British Raj: Merchants, Rulers, and the British in the Nineteenth Century Gulf* (Oxford: Oxford University Press, 2007).
60 Hay, "Persian Gulf States."
61 Fatma Al-Sayegh, "Merchants' Role in a Changing Society: The Case of Dubai, 1900–90," *Middle Eastern Studies* 34, no. 1 (1998): 87–102.
62 Husain M. Al-Baharna, *The Legal Status of the Arabian Gulf State: A Study of Their Treaty Relations and Their International Problems* (Manchester: Manchester University Press, 1968); J. B. Kelly, "The Legal and Historical Basis of the British Position in the Persian Gulf," in *St. Anthony's Papers: No. 4; Middle Eastern Affairs*, Vol. 1 (London: Chatto & Windus, 1958); James Onley, "The Politics of Protection: The Arabian Gulf Rulers and the Pax Britanica in the Nineteenth Century," *Liwa: Journal of the National Center for Documentation and Research* 1, no. 1 (2009): 25–46; Malcom C. Peck, *The United Arab Emirates: A Venture in Unity* (Boulder: Westview Press, 1986); David B. Roberts, "The Consequences of the Exclusive Treaties: A British View," in *The Arab Gulf and the West*, ed. Brian R. Prindham (London: Croom Helm, 1985).
63 Mitchell, *Carbon Democracy*; Yergin, *The Prize*.
64 Soja, "The Political Organization of Space."
65 Nelida Fuccaro, "The Making of Gulf Ports Before Oil," *Liwa: Journal of the National Center for Documentation and Research* 2, no. 3 (2010): 19–32; Julian Walker, "Social and Economic Developments in Trucial Oman in the First Half of the Twentieth Century," *Liwa: Journal of the National Center for Documentation and Research* 2, no. 3 (2010): 33–47.
66 Nelida Fuccaro, *Histories of City and State in the Persian Gulf: Manama Since 1800* (Cambridge: Cambridge University Press, 2009).
67 Sulayman Khalaf, "The Evolution of the Gulf City Type, Oil, and Globalization," in *Globalization and the Gulf*, eds. John Fox, Nada Mourtada Sabbah, and Mohammed Al-Mutawa (Oxford: Oxford University Press, 2006); Onley, "Politics of Protection"; Henry Rosenfeld, "The Social Composition of the Military in the Process of State Formation in the Arabian Desert," *Journal of the Royal Anthropological Institute of Great Britain and Ireland* 95 (1965): 75–86; Richard N. Schofield and Gerald Blake, *Arabian Boundary Disputes on the Arabian Peninsula* (Cambridge: Archive Editions, 1988), 25; Rosemarie Said Zahlan, *The Origins of the United Arab Emirates: A Political and Social History of the Trucial States* (New York: St. Martin's Press, 1978).
68 Zahlan, *United Arab Emirates*, 6.
69 J. P. Tripp, "Political Agency, Trucial States. February 15, 1957 No. 6 Dispatch to Sir Bernard Burrows," in *Arabian Boundary Disputes on the Arabian Peninsula*, Vol. 22, ed. Richard N. Schofield (Cambridge: Archive Editions, 1991), 77.
70 J. C. Wilkinson, "Britain's Role in Boundary Drawing in Arabia: A Synopsis," in *Territorial Foundations of the Gulf*, ed. Richard N. Schofield (New York: St. Mark's Press, 1994).
71 Wilkinson, "Nomadic Territory," 45.
72 Wilkinson, "Britain's Role," 97.
73 Onley, "Politics of Protection," 29; Peter Lienhardt, "The Authority of Shaykhs in the Gulf: An Essay in Nineteenth-Century History," *Arabian Studies* 2 (1975): 61–75; James Onley and Sulayman Khalaf, "Shaikhly Authority in the Pre-Oil Gulf: An Historical–Anthropological Study," *History and Anthropology* 17, no. 3 (2006): 189–208.
74 Onley, "Politics of Protection."
75 Ramos, "Historical Examination."

76 Fuccaro, "Between Imara, Empire and Oil."
77 Jones, *Desert Kingdom*.
78 J. B. Kelly, "The Buraimi Oasis Dispute," *International Affairs*, 32, no. 3 (1956): 318–26.
79 Michael Quentin Morton, "The Buraimi Affair: Oil Prospecting and Drawing the Frontiers of Saudi Arabia," *Asian Affairs* 46, no. 1 (2015), 1–17.
80 Ibrahim, Ibrahim, "Sovereign States and Borders in the Gulf Region: A Historical Perspective," in *The Gulf Crisis: Background and Consequences*, ed. Ibrahim Ibrahim (Washington, DC: Center for Contemporary Arab Studies, 1992); Krista E. Wiegand, "Resolution of Border Disputes in the Arabian Gulf," *Journal of Territorial and Maritime Studies* 1, no. 1 (2014): 33–48.
81 Michael Quentin Morton, *Buraimi: The Struggle for Power, Influence and Oil in Arabia* (London, New York: I. B. Tauris, 2013).
82 Richard N. Schofield, "Negotiating the Saudi-Yemeni International Boundary," Lecture delivered to the British Yemeni Society, March 31, 1999.
83 Balfour Paul, *End of Empire*.
84 Morton, *Buraimi*; Wiegand, "Resolution of Border Disputes."
85 Jones, *Desert Kingdom*.
86 Mitchell, *Carbon Democracy*.
87 Balfour Paul, *End of Empire*; Wilkinson, "Nomadic Territory"; J. B. Kelly, *Arabia, the Gulf and the West*, (London: George Weidenfeld & Nicolson, 1980).
88 Wilkinson, "Nomadic Territory."
89 R. Seddiq, "Border Disputes on the Arabian Peninsula," in *Policy Watch # 525* (Washington, DC: Washington Institute, 2001); Wiegand, "Resolution of Border Disputes"; Wilkinson, "Nomadic Territory."
90 Julian Walker recounts that their British camp almost comes in conflict with an ARAMCO delegation near Buraimi in what he considered Abu Dhabi territory. Julian Walker, *Tyro on the Trucial Coast* (Durham: Memoir Club, 1999), 103.
91 Gwenn Okruhlik and Patrick J. Conge, "The Politics of Border Disputes on the Arabian Peninsula," *International Journal (Toronto, Ont.)* 54, no. 2 (1999): 230–48.
92 G. Balfour Paul, *Bagpipes in Babylon: A Lifetime in the Arab World and Beyond* (London: I.B. Tauris & Co, 2006).
93 Walker, *Tyro on the Trucial Coast*.
94 Hay, "Persian Gulf States"; Ali Mohammed Khalifa, "The United Arab Emirates: Unity in Fragmentation—A Study in Ministate Integration in a Complex Setting, 1968–1976" (PhD diss., University of California, Santa Barbara, 1978).
95 Walker, *Tyro on the Trucial Coast*, 109; Tahira Yaqoob, "A Return to the Country He Drew," *The National* (November 25, 2008), https://www.thenational.ae/uae/a-return-to-the-country-he-drew-1.484490
96 Zahlan, *United Arab Emirates*, 148; George Joffé, "Concepts of Sovereignty in the Gulf Region," in *Territorial Foundations of the Gulf*, ed. Richard N. Schofield (New York: St. Mark's Press, 1994).
97 Fuccaro, "Knowledge at the Service of the British Empire"; Joffé, "Concepts of Sovereignty.".
98 Julian Walker, "Practical Problems of Boundary Delimitation in Arabia: The Case of the United Arab Emirates," in *Territorial Foundations of the Gulf*, ed. Richard N. Schofield (New York: St. Mark's Press, 1994), 114.
99 Ibid., 116.
100 Edney, *Mapping the Empire*; Fuccaro, "Knowledge at the Service of the British Empire"; Mitchell, *Rule of Experts*.
101 Walker, *Tyro on the Trucial Coast*, 121.
102 Ibid.
103 Ibid., 98.
104 Stephen Graham and Simon Marvin, *Splintering Urbanism: Networked Infrastructures, Technological Mobilities and the Urban Condition* (London: Routledge, 2001); Jones, *Desert Kingdom*; Brian Larkin, *Signal and Noise: Media, Infrastructure, and Urban Culture in Nigeria* (London: Duke University Press, 2008); Mitchell, *Rule of Experts*, 201.
105 Michael Mann, "The Autonomous Power of the State: Its Origins, Mechanisms and Results," *Archives Européennes de Sociologie* 25 (1984): 185–213.
106 Ramos, "Historical Examination."

107 See chapters by Fuccaro and Hindelang, this volume; Owen, "Cities of the Persian Gulf."
108 Balfour Paul, *End of Empire*; Antoine Benjamin Zahalan, *The Arab Construction Industry* (New York: St. Martin's Press, 1984).
109 Ramos, *Dubai Amplified.*
110 Ibrahim, "Sovereign States."
111 Kelly, *Arabia, the Gulf and the West*, 283.
112 Raffestin, "Space, Territory, and Territoriality," 129
113 Boodrookas and Keshavarzian, "Forever Frontier"; Fuccaro, "Knowledge at the Service of the British Empire"; Khalaf, "The Evolution of the Gulf City Type."
114 William Cronin, *Nature's Metropolis Chicago and the Great West* (New York: W.W. Norton & Co., 1992).

5

BETWEEN VISIBLE AND INVISIBLE

ENI and the Building of the African Petroleumscape

Giulia Scotto

In 1955, the Italian national hydrocarbon agency Ente Nazionale Idrocarburi (ENI) launched what the coordinator of ENI's activities in North and West Africa later described as its "grande disegno africano," the company's exploratory and commercial expansion into the African continent.[1] Despite Italy's colonial past, ENI managed to establish itself as a seemingly neutral actor and, within ten years, it entered the oil market of twenty-five newly founded African countries—including the former Italian colonies and many newly independent countries of the sub-Saharan region. Through diplomacy, advertising, and the construction of material artifacts such as refineries, pipelines, motels, and gas stations, ENI (or "ENI" as it was renamed after privatization in the 1990s) developed an incremental network of more and less visible lines and nodes through which oil was extracted, transported, refined, and commercialized.

In order to understand the impact of the oil industry on spatial and social realities, we need to consider, as suggested by this volume, the entire petroleumscape: "the layered physical and social landscape" created by petrocapitalism.[2] The Lefebvrian triad of represented, material, and experienced spaces linked by the notion of the petroleumscape is particularly useful as a way of questioning the relationship between the spatial and the propagandistic operations of ENI and the lived experience of the infrastructural artifacts built by its subsidiaries.[3] Adopting this holistic approach, this chapter interrogates actors, typologies, spaces, and scales often neglected by architectural historians, and combines the study of the artifacts' material presence with an analysis of how they have been represented and appropriated.

Oil infrastructures have played a key role in defining the dynamics and aesthetics of petrocapitalism. In *Extrastatecraft: The Power of Infrastructure Space*, Keller Easterling compares infrastructures to the medium of Marshall McLuhan's famous expression: "the medium is the message." Following McLuhan's suggestion, this chapter considers ENI's oil infrastructures as both messages and media in order to reveal what they are "doing" beyond what they are "saying."[4] In other words, it reads the difference between the "declared intent" and the "underlying disposition" of gas stations, refineries, and pipelines disseminated by ENI across the African continent.[5]

DOI: 10.4324/9780367816049-7

In postcolonial Africa, gas stations and motels might convey a sense of freedom, refineries might evoke a narrative of national self-reliance, but what was ENI's intention when they designed and constructed them? Were they, as claimed by ENI, decolonizing tools enhancing energy independence or rather instruments to establish a neocolonial petrol-based empire? These questions have no simple answers, but by examining spatial, architectural, and aesthetic aspects of ENI's projects, their "dispositions," and the multiple ways in which they were communicated, this chapter explores the unspoken territorial and social vision that the Italian company pursued on the African continent and the consequences of their materialization. This chapter offers an alternative to the official narrative of the pioneering and developmental mission proposed by ENI in Italy and abroad. It first describes ENI's role in the Italian postwar context. It then traces ENI's African expansion through three projects, representing the different phases of ENI's oil supply chain: production, circulation, and commercialization. The three projects are the Tema refinery in Ghana, the Tanzania-Zambia Mafuta (TAZAMA) oil pipeline linking Zambia and Tanzania, and Agip (originally the acronym for Agenzia Generale Italiana Petroli) service stations and motels (especially the Agip Motel of Dar es Salaam). In order to understand the role played by these infrastructures within ENI's broader expansionist project, I consider their material and spatial impact, but also how they have been represented and narrated, mostly by ENI and its local partners—the African governments—as well as by African artists and citizens.

The material presence of oil infrastructures ranges from the invisible to the hyper-visible, from the mundane to the spectacular. As we will see, these seemingly neutral objects can play an important role in an oil company's geopolitical and territorial strategies and in the definition of urban and territorial patterns. They can be strategically concealed, located in remote sites, buried underground, enclosed in inaccessible industrial areas, advertised in postcards and stamps, or they can be too ubiquitous to notice. Philosopher Jacques Ranciere argues that the distribution or partition of the sensible, far from being merely phenomenological, is an inherently political gesture.[6] Exploring how "(in)visibility is mobilized and why"[7] allows us to read the complex system that goes from refinery to filling station and beyond as a cultural, political, and historical phenomenon. ENI's "disegno," I argue, should not be understood as an assemblage of utilitarian and neutral technological artifacts but as the product of power ideologies entangled with historical, geographic, and social circumstances that radically affected lived experiences and geopolitics.

ENI and Italy

In 1953, the Italian government established ENI to support Italy's economic development through the production and supply of oil and gas. Within a few years, under the guidance of its first president Enrico Mattei, the public holding evolved into a powerful entity that has been described as "a parallel state" capable of influencing Italy's foreign politics and diplomacy.[8] During the 1960s, thanks to the support of the US through the Marshall Plan, Italy underwent a phase of transformation: a predominantly rural nation was converted to a proto-industrialized and consumption-oriented society. ENI did not restrict its activities to extracting and distributing gas and oil derivatives: it soon expanded its scope into the petrochemical industry, engineering, and construction while Agip, the group's commercial branch, began promoting an oil-consuming lifestyle based on individual mobility.[9]

The image of a seductive and motorized lifestyle proposed by Agip was carefully created through tailored advertising campaigns, including television shows, documentaries, and sponsored appearances of Agip gas stations in Italian movies.[10] The company adopted the ideal of individual freedom and automobility to support narratives of national independence. Stopping to fill your car at the Agip "was much more than choosing a brand; it meant choosing Italy and the company that represented the nation."[11]

The Agip logo, the six-legged fire-spitting dog, became a familiar presence across the Italian landscape.[12] The mythological animal combined the power of the fire-spitting

FIGURE 5.1 ENI's industrial and commercial facilities distribution in Italy in 1959. The map shows the locations of multiple refineries, petrochemical and chemical plants, mechanical factories, power plants, hydrocarbon deposits, gas bottling plants, offices, laboratories, and gas stations. *Source*: ENI Ente Nazionale Idrocarburi, Relazioni e Bilancio al 30 Aprile 1960, ENI Historical Archive.

dragon and the devotion of the dog to symbolize the power of Agip gasoline and its reliability. Celebrated by Agip's motto, the "six-legged dog, loyal friend of the four-wheeled man," the yellow sign accompanied Italian drivers across the pENInsula making them feel at home on the go (Figures 5.1–5.3).[13]

FIGURE 5.2 An Agip gas station along the Italian Highway. *Source*: ENI Historical Archive.

FIGURE 5.3 Agip Motel in Montalto di Castro, Italy. *Source*: ENI Historical Archive.

Architecture was one of the main instruments of ENI's propaganda. During the 1950s, Enrico Mattei invited the most prominent Italian architects and designers to envision its headquarters, factories, company towns, and filling stations.[14] The goal of the newly designed gas stations, usually associated with a bar, a restaurant, and a shop for Agip products, was to be recognizable and to invite the customers to stop, relax, and consume. The elegant and sober corporate design transformed the mundane activity of fueling the car into an aesthetic experience.

To complement the distribution network, Agip implemented Italy's first chain of motels. These Italian versions of an American typology, reinterpreted as small suburban motor-hotels, were located along the highways and main roads in proximity to urban areas and tourist attractions and offered relatively cheap accommodation to motorists, tourists, and truck drivers. In 1962, the *Architectural Review* described Agip's corporate image as the "public face of an Italian economic miracle." The article compared Agip's house style to the work of renowned designers hired by the Olivetti and Braun companies and explained its success in the masterly play of subtle variations and alternative compositions of basic elements. The author concludes that "entering a highly competitive field later than its rivals, Agip has established itself by offering a better service than the other companies, better in the double sense of being more comprehensive and better presented" (Figures 5.2 and 5.3).[15]

The "Grande Disegno Africano"

ENI's initial attempts to find oil within Italy were partially successful, but not sufficient to satisfy the demand resulting from the rapid industrialization of the Po Plain (see Geroldi and Pessina, this volume). Frustrated by the poor hydrocarbon discoveries on the Italian territory and by the constraints on crude oil import conditions imposed by the US and the international oil cartel of the so-called "Seven Sisters,"[16] in 1955 ENI's president Enrico Mattei launched what was later described as the "grande disegno africano," which can be translated as the "big African project" or "big African scheme." Through its African expansion, ENI expected to discover rich oil deposits to supply Italian factories and enter new and rapidly growing markets for Agip's refined products.

Between 1914 and 1943, Italy had a substantial colonial presence in Africa with colonies in Eritrea, Somalia, Tunisia, and Libya. After World War II, it hoped to regain control of its former possession. The national neocolonial ambition was precluded by the official denial of the UN, which forced Italy to adjust its geopolitical attitude.[17] The denial of the UN led to the strategic adoption of an "anticolonial" position in support of African leaders' struggle for independence and against their former colonial masters. In those years, after decades of colonialism and centuries of exploitation, independence movements were reaching the point of negotiating their formal independence from France, Britain, and Belgium. ENI's parallel diplomacy understood the decolonization of Africa as a moment of possibility in which global power relations and alliances could be redefined, but also as the right moment for Italy to regain access to Africa's wealth and emancipate itself from the international oil lobby.[18] Italy's access point to Africa's postcolonial politics was the notion of "development," a Western-forged notion that, since the late colonial era, became the primary aspiration of independent African states. The importance of fossil fuels to activate—both literally and metaphorically—the engine of development was clear to both African leaders and to ENI.

In the draft of a speech to give in Tunis in 1961, ENI's president Enrico Mattei stated:

> I am here to (...) help you in the fight against underdevelopment.
> I do not only believe in decolonization for moral reasons of human dignity but for economic and productivity reasons too.
> Before all this, I had to decolonize Italy because many sectors of the Italian economy were colonized (...). Colonialism is not only political; it is first and foremost economic; the colonial condition is given when the infrastructure to transform raw materials are missing and when the demand and supply of a fundamental resource are defined by a hegemonic power (...)
> (...) I offer parity, co-management, the education of a technological elite that will make you economic subject, rather than the object or passive receiver of a foreign initiative.[19]

ENI's president underlined the economic and material aspect of imperial domination and depicted both Italy and newly independent African countries as victims of the economic colonialism of the oil cartel. By providing "missing infrastructures," such as refineries, pipelines, and gas stations, he offered to help break this energy dependence and spur self-reliance (Figure 5.4). All this, with better conditions than oil multinationals would offer, thanks to the support of the Italian government.

Playing the dual role of the caring national agency and the profit-oriented multinational, ENI's operations in Africa fluctuated between foreign investment and cooperation aid. This commercial and diplomatic strategy granted ENI a friendly image and paved its economic penetration with paternalistic good intentions. What ENI offered was actually a paradoxical "oil-based energy independence" sold as a key step toward development and modernization.

The vision of the "grande disegno africano" materialized into an incremental network of pipes and nodes that would guarantee ENI's presence in a growing number of African countries. Despite ENI's "anticolonial" stance, its first African agreements were concluded with former Italian colonies: with Eritrea in 1956 and with Ethiopia and Somalia in 1958. In 1957, ENI signed its first contract with Egypt and, in 1959, in partnership with local governments, it established Agip Casablanca and Agip Tunis. Contracts followed in the next few years to form Agip Congo, Agip Ghana, Agip Tanzania, Agip Zambia, and many others. Within ten years, ENI and its engineering subsidiary (Societa Nazionale Metanodotti [SNAM] Progetti) built six refineries (in Morocco, Ghana, Tunisia, Tanzania, Congo, and Zambia) and fifteen coastal deposits from which Agip local branches were distributing gas and gasoline through over 1,300 gas stations in over twenty-five countries.[20] The administration of ENI in Africa was organized into four geographic sectors: North Africa was managed from ENI's headquarters in Rome, East Africa from Nairobi, West Africa from Accra, and Central Africa from Brazzaville.[21] Here, and in many other capital cities of newly founded African states, ENI acquired prestigious representative buildings in prominent positions to host its regional and national headquarters. Often, ENI shared these spaces with other Italian companies (mainly FIAT, the Italian car manufacturer), Italian schools, and cultural centers. In Kampala, Uganda, ENI shared space with the Italian embassy. This administrative structure was supported by a more widespread network of local Agip offices, where ENI's employees studied the local market, made geological surveys, and continuously planned the growth of ENI's empire. This diffuse territorial presence made ENI's "parallel diplomacy" more present and locally engaged than Italy's official diplomacy.[22]

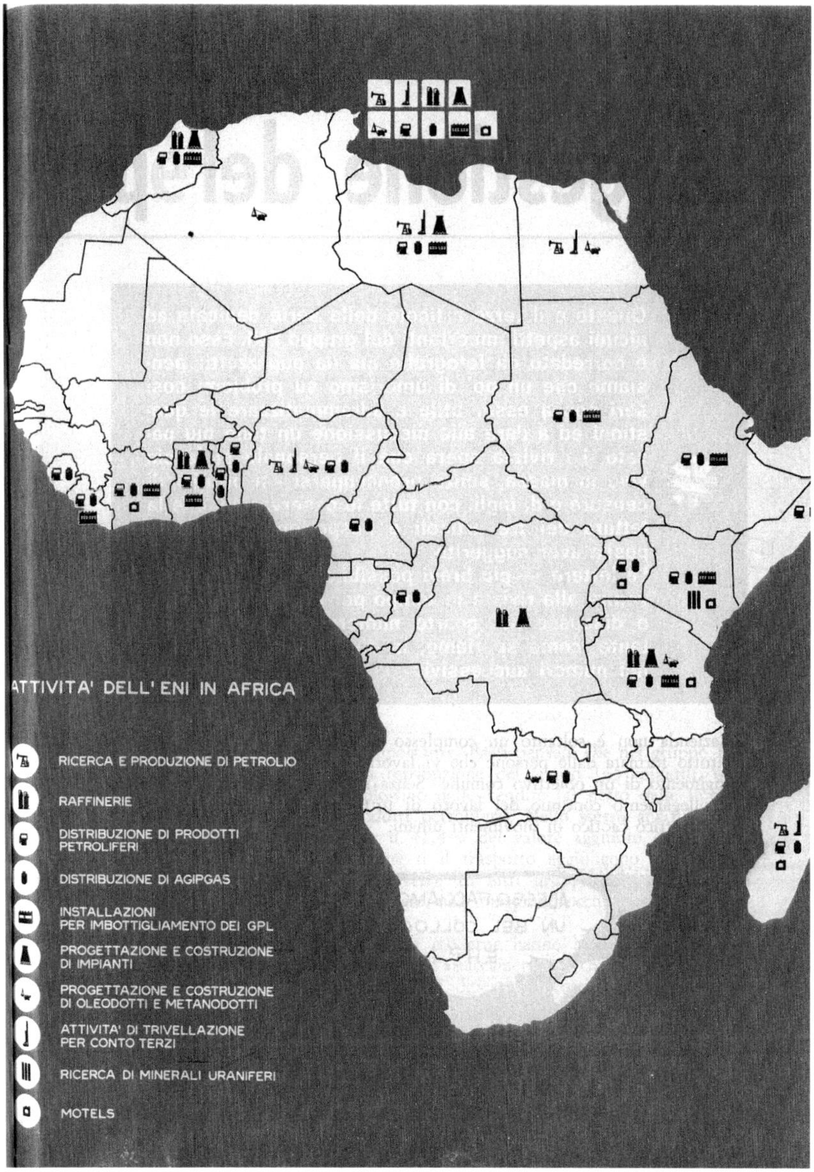

FIGURE 5.4 ENI's activity in Africa in 1968. The map shows the countries where ENI's subsidiaries were active in extracting, refining, and distributing petroleum products, in searching for uranium, and in the engineering and construction of pipelines and other industrial plants. ENI n.3, 1968. *Source*: ENI Historical Archive.

The Tema Refinery, Ghana, 1960–1963

In 1957, Ghana became the first country in sub-Saharan Africa to achieve political independence. Kwame Nkrumah, Ghana's first president, was a committed leftist, anti-imperialist, and Pan-Africanist, but also a skillful diplomat able, in the context of the cold war, to gain foreign aid and investments from both the capitalist and communist blocs.[23] He firmly believed that economic independence would be achieved through the development of the industrial sector and radical social modernization. His ambition took shape in the Volta River Project, a territorial infrastructural scheme that consisted of a huge hydroelectric dam and a new industrial area served by an international maritime port in Tema, a former fishing village located a few kilometers from the capital city, Accra. Nkrumah's developmental narratives depicted the Akosombo dam and the Tema port as a step toward self-reliance: they were expected to satisfy the energy requirements of a flourishing industrial sector and provide the basic infrastructures to independently process Ghana's natural resources, which in the colonial era had been exported as raw material and manufactured abroad. The Tema industrial area hosted factories for the processing of timber, bauxite, cocoa, and crude oil (Figure 5.5).

The rapid development of postcolonial Ghana inspired many African leaders and showcased Africa's development potential. Envisioning multiple benefits of participating in Ghana's economic growth, Enrico Mattei decided to start investing in the country. In 1960, ENI registered Agip's local branch as the Ghanaian Italian Petroleum Company (GHAIP), and in the following year, proposed to President Nkrumah that ENI's engineering department build Ghana's first refinery. After ENI's proposal was made public, other oil multinationals rose up against this initiative. Mobil, the American oil multinational, made a counter offer, but after evaluating both proposals, the Ghanaian government assigned the construction to SNAM Progetti, ENI's engineering subsidiary, which offered better conditions and more space to local actors. The agreement between ENI and the Ghanaian authorities stated that GHAIP would run and own the refinery for ten years, after which the Ghanaian government would get 50 percent of its shares. To ensure control over the plant's operation, the Ghanaian government would own the pipeline connecting the tanker harbor to the refinery, through which crude oil was delivered.[24]

One of ENI's newsreels recorded the opening ceremony of the refinery, which was inaugurated in 1963 in the presence of the Ghanaian and Italian authorities, local chiefs (owners of the land on which the plant stands), and the first group of GHAIP workers.[25] During his speech, President Nkrumah declared: "Since we reached independence we wanted Ghana to be a modern industrialized state. Only in this way could we survive as an independent nation. The oil-product made in Ghana will reduce our dependence on the import of energy resources."[26] Thanks to the new plant, Ghana could negotiate and buy crude directly from oil-producing countries and independent producers (like ENI) instead of importing refined products at the fixed prices imposed by the oil cartel. The shared struggle for energy independence united the interests of ENI, Ghana, and many newly independent nations. During the 1960s, ENI and other oil companies successfully marketed refineries as a step toward energy self-reliance, to the point that in 1963 thirteen new plants were under construction across the continent.[27]

The GHAIP facilities consisted of two parts: the campus and the plant itself. The most representative space of the campus, situated along the main access road, was a green and tidy compound where a composition of low, white modernist buildings hosted the offices, canteen, clinic, and laboratory. Behind the campus, the plant was organized as an ordered labyrinth of pipes, conduits, machinery, and tanks from which emerged the distillation tower.

The plant started operating as a tolling refinery, processing crude oil for third parties who paid a fee for the service. The oil companies could ship their tankers filled with crude to the port of Tema, where the oil would enter the refinery through the state-controlled

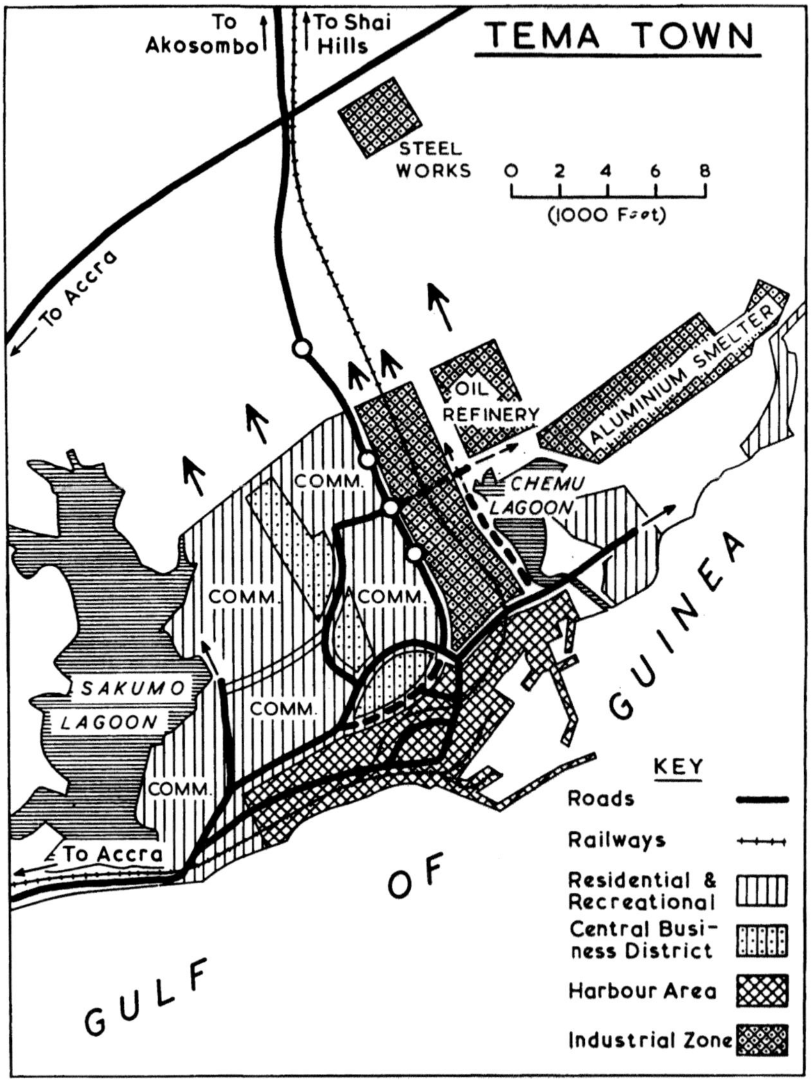

FIGURE 5.5 Tema, new town and industrial area master plan. *Source*: David Hilling, "Tema—The Geography of a New Port," *Geography* 51, no. 2 (1966): 111–25.

pipeline, and once processed, the by-products could be stored or loaded on gas or fuel trucks and distributed in Ghana and in neighboring countries.[28] Like other ambitious facilities built during the Nkrumah era, the refinery's capacity was based on expected domestic demands for fossil fuel that did not grow as rapidly as anticipated. Consequently, the refinery only operated intermittently.[29]

After the contract was signed, in 1961, ENI started recruiting a local workforce. The selection process led to the employment of around one hundred Ghanaians with different levels of education and experience. Before the opening of the refinery, the technical staff with responsible positions were sent to Italy for training in ENI's factories, refineries, and laboratories.

GHAIP workers were offered a series of benefits: they and their close families could enjoy the company's medical facilities, they were offered favorable loans to buy Italian cars, and they had the right to a monthly allotment of free gasoline. GHAIP employees were also assigned flats in the newly developed settlement of Tema, just a few hundred meters from the plant.[30]

The master plan for the new town was developed by the Greek firm Doxiadis Associates to accommodate the workers of the industrial areas and port of Tema (Figures 5.6 and 5.7). Interlocking communities or neighborhoods were divided into low-, middle-, and high-income levels. The different communities were characterized by different densities and housing typologies, all based on the needs of the nuclear family supported by one—usually male—wage laborer employed at the industrial park.[31] Doxiadis's approach to planning complemented Nkrumah's desire to modernize all aspects of Ghanaian society and ENI's desire to transform citizens into consumers.

During the construction of the refinery and the first decade of joint operation, ENI's geologists were searching for potential deposits in Ghana and offshore, while GHAIP consolidated a growing distribution network. Even though the Italo-Ghanaian company entered a market sector where other oil multinationals were already established, it benefited from the positive image gained by constructing the refinery. Through its paternalistic politics and the advantageous conditions of its contracts, ENI managed to convey the image of a friendly public company of a Western but neutral country that supported Ghana's struggle for independence.[32] In reality, large-scale infrastructures like the GHAIP refinery operated as modern Trojan horses, allowing the company to access new territories while the plant's constant requirement of maintenance and improvement generated a new dependence on

FIGURE 5.6 Ghana's President Kwame Nkrumah and ENI's representatives at the opening ceremony of the Tema refinery. *Source*: ENI Historical Archive.

FIGURE 5.7 A page from the Ghanaian government publication *Accra-Tema in Pictures*. Accra, Ghana: Advanced Press Ltd., 1969. *Source*: Balme Library, University of Ghana.

ENI's skills and spare parts, as the long-lasting cooperation between ENI and the Ghanaian government demonstrated.

Another tool for ENI's successful penetration of the Ghanaian market was advertising. ENI's activity in Ghana was celebrated in 1974 with the company's publication of a calendar featuring a series of pictures taken by James Barnor, a Ghanaian photographer and photojournalist. Barnor collaborated with various Ghanaian and British newspapers and with the influential South African anti-apartheid magazine *Drum*. For the calendar, which was distributed to GHAIP clients, Barnor portrayed elegantly dressed women surrealistically posing in front of Agip barrels and gas tankers. The minimal compositions of the black-and-white images of shiny clean oil barrels are highly aestheticized. Throughout this set of pictures, oil is never visible but always present. Taking a synecdochic approach, Barnor creatively photographed oil without directly showing it, while the emancipated, fashionable female models suggest oil's broader "economic and social benefits."[33] Barnor depicted the promise of wealth and modernity that access to oil was expected to make available to Ghanaian citizens and showed that oil not only embodied Nkrumah's vision of industrialization and self-reliance, it also carried expectations of individual freedom. Like Ross Barrett and Daniel Worden observe in their volume *Oil Culture*, such representations fetishize oil's value as a magical substance able to bring development and freedom while detaching it from dirt and labor (the latter appears in one of the pictures as a blurry presence in the figure of a tanker driver). Images like the ones Barnor produced for Agip, particularly popular in the 1960s, reflected the shared enthusiasm for oil power and played a key role in the expansion of global "petrocapitalism" and its aesthetic (Figures 5.8 and 5.9).[34]

FIGURE 5.8 Agip Ghana Calendar, 1974. Model with oil drums. Picture: James Barnor. Courtesy of the artist and October Gallery, London.

FIGURE 5.9 Agip Ghana Calendar, 1974. Model with Tank and Driver. Picture: James Barnor. Courtesy of the artist and October Gallery, London.

The "Freedom Pipeline": Zambia and Tanzania, 1966–1968

The TAZAMA[35] Pipeline is probably the most ambitious of ENI's African projects, not only because of its size (at 1,700 km, it was the longest African oil pipeline at the time), but also for its geopolitical relevance (Figures 5.10 and 5.11). In 1966, the Zambian government invited ENI—which had gained a reputation as a reliable oil multinational by constructing refineries in Ghana, Morocco, Tanzania, and Tunisia—to present a proposal for the construction of a pipeline connecting the Zambian Copperbelt with the Tanzanian port of Dar es Salaam.

Zambia, known as Northern Rhodesia at the time of British rule, was facing an oil crisis as a consequence of the embargo the UN imposed on Southern Rhodesia after its white

Oil goes to Mikumi by rail then the African lorry driver takes over.

Americans fly the oil in Boeings to Elizabethville from there it is taken by road to Zambia.

PICTURES BY MOHAMED AMIN

At Mikumi you could walk on oil drums from one end of the town to the other.

GETTING THE OIL TO ZAMBIA

They all said it would be impossible to get enough oil through to Zambia. They said Zambia would be brought to its knees before Ian Smith cracked. They said it was a hopeless venture. But the efforts of all concerned in the Great Zambian Oil Lift have proved the pessimists wrong. The R.A.F. pilots, the tough Americans and Canadians the African dockers, mechanics and railwaymen, and thousands of African lorry drivers have got so much oil through that Zambia has gradually eased rationing and is now stockpiling the stuff !

FIGURE 5.10 *Drum East Africa* article about the Zambian oil crisis and the oil lift, March 1966. *Source*: British Library.

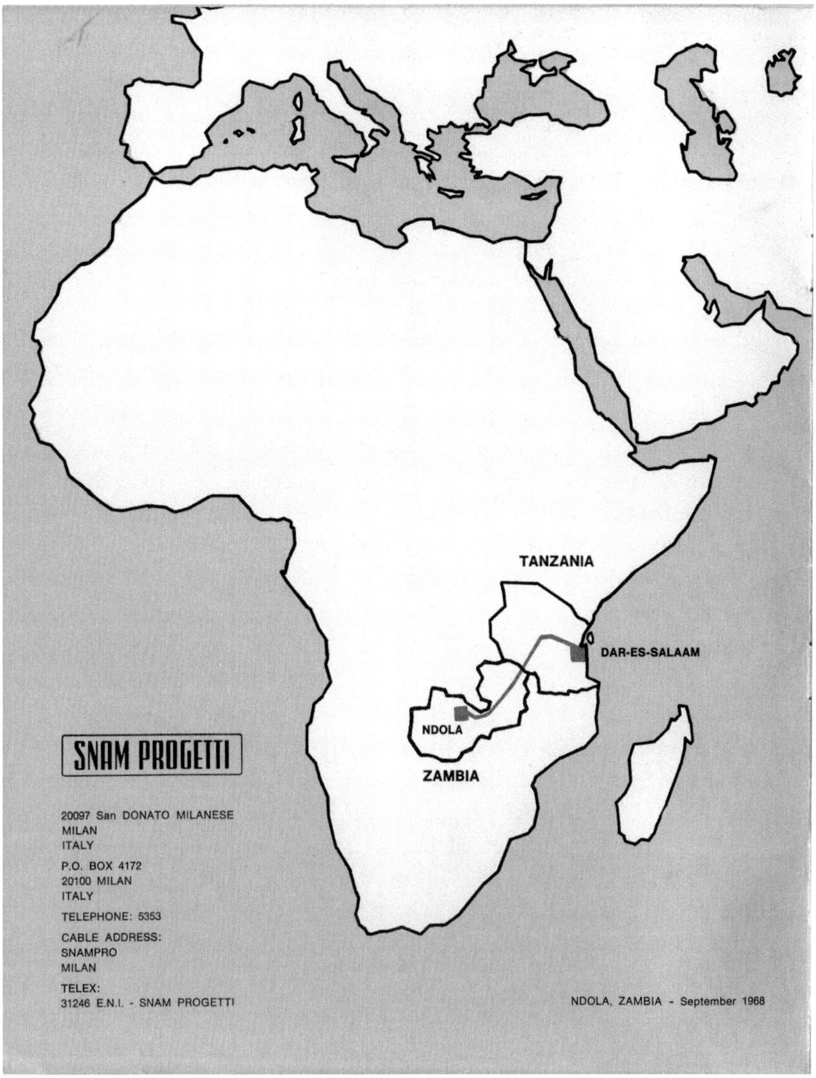

FIGURE 5.11 SNAM Progetti's brochure of the TAZAMA pipeline. *Source*: ENI Historical Archive. Location: Relazioni esterne, b.155 f.2FC0.

minority's Unilateral Declaration of Independence. Zambia was the world's third-largest copper producer, and its highly specialized economy was vulnerable to copper price fluctuations and depended on the availability of fossil fuels to support mining complexes. During the years of colonial domination, coal, oil, and copper were shipped through Southern Rhodesia and the closure of the Zambia-Rhodesia border left Zambia almost completely isolated and surrounded by white-controlled territories. The oil crisis forced the landlocked country to adopt severe oil rationing, only partially eased when British, American, and Canadian air forces organized an airlift to bring oil in from Dar es Salaam and Nairobi.[36]

ENI's pipeline proposal unexpectedly defeated the proposal made by the British "London and Rhodesian Mining and Land Company" (LONRHO), the industrial partner of the two

Rhodesias during the colonial era. A 1969 *Financial Times* article, entitled "Italian penetration of the Zambian economy," explains:

> The ascendancy which has been gained by Italian interests in Zambia in recent years has caused considerable heart-burning among advocates of British interest who have tended to regard Zambia as a traditional market because of its former British connections.
>
> The reasons for this development are two-fold: The first is that the Zambians regard the Italians as politically more acceptable than the British in view of their lack of a colonial past in Central or Southern Africa.
>
> Going to Italy in preference to Britain is also helpful in demonstrating to the public that the new independent Government is no longer tied to Britain's apron strings. Such a demonstration is always politically popular.
>
> But there is a second reason which is equally valid and perhaps in the long run more important. In whatever sphere the Italians have come in they have done an efficient job at reasonable prices and rates of interest with a minimum of bureaucratic red tape hold-ups.[37]

The pipeline contract indeed involved SNAM Progetti, ENI engineering subsidiary, as the engineering company and a consortium of Italian banks as funding institutions granting a soft terms loan to the Zambian government payable in fifteen years. This advantageous package was made possible by ENI's legal status as a state-owned company.

The 8.5-inch looped pipeline with its five pumping stations delivered the first shipment of refined products on September 2, 1968, thirteen months after the beginning of construction. The refined liquid had departed from the TIPER (Tanzanian and Italian Petroleum Refining Company Limited) refinery of Dar es Salaam, built by ENI and managed in partnership with the Tanzanian government, thirty days earlier. Owner and manager of the pipeline was the newly established TAZAMA Authority Ltd., a joint venture between the Zambian and Tanzanian governments, but ENI could still control the flow of diesel fuel departing from Dar es Salaam.

The pipeline carried refined oil products until 1975 when, based on the trust earned from the pipeline construction, the ENI group was appointed to build Zambia's first refinery. After the construction of the Zambian refinery, SNAM Progetti was contracted to modify the pipeline to allow the flow of crude oil and convert the pumping stations' engines to the new propellant. In the following years, SNAM was asked to increase the pipeline capacity with an additional outside pipe of 769 km; two pumping stations were also added and the existing ones were provided with additional engines.[38]

The pipeline and the refinery were perceived by Zambians as an important step toward real economic and energy independence. Even though the main goal of the pipeline was to supply fossil fuels to the mining industry of the Copperbelt, the postcolonial regime inserted the new infrastructure into narratives of nation-building, Pan-African cooperation, and decolonization in relation to the white-controlled south and the emancipation from colonial contractors. The new geography defined by the pipeline did indeed alter the territorial structure that was subjecting Zambia to southern Africa, but it reproduced a network based on resource extraction and control where new foreign powers like ENI would play a central role.

In a letter to one of his collaborators, Marcello Boldrini, ENI's president after Mattei's death in 1962, described the project as "an operation that opens new possibility for the Italian industry and marks a material and long-lasting presence in a territory, that of East Africa, beyond which Mozambique, Rhodesia and South Africa represent the 'Hercules pillars' of the free Occident."[39] The pipeline embodied ENI's strategy of progressive territorial conquest through "material and long-lasting" infrastructural artifacts justified as a "decolonizing" mission in support of independent governments and based on the belief that modernity and development were a direct consequence of easy access to oil (Figures 5.12 and 5.13).

In 1968, the construction of the pipeline was epically narrated in a video documentary *De l'océan indien au coeur de l'Afrique: L'oléoduc Dar es Salaam–Ndola* (*From the Indian Ocean to the heart of Africa: The oil pipeline Dar es Salaam–Ndola*), directed by Raoul Brunlinger for SNAM Progetti.[40] This was one of many industrial documentaries that ENI commissioned from well-known directors to shape the company's public image and communicate its often invisible work in a highly aestheticized manner. ENI documentaries directed by Bernardo Bertolucci, Gilbert Bovay, Joris Ivans, and many others appeared at international industrial film festivals and often screened (in shortened versions) on Italian TV and in movie theaters.[41]

Brunlinger's film follows the 45,000 tons of steel pipes from their arrival in Dar es Salaam to the construction site of the terminal tank in the Zambian Copperbelt. The steel pipes, through which oil normally flows unseen, serve as the main character of the movie, which shows how the laying of the pipes progresses, despite technical and contextual difficulties, thanks to the hard work of Italian, Zambian, and Tanzanian workers. The narration underlines the workers' effective cooperation, the advanced technological means offered by SNAM, and simultaneously, the creative solutions deployed to overcome Tanzania's

FIGURE 5.12 Construction of the TAZAMA pipeline, Tanzania 1966. *Source*: ENI Historical Archive.

TIMES OF ZAMBIA

FORWARD WITH THE NATION

MONDAY, March 15, 1971

RAY-O-VAC

It's fuel independence

K24 million refinery to pump out all we need

By Times Reporter

ZAMBIA will be self-sufficient in fuel-oil supplies by 1972, permanent secretary in the Ministry of Trade and Industry, Mr Andrew Kashita, said in Lusaka yesterday.

Religious group calls for anti-racist 'fighters'

Put the screws on thieves, UNIP meeting demands

FLOOD FUND CASH IS POURING IN–D.G.

Rose gets life for murder

Congo copper gets K11.5m shot in arm

Rabies reported in four districts

New 'squad' killings?

Shipping lines raise charges

Crude oil

Shortage

Freedom men winning the war in Portuguese Guinea

Meeting that never was

LATEST

TERBLANCHE & CO LTD
Quality Products
TELEFUNKEN
HI-FI, TAPE RECORDERS, TELEVISION, RADIOGRAMS, PORTABLE RADIOS, SPEAKERS
ZANUSSI
DEEP FREEZERS, WASHING MACHINES, GAS STOVES, REFRIGERATORS
DEFY
ELECTRIC STOVES, WASHING MACHINES
KDK
ELECTRIC FANS
ONLY THE BEST FOR COMFORT
at
TERBLANCHE CO. LTD.
Ndola.

FIGURE 5.13 The *Times of Zambia* celebrating the completion of the TAZAMA Pipeline. *Source*: Zambia National Archive.

topographic complexity. The visualization of oil-related labor here, as in other ENI's neorealist documentaries, is showcased to honor workers' epic endeavor to rescue Zambia from its energy isolation. Indeed, the documentary's voice-over proudly announces that despite multiple difficulties, the construction of Africa's longest pipeline was heroically achieved ahead of schedule because "the needs of modern life do not allow delay."[42] "From the Indian Ocean to the heart of Africa" displays what Georgiana Banita defines as "naive petro-progressivism," through which ENI depicted itself as a godsend satisfying Zambia's craving for energy independence. The pipeline was primarily intended to serve the company's profit and expansion.[43]

Agip Architecture on Africa's Roads in the 1960s and 1970s

With the end of formal colonialism and of urban and territorial racial segregation measures, Africans regained freedom of movement. After many years of colonial segregation, mobility and unrestrained flows were perceived as an important part of decolonization. This newly achieved freedom went hand in hand with continental, regional, and national developmental policies, aiming, through the construction of roads and highways, to foster trade, tourism, and Pan-African integration.[44] But circulation and automobility required another kind of oil-related infrastructure: gas stations. Despite their colorful and decorated architecture, these small-scale roadside facilities, more mundane than other oil infrastructures, tend to escape both everyday and scholarly attention. Nonetheless, they played a fundamental role in increasing mobility and expanding ENI's empire.

In the African landscape, Agip gas stations not only advertised ENI's refined products, they also served as quasi-military outposts marking ENI's "conquered territories" and granting the company a widespread and enduring territorial presence. Agip was in charge of this last segment of the oil supply chain, the most visible and recognizable of the company. As an indication of its importance, Enrico Mattei decided to retain the presidency of Agip and direct control of its commercial activities, which he considered the "facade of the group."[45]

In the early 1960s, the "facade of the group" made its appearance along the expanding road network of sub-Saharan Africa. In East Africa, the Agip logo became so well known that in his 1977 novel *Petals of Blood*, Kenyan novelist Ngũgĩ wa Thiong'o included a scene where boys and girls of Ilmorog village sit along the road and playfully "spell out

FIGURE 5.14 Agip service station in the Republic of the Congo, 1967. *Source*: ENI Historical Archive.

FIGURE 5.15 Agip Italian advertising campaign showing various Agip activities in Africa, including a gas station in Ibadan, Nigeria. *Source*: ENI Historical Archive.

LONRHO, SHELL, ESSO, TOTAL, Agip beside the word DANGER on the side belly of the tankers."[46] Ngugi's novel offers a nuanced perspective on infrastructural development in early postcolonial Kenya and across the African continent. While he criticizes the neocolonial attitude of international capital and the collusion of the postcolonial state, he also reminds us, through the joyful song of Ilomorog's children, that infrastructures have a generative social and cultural power. ENI and other oil companies understood this power and invested in infrastructures to promote travel, leisure, and consumption.[47]

Agip's African gas stations were based on the prototypes "59" and "61" (Figures 5.2 and 5.14) designed by ENI's internal engineering office (the Agip Ufficio tecnico) and were composed of one or more pumps and a small building hosting the gas station manager's office, a toilet, and a store for Agip lubricants and bottled gas. In the bigger service stations, these standard elements could be integrated with a snack bar and a mechanical workshop offering technical support and a car washing service. Despite the specificities of the different contexts, all stations were made of basic standardized elements united under the steel projecting roof with the Agip sign (translated in Aramaic or Arabic where needed) and by the omnipresent six-legged dog logo. The metal structure of the canopy, the minimal modular construction of the building, and the pumps themselves gave the gas stations the image of a perfectly functioning engine.

In the African landscape, Agip stations emerged as enclaves where every detail was designed and produced in Italy, including the pumps, workers' uniforms, the yellow furniture of the snack bar, and the trucks delivering the refined products. Agip's Italian advertising campaigns often played with the contrast between the informal activities happening along African roads and the modern, controlled, and tidy space of the service station. Like Foucauldian heterotopias of compensation, the stations were depicted as "perfect, meticulous and well arranged" "other spaces" juxtaposed with the "disordered, ill-conceived" outside[48] (Figures 5.14 and 5.15). These images of Agip African facilities produced by ENI's photographers to document and advertise the company achievements show how the story of ENI's involvement in postcolonial Africa is also a story of "the familiar" versus "the strange," through which the first one defines itself in relation to the latter, serving as the "other."[49] For postwar Italy, a poor but industrializing country recovering from defeat in World War II, Africa was not only a geographical entity, but also a means for Italy to develop its own Western identity in opposition to an unknown and schematized "other"—represented by African citizens depicted in their traditional costumes and activities. Making Africa the "other" allowed Italy—despite its poverty relative to countries like France, Germany, and the US[50]—to represent itself as a modern and developed country at the heart of the hegemonic cultural imaginary of the West.[51]

Aiming at keeping people on the road and increasing circulation and consumption, in Italy ENI built a chain of over forty Agip motels.[52] Only a few were built outside Italy. Some motels operated as Agip showcases in prominent urban areas like those of Nairobi and Dar es Salaam, others were strategically located at relevant highway intersections like in Mbarara, Uganda, or in proximity to ENI's extraction or refining sites, like in Mohammedia, Morocco.[53] The Motel Agip of Dar es Salaam was inaugurated in 1964, the year of the birth of independent Tanzania as the union of Tanganyika and Zanzibar.[54] The motel was located in the center of the city, between the commercial neighborhood or Indian Bazaar and the administrative district originally planned as the German colonial quarter. The structure hosted the motel itself, a large service station offering full mechanical service, and a showroom for FIAT, the Italian automobile manufacturer (Figures 5.16 and 5.17).

In 1970, ENI commissioned Lionello Massobrio to film the development of Africa's local tourism and increased automobility. In "Africa on the Road," the director collected a series of staged episodes that illustrate the experiences of various African motel customers: an elegantly dressed family with two kids on a road trip to Kenya's national parks, a young couple on a romantic honeymoon in North Africa, and a businessman driving his car from the Tanzanian countryside to the busy city of Dar es Salaam. In Dar es Salaam, the businessman parks his car at the gas station where a diligent group of Agip workers takes care of his vehicle while he checks in to the motel. In the following scene, the hotel's concierge

FIGURE 5.16 A postcard featuring the motel Agip and the Agip service station in Dar es Salaam, Tanzania, ca. 1970. *Source*: delcampe.net.

FIGURE 5.17 Workers at the Agip service station in Dar es Salaam. In the background is the Agip Motel. Dar es Salaam, Tanzania, ca. 1974. *Source*: ENI Historical Archive.

collects the car from the gas station, drives up the ramp and parks the car in front of the motel entrance so that on the following day the businessman can quickly leave the building and proceed with his busy schedule.[55]

The car-oriented architecture of the motel and its location shaped its clientele, which consisted mainly of white settlers driving to the city and to "the Agip" for short stays in the capital, foreign workers of nongovernmental organizations and international companies who stayed as long-term motel guests, and members of the government who worked in the proximity and drove to "the Agip" for business lunches and meetings.[56] On the motel's opening day, the Tanzanian government-owned newspaper *The Nationalist* reported that "from the car showroom and petrol station at ground level to the peaceful roof garden the motel is every inch modern," underlining the role that automobility played in Tanzanians' perception of modernity.[57] Indeed, to the Tanzanian members of the upper class who could afford it, automobility came to symbolize a "drive into the modern."[58] Cruising up the concrete ramp of the motel and enjoying their meal at "the Agip" restaurant, with its direct view of the parking lot, offered a stage to perform their newly achieved aspirations. Beyond their technical functioning indeed, AGIP roadside facilities operated "on the level of fantasy and desire" and stimulated a "deeply affectual relation" that "is an important part of their political effect"[59] and of the successful expansion of the petroleumscape.

The Dar es Salaam motel was one of over thirty motels built by Agip as a variation of motel prototype "59" developed by Agip's in-house design office.[60] Unlike the site-specific, singularly crafted buildings of the early 1950s, these standardized typologies were developed in order to optimize the design and construction process and to improve the company's marketing performance, as suggested by studies of branding recognition and consumer trust. Through prototyping and prefabrication, Agip technicians sought to increase control over construction time, costs, and quality and to minimize accidents caused by unskilled workers and extreme atmospheric conditions. The need to control processes despite ENI's intercontinental expansion affected construction processes as well as Agip building aesthetics.

In the Tanzanian postcolonial context, like in other newly independent countries, Agip's sober modernist architecture was welcomed as a sign of newness and modernity. For its ability to materialize a visual rupture with the colonial past, modern architecture became the preferred style of independent Tanzania, through which Julius Nyerere, Tanzania's first president, sought to develop a new and inclusive national identity. Furthermore, as proudly underlined by the *The Nationalist* article on the Agip Motel, the new facility was a sign that Tanzania was increasingly attracting foreign tourists and investors.[61]

Thanks to ENI's diplomatic and propagandistic efforts, Agip's modernist architecture and its aesthetic principles broadly circulated along its extensive supply chain and became a common presence in the African landscape. For this reason, the Italian oil company should be considered one of the main actors responsible for the globalization of the "aesthetic regime of modernism."[62] Shifting the focus from accounts of a few well-known master architects (often of Western origins) designing institutional buildings of newly independent states to the work of a national and multinational oil company, global modernism can be seen as a complex phenomenon entangled with the geopolitical and "geoeconomical" dynamics of nation-building, development, and petrocapitalism.

Conclusion

The final goal of the "grande disegno africano" was to broaden ENI's market and acquire rights for oil exploration in an ever-growing number of territories in order to sustain Italy's industrial development. Looking at ENI's empire from its "margins" and reading its operations "against the grain" reveals a series of inconsistencies in ENI's narratives and exposes the violence and the impact of its seemingly neutral operations on African territories and societies.

The end of Africa's formal colonization offered the precondition for ENI's penetration that was achieved thanks to diplomatic, propagandistic, and infrastructural effort. ENI's narratives hinged on promises of modernity and development and on the reputation of the Italian company as a benevolent national agency different from rapacious and profit-oriented oil multinationals. Indeed, unlike other private oil companies, ENI offered advantageous contracts, soft terms loans, and 50-50 partnerships with local governments, paving its expansion with paternalistic good intentions.

Infrastructures such as refineries, pipelines, and gas stations were fundamental to ensure oil production, circulation, and commercialization, but unlike their Italian counterparts, ENI's infrastructures in the African context operated as quasi-military outposts for its logistic expansion. The same infrastructures were successfully marketed by ENI's propaganda as tools to transcend colonial dependencies and achieve energy independence at the national level and as a means of acquiring individual modernity and freedom. With the support of local governments, ENI reshaped the African territory and—by promoting an oil-based lifestyle—affected people's everyday lives and imaginaries, contributing to the flourishing of an oil-based consumerist civilization.

Notes

1 Giuseppe Accorinti, *Quando Mattei era l'impresa energetica io c'ero*, 3rd ed. (Matelica: Hacca, 2008), 160.
2 The notion of the petroleumscape was coined by Carola Hein. Carola Hein, "Oil Spaces: The Global Petroleumscape in the Rotterdam/The Hague Area," *Journal of Urban History* 44, no. 5 (September 2018): 887–929, https://doi.org/10.1177/0096144217752460; Carola Hein and Mohamad Sedighi, "Iran's Global Petroleumscape: The Role of Oil in Shaping Khuzestan and Tehran," *Architectural Theory Review* 21, no. 3 (September 2016): 349–74, https://doi.org/10.1080/13264826.2018.1379110.
3 Hein, "Oil Spaces."
4 Keller Easterling, *Extrastatecraft: The Power of Infrastructure Space* (London: Verso, 2014), 13.
5 Ibid., 13.
6 Jacques Ranciere, *The Politics of Aesthetics*, trans. Gabriel Rockhill (London, New York: Bloomsbury Academic, 2006).
7 Brian Larkin, "The Politics and Poetics of Infrastructure," *Annual Review of Anthropology* 42, no. 1 (2013): 327–43, https://doi.org/10.1146/annurev-anthro-092412-155522.
8 Andrea Greco and Giuseppe Oddo, *Lo Stato parallelo: La prima inchiesta sull'ENI tra politica, servizi segreti, scandali finanziari e nuove guerre. Da Mattei a Renzi* (Milan: Chiarelettere, 2016).
9 Elisabetta Bini, *La potente benzina italiana: Guerra Fredda e consumi di massa tra Italia, Stati Uniti e Terzo mondo (1945–1973)* (Roma: Carocci, 2013), 142.
10 See, e.g., Dino Risi's "Il sorpasso," 1962; Mario Missoroli's "La bella di Lodi," 1963.
11 Francesca Carnevali, "State Enterprise and Italy's 'Economic Miracle': The Ente Nazionale Idrocarburi, 1945–1962," *Enterprise & Society* 1, no. 2 (June 2000): 249–78https://doi.org/10.1093/es/1.2.249.

12 Venturi, Scott Brown and Izenour wrote: "The Italian landscape has always harmonized the vulgar and the Vitruvian: the contorni around the Duomo, the portiere's laundry across the padrone's portone, Supercortemaggiore against the Romaneques apse." "Supercortemaggiore" was the name of Agip gasoline. Its name came from Corte Maggiore, the town where Italian gas deposits were first discovered, while the gasoline was refined in Italy and extracted mainly in Iran, and later also in Egypt and Tunisia. Robert Venturi, Denise Scott Brown, and Steven Izenour, *Learning from Las Vegas* (Cambridge, MA: MIT Press, 1972); Carnevali, "Economic Miracle."
13 Author's translation of Agip's motto: "Il cane a sei zampe, amico fedele dell'uomo a quattro route."
14 Bacciocchi, Baccigalupo and Ratti, Gellner, Mollino, BBPR, and Castiglioni among others. For an account of ENI's architectural interventions in Italy, see Dorothea Deschermeier, *Impero ENI: l'architettura aziendale e l'urbanistica di Enrico Mattei* (Bologna: Damiani, 2008).
15 "The Face of the Miracle," *Architectural Review* CXXI, no. 783 (May 1962).
16 The term "Seven Sisters" (for some, a creation of Enrico Mattei himself) referred to the seven international oil companies forming the "Consortium for Iran" oligopoly that included the Anglo-Iranian Oil company (now BP), Royal Dutch Shell, Standard Oil California (later Chevron), Gulf Oil, Texaco, Standard Oil of New Jersey (later Esso, Exxon, and now ExxonMobil), and Standard Oil New York.
17 Elisabetta Bini, *La potente benzina italiana: Guerra Fredda e consumi di massa tra Italia, Stati Uniti e Terzo mondo (1945–1973)* (Roma: Carocci, 2013), 142.
18 ENI strategically supported the leaders of independent states, but also the leaders of independence movements still struggling for decolonization like in the case of Algeria. Bini, La potente benzina italiana: Guerra Fredda e consumi di massa tra Italia, Stati Uniti e Terzo mondo.
19 Enrico Mattei's speech, Tunis, 9–10 June 1960. Translated by the author from Italian. Enrico Mattei and Paolo Mieli, *Scritti e discorsi 1945–1962: raccolta integrale dall'archivio storico Eni*, ed. speciale per Eni (Milano: Rizzoli, 2012), 730.
20 Daniele Pozzi, *Dai gatti selvaggi al cane a sei zampe: tecnologia, conoscenza e organizzazione nell'Agip e nell'Eni di Enrico Mattei* (Venezia: Marsilio, 2009).
21 Pozzi, *Dai gatti selvaggi al cane a sei zampe: tecnologia, conoscenza e organizzazione nell'Agip e nell'Eni di Enrico Mattei*; Accorinti, *Quando Mattei era l'impresa energetica io c'ero.*
22 Author's interview with A.M. and M.C. in Accra, Ghana, November 2019.
23 Stephan F. Miescher and Dzodzi Tsikata, "Hydro-Power and the Promise of Modernity and Development in Ghana: Comparing the Akosombo and Bui Dam Projects," *Ghana Studies* 12/13 (2011): 15–53.
24 See Ghana Oil Refinery Gh/Praad/RG.7/1/105, Tema Industrial Developement Gh/Praad/RG.7/1/368, Agipt Gh/Praad/RG.7/1/391, Ghaip Gh/Praad/RG.7/1/1840-43. Public Records And Archives Administration Department Accra, Ghana.
25 Lavoro italiano in Africa—insediamenti ENI in Congo Belga e Ghana, Cinegiornali C, ENI Historical Archive. "Archivio Storico ENI", https://archiviostorico.eni.com/aseni/it/explore/audiovideo/IT-ENI-AV0001-000139?r=search#/search/ghana, accessed June 19, 2020.
26 Gatto Selvatico, Anno IX, N10, 1963.
27 B. S. Hoyle, "New Oil Refinery Construction in Africa," *Geography* 48, no. 2 (1963): 190–94.
28 The plant produced gasoline, liquefied petroleum gas (LPG), diesel fuel, kerosene, and other refined products with a capacity of 28,000 barrels per stream day.
29 Interview with Mr. Forson, former employee and managing director of the Tema Refinery (1970–2004), Tema, November 2019.
30 Interview with Mr. Addoo, former employee and chief manager of the Tema Refinery, Tema, November 2018.
31 Viviana d'Auria, "From Tropical Transitions to Ekistic Experimentation: Doxiadis Associates in Tema, Ghana," *Positions*, no. 1 (2010): 40–63.
32 This perception emerged through various interviews with former workers of the refinery.
33 Barrett and Worden, *Oil Culture*, (Minnesota: Minnesota University Press, 2014): XXV.
34 Ibid.
35 TAZAMA is the acronym of Tanzania Zambia Mafuta (oil in Swahili).
36 Andrew Cohen, "Britain and the Breakdown of the Colonial Environment: The Struggle over the Tanzam Oil Pipeline in Zambia," *Business History Review* 88, no. 4 (2014): 737–59, https://doi.org/10.1017/S0007680514000749.

37 "Italian Penetration of the Zambian Economy," *The Financial Times*, August 14, 1969.
38 Author's interview with Davison Thawethe, managing director of the TAZAMA Pipeline Ltd., Dar es Salaam, August 2019, and Eni Historical Archive, Fondo ENI: Relazioni Esterne, b.155 f.2FC0; Estero, Iniziative industriali, Zambia, b.69 f.1; Estero, Iniziative industriali, Tanzania, b. 384 f.1CCD; Estero, Coordinamento regioni estere,Tanzania, b.30 f.12.
39 Letter of ENI President M. Boldrini to Stammatti, 7 October 1966: "Nella mia valutazione, si tratta di realizzare un 'opera che, oltre ad essere economicamente valida e ad aprire nuove possibilita al lavoro italiano, costituirà una presenza concreta e duratura nell'ultimo lembo dell'Africa Orientale, oltre al quale, Mozambico, Rhodesia e Sud Africa, rappresentano un po' le Colonne d'Ercole dell'Occidente libero." ENI Historical Archive, Fondo ENI. Presidenza, Presidenza Marcello Boldrini, 143-32E832E8.
40 "From the Indian Ocean to the Heart of Africa. The oil pipeline Dar es Salaam – Ndola," Raoul Brunlinger, 1968. ENI Historical Archive, Rome.
41 See, for example, Joris Ivens's 1960 documentary *L'Itala non è un paese povero* (*Italy is Not a Poor Country*), Gilbert Bovay's *Gli uomini del petrolio* (*The People of Petroleum*) of 1965 and *Africa, nascita di un continente* (*Africa: The Birth of a Continent*) of 1968.
42 Author's transcription and translation from French from "De l'océan indien au coeur de l'Afrique. L'oléoduc Dar es Salaam – Ndola," Raoul Brunlinger, 1968, ENI Historical Archive, Rome.
43 Georgiana Banita, "From Isfahan to Ingolstadt," in *Oil Culture*, eds. Allan Stoekl, Ross Barrett, and Daniel Worden (Minneapolis: University of Minnesota Press, 2014), 145–68, https://doi.org/10.5749/minnesota/9780816689682.003.0008.
44 See, for example, the project of the Trans-African Highway promoted in the late 1960s by the UN Economic Commission for Africa.
45 Interview with Eng. Ennio Ghellini Sarqenti, Milano, 28 marzo 1987, ENI Historical Archive.
46 Ngugi wa Thiong'o, *Petals of Blood* (Oxford: Heinemann International Literature and Textbooks, 1986).
47 For an in-depth analysis of *Petals of Blood* and road infrastructure development in postcolonial Africa, see Kenny Cupers and Prita Meier, "Infrastructure between Statehood and Selfhood," *Journal of the Society of Architectural Historians* 79, no. 1 (March 1, 2020): 61–81, https://doi.org/10.1525/jsah.2020.79.1.61.
48 Michel Foucault, "Of Other Spaces: Utopias and Heterotopias," in *Rethinking Architecture: A Reader in Cultural Theory*, ed. Neil Leach (New York: Routledge, 1997), 330–36.
49 Edward W. Said, *Orientalism* (London: Penguin, 2003).
50 "In 1956, for example, Italy had one car for every 47.4 people, compared to 1 to 3 in the United States, 1 to 11.8 in France and 1 to 20.7 in West Germany." Elisabetta Bini, "Selling Gasoline with a Smile: Gas Station Attendants between the United States, Italy, and the Third World, 1945–1970," *International Labor and Working-Class History*, no. 81 (2012): 69–93.
51 Walter D. Mignolo, *Local Histories, Global Designs: Coloniality, Subaltern Knowledges, and Border Thinking* (Princeton, NJ: Princeton University Press, 2012).
52 Accorinti, *Quando Mattei era l'impresa energetica io c'ero*, 36.
53 Fernanda De Maio et al., *e169 | EniWay* (Edizioni Engramma, 2019).
54 Tanganyika achieved formal independence in 1961.
55 Massobrio Lionello, *Africa on the Road/ Al servizio dell'Africa*, 1970, ENI Historical Archive.
56 Author's telephone interview with N. M., manager of the motel from 1964 to 1999. August 2019. Author's interview with Mr. Mkindi, former employee of the Agip Motel, Dar es Salaam, July 2019.
57 Jordan Lane, "Agip Motel Adds Beauty to Dar," *The Nationalist*, July 31, 1964.
58 Laura Fair, "Drive-In Socialism: Debating Modernities and Development in Dar es Salaam, Tanzania," *American Historical Review* 118, no. 4 (2013): 1077–104.
59 Larkin, "Politics and Poetics of Infrastructure."
60 Laura Greco and Stefania Mornati, *Architetture Eni in Italia (1953–1962)* (Roma: Gangemi, 2018).
61 Lane, "Agip Motel."
62 Tom Avermaete, Serhat Karakayali, and Marion Von Osten, *Colonial Modern: Aesthetics of the Past Rebellions for the Future* (London: Black Dog Publishing, 2010).

6

THE OFFSHORE PETROLEUMSCAPE

Grids, Gods, and Giants of the North Sea

Nancy Couling

The first major discovery of oil in Europe was made in the North Sea in December 1969 by the American company, Philipps. This historic event confirmed that the North Sea contained hydrocarbons in promising commercial quantities and placed it at the center of international interest, reconfiguring the global geography of oil. Since that discovery and reaching a peak in the 1980s, the existing landside petroleum infrastructure has been extended and diversified, creating an important offshore petroleumscape in the North Sea. While oil had previously been extracted close to shore in the Gulf of Mexico and in the shallow waters of the Persian Gulf and the Caspian Sea, the depths and wave heights of the North Sea represented new challenges. Europe was eager to reduce its dependence on oil, which was imported primarily from the Gulf.

Similar to the one on land, the offshore petroleumscape is complex and palimpsestic, with the critical difference being that it is not entangled with other fixed structures. It has established completely new sets of relations to the established petroleum nodes as well as to new geographic places on- and offshore. Offshore, the petroleumscape is made up of unprecedented, tailor-made, extended territorial frameworks—the sea needed to be prepared for such an extension of industrial activities as it had previously been a fluid space held open in the interests of trade. The establishment of the offshore petroleumscape gave companies and corporations temporary rights to inhabit sea space, install fixed structures, and carry out a new type of production at sea.

The North Sea is a particular example of this sea-based petroleumscape. Organized around these abstract regulatory frameworks, the offshore petroleumscape became a physical space constructed to high technical requirements with vast amounts of concrete and steel. Over time, and in particular at the 1980s peak, offshore fields developed into complexes with diversified uses and large rotating workforces—assuming the characteristics of artificial urban archipelagos in their own right. "Ekofisk City,"[1] on the Norwegian continental shelf, is exemplary of this urban dimension. More intangible are the cultural dimensions of the offshore petroleumscape, in particular within the Norwegian context: it is

DOI: 10.4324/9780367816049-8

a mythological space, a legendary space of everyday culture, and in selected cases, a *historic artifact* which is digitally preserved.

The combination of specific state-led territorial organization, physical production of space, and everyday working life are indicators of specific processes of urbanization, first proposed by Henri Lefebvre[2] and more recently further articulated in Neil Brenner and Christian Schmid's theory of planetary urbanization.[3] The petroleum industry was largely responsible for the creation of a new offshore production space at the outset, and the resulting petroleumscape, as explored in the examples outlined in this chapter, can be understood as a powerful vehicle of the urbanization of the sea.[4] Now the most industrialized sea in the world,[5] the North Sea has produced a particular type of offshore urbanity in conjunction with the construction of its petroleumscape. The offshore petroleumscape leaves a formidable territorial legacy: it has set a precedent in transforming the North Sea into an *energy seascape*, which is being appropriated, renovated, and redirected toward renewables at the same time as processes of optimization are extracting the last hydrocarbons from existing fields and finding other substances to inject into their hollowed-out subsea spaces.

The first part of this chapter gives a historical overview of interactions that have determined and formalized spatial definitions at sea and it describes the context within which the postwar offshore petroleumscape emerged as a transnational extraction grid in the North Sea. The prolific production, longevity, and widespread sociocultural impact of North Sea oil and gas make it a valuable case study of the offshore petroleumscape discussed in the chapter's second part. Here, its physical and cultural layers are examined more closely: the Ekofisk City complex located in the middle of the North Sea, the political influence exerted in naming newly estabished geographic places offshore, and the absorption of the oil industry into specialized strands of Norwegian maritime culture. The third part considers how emerging post-oil energy strategies are perpetuating the extractivist practices and established petroleumscape of previous generations. Rather than wholesale post-oil abandonment, current developments demonstrate a range of scenarios set to maintain the unprecedented, systematic reprogramming of sea space that has taken place on the North Sea continental shelf.

Preparing the Offshore Petroleumscape of the North Sea: A Brief Historical Overview Before 1964

Seas are tangible historical spaces that have harbored and facilitated continuous exchanges. In Europe, human interaction with the sea has been accompanied by instruments of measure and control, both of which are a prerequisite to planning. Formalization of these spaces into legally bound units is a political act involving either negotiation or the execution of power. Early alliances and rivalries between tribes, political units, or confederations at sea were fluid and dynamic, similar to the situation that prevailed on land. Initially a platform for fluid material and cultural exchange between important urban centers, the sea space—such as notably the North Sea—has become a fixed production site serving—but increasingly severed from—landside hubs. The offshore petroleumscape has played a significant role in this transformation.

The North Sea is exemplary. The story of its rich history of trade and cross-cultural exchange includes the Frisians (from the seventh to the eighth centuries), followed by the Vikings (from the mid-eighth to the twelfth century), and the Hanseatic League (from the twelfth to the fifteenth centuries). Before mobility infrastructure on land was well developed, because it was easier to travel by freshwater or saltwater routes than on land, the North Sea became the major logistical space of the region. The history of sea space was intimately tied to that of urban culture in the major cities of London, Antwerp, and Amsterdam, because it was through the North Sea that critical connections were established and maintained between centers whose wealth frequently depended more on international trade connections than on hinterlands.[6] This was particularly true for the Viking capital Hedeby and ports such as Bergen, Bruges, and Aberdeen. Likewise, people were frequently on the move around the North Sea region, profoundly influencing local languages, traditions, music, and creating a shared culture.[7]

Human interactions across the land-sea threshold result in territorial practices which shape the sea, both physically and conceptually. In northwestern Europe, the way in which we understand sea space in relation to urban space has changed dramatically since the late seventeenth and early eighteenth centuries. Traditionally, the sea provided a medium through which urban systems were loosely extended and a space within which to capture resources, demonstrate power, and carry out free trade. With the rise of the nation-state, there was less recognition of the fluid, extended forms of urbanity sustained by the sea and a process of increased appropriation and enclosure began to unfold around the emerging offshore petroleumscape.

After World War II, it was largely because of pressure from the petroleum industry that the first comprehensive legal framework for the ocean was formulated—historical legislation that opened up a vast new territory to a new type of industry for the North Sea, executed through urban devices and ultimately leading to the construction of specific new geographic places of urban dimensions. The concept of mare liberum was the doctrine of the freedom of the high seas in the general interests of humanity, defended by the Dutch jurist and philosopher Hugo Grotius in 1609 in the book of the same name.[8] This book was a response to the threat of a Portuguese monopoly on the East India trade route. Mare liberum was well suited to the ambitions of maritime trading nations, and was therefore upheld alongside the widely accepted notion of a coastal strip of territorial sea about the width of a cannon shot, which, in the early eighteenth century, was formalized by the Dutch to a width of three nautical miles.[9] Hence, while the sea was an essential space of connection and full of maritime activity, it was also conceived and maintained as a fluid space distinct from land and free from territorial domination.

The extension of sovereign exploitative rights into the sea was inextricably linked to the search for oil. The delineation of the territorial sea remained relatively unchanged until after World War II, by which time Western Europe had shifted its energy source from indigenous sources such as coal to become heavily dependent on oil, mostly imported from the Middle East.[10] Apart from initial offshore drilling in shallow waters of 3–7 m off the California coast at the end of the nineteenth century, the ocean remained a boundary for oil exploration until the 1930s.[11] By this time, onshore fields had become difficult to find around the Gulf of Mexico and large geological surveys predicted offshore finds; the refinery infrastructure was already in place along the coast due to the importance of shipping to transport oil. In 1938, the Creole field produced the first oil from "open waters" 4.5 m deep and 1.6 km into the gulf.[12] These explorations were well within the three-nautical-mile limit, but after steel became more readily available at the end of the World War II, the oil and gas industries experienced a boom and companies raced to find oil farther out in the Gulf of Mexico.

In 1945, under pressure from the petroleum industry, US President Harry Truman declared all natural resources on the continental shelf under US jurisdiction.[13] This was an unprecedented expansion of national territory into the sea and unleashed a series of similar claims from other nations. Under these circumstances, and in combination with other pressures on the world's oceans such as the threat of pollution and the need to regulate the expanding shipping trade, the UN initiated the first comprehensive legal framework for ocean space. This process began in 1949, resulting in the first UN Conference on the Law of the Sea (UNCLOS) in Geneva in 1956 and the subsequent finalization of four conventions relating to the Territorial Sea (1964), the Continental Shelf (1964), the High Seas (1962), and Fishing and Conservation of Living Resources of the High Seas (1966). The continental shelf itself was here defined according to the possibility of resource exploitation in terms of what was possible at the time:

> the term "continental shelf" is used as referring (a) to the seabed and subsoil of the submarine areas adjacent to the coast but outside the area of the territorial sea, to a depth of 200 meters or, beyond that limit, to where the depth of the superjacent waters admits of the exploitation of the natural resources of the said areas.[14]

These initial four conventions were subsequently replaced by the UNCLOS in 1982, which included a more precise definition of the extent of national offshore resource extraction within a littoral nation's Exclusive Economic Zone.[15] It is imperative to note the implicit underlying liaison between petroleum and the formalized territorialization of ocean space from the outset. UNCLOS 1982 was the first fundamental legal territorial concept to operate on an international scale. Its implementation has resulted in ongoing spatial reconfiguration of the planetary ocean and seas; according to EU directives, all coastal nations are required to have Maritime Spatial Plans for their Exclusive Economic Zones in place by March 31, 2021.[16] UNCLOS laid the foundations for the establishment of large-scale planned seascapes and the emergence of a radical new urban realm within the unfolding history of planning.[17]

Prior to the UN regulations outlined above, the existing petroleumscape had utilized the sea only as a transport surface to connect the global oil supply and delivery networks. Extraction was only taking place offshore in a tentative way in areas close to shore. However, in addition to the feedback loops, described by Carola Hein, as imperative to the ongoing reinforcement of the palimpsestic petroleumscape, this fundamentally capitalist-driven operation relies on expansion to secure the ongoing accumulation of capital. Both David Harvey[18] and Neil Brenner have discussed the de-territorializing effect of capitalist socio-spatial dynamics as, in Brenner's words, "capital's drive toward spatial expansion, temporal acceleration and relentless spatiotemporal restructuring,"[19] which is accompanied by dependence on territorially rooted, built infrastructure to achieve the same objective. I argue that the extension of the petroleumscape into the offshore exemplifies these processes. The following section explains how this expansion took place in the North Sea.

Subdivisions of the North Sea Since 1964

In 1964, several North Sea countries passed acts that effectively released the North Sea from its status as a common space and turned it into a vast, gridded site of petroleum exploration and production which marked a watershed in Northern European urbanization processes. Firmly controlled by governments and industrialists, the implications were neither thoroughly discussed in terms of planning nor, until recently, theorized. Immediately subsequent to the establishment of the first legal regulation of the seabed, an entire sea, 600 km wide, became a territory opened up for development. Previously not considered a potential source of hydrocarbons, in the 1960s the North Sea rapidly developed into the focus of international exploration activity with competition between neighboring countries, organized around newly determined offshore borders.

Within the hitherto "undeveloped" space of the sea, the petroleumscape could freely expand without concern for, or conflict with, existing urban layers, yet it is inextricably linked to the physical networks and energy demands of landside urban agglomerations. The theory of planetary urbanization, in particular "extended urbanization," provides a useful way to think about such remote and unevenly dispersed sites that operate within a stretched, dialectical relation to urban centers.[20] Within the reciprocal urban processes of "implosion and explosion,"[21] extended urbanization refers to the far-reaching systems and mechanisms set up to support urban agglomerations. These systems selectively cross and erase historical borders between inherited categories such as the urban and the rural or land and sea, sometimes densifying into extreme and unfamiliar settlement formations charged with circulating goods and data, extracting resources, and engaging in specialized production. In contrast to the earlier fluidity of extended commercial and cultural connections across the North Sea, the strategic instruments applied in the process of extended urbanization fundamentally changed the nature of the seascape into a fixed set of Cartesian coordinates designed to produce a seascape "operationalized"[22] for the extraction and production of oil and gas.

The giant Slochteren gas field near Groningen in the Netherlands, discovered in 1959 and proving to be the second-largest natural gas field in the noncommunist world, stimulated further exploration on the continental shelf, leading to the realization that gas formations extended seawards.[23] Legislation for drilling outside of Dutch territorial waters had not yet been set in place and hindered further exploration until 1967. But in the UK, the government was keen to capitalize on international oil companies' intense interest in the North Sea and quickly prepared to begin issuing licenses. Before the UK Continental Shelf Act of April 1964 was passed and the "median line" separating UK and Norwegian waters agreed upon (May 15, 1964), the UK Ministry of Power prepared the first licensing areas by dividing the global geographical baselines of 1 degree latitude by 1 degree longitude into 30 blocks of 100 mi^2.[24] In total, the offshore area comprised 960 such blocks or 86,000 square miles. A notice inviting applications for production and exploration licenses was issued on the same day as the Act and Regulations became effective, and by September of the same year, 52 licenses covering 394 of the 960 blocks were issued to 22 applicants and a total number of 51 companies.[25]

The wholescale subdivision of the North Sea into developable blocks, comparable to greenfield development sites on land, was not questioned by the UK's Ministry of Power. Likewise, according to geographer Keith Chapman, legal questions about the ownership of the seabed were mere details for the oil industry: "The activities of the oil industry made these issues a matter of practical importance rather than philosophical debate."[26] The UK government was well aware of their advantage over other North Sea countries in pushing through rapid legislation and offering blocks for exploration. Their goal was to "incite the maximum of activity,"[27] encouraging oil companies to initiate and maintain operations in the UK part of the North Sea.

As a device of preliminary appropriation around which hydrocarbon infrastructure is organized, the grid represents a powerful vehicle of extended urbanization. While familiarly banal, the grid belies the spatial power executed by the North Sea littoral states. With the exception of Germany, which awarded rights to a consortium for the entire area, the grid system was subsequently adopted by neighboring North Sea countries with slight variations in the interest of promoting rapid exploration (Figure 6.1).[28] Norway, for example, adopted a 500 km^2 or 200 mi^2 subdivision of the block in 12 units rather than the 30-unit UK division and released 278 blocks for licensing in the first round in 1965, including the whole

FIGURE 6.1 North Sea petroleum grid. Map by Nancy Couling.

Norwegian continental shelf south of the 62 degree parallel.[29] The vast and abstract grid was circulated as a plan only within government and petroleum industry circles. Although not visible per se, the grid established an immutable, orthogonal, and highly resilient referential layer, presupposing the possibility of seafloor development anywhere within its range, while in theory the water column still remained a dubious "commons" effectively rented out to transient oil rigs.

In the 1960s, over 200 exploratory wells were drilled in the North Sea. Gas was discovered by Conoco/National Coal Board in the UK sector at the Viking field in 1965 and the Hewitt (Arpett Group) and Leman Bank fields (Shell/Esso) in 1966, the latter proving to be the world's sixth-largest offshore gas field, and it transformed the British gas supply industry.[30] The abstract grid was therefore activated into a series of nodes of intense technological and human activity as floating or jack-up drilling rigs were relocated and accommodation platforms were installed, while workers circulated by helicopter and service vessels delivered supplies and equipment. As gas fields came online, a system of delivery pipelines and receiving terminals at the coast was also required. At this time, North Sea waters presented the most demanding offshore environment, surpassing previous technical limits of depth and wave height. Previously, depths of over 400 feet (ca. 122 m) had been considered "deep." In the central North Sea, "one of the world's roughest bodies of water,"[31] depths were up to 500 feet (152 m) and waves reached heights of 75–100 feet (22–30 m).[32] Construction systems inherited from more benign seas such as the Persian Gulf had to be adapted to these new conditions. In 1968, the Cod gas field was discovered in the Norwegian sector, followed by the 1969 discovery of Ekofisk, the first major commercially viable oil field in the North Sea. Ekofisk was a giant find and due to the density and complexity of its offshore infrastructure, its early discovery, and subsequent longevity, it irrevocably transformed the North Sea into a dedicated petroleumscape and made Norway a society based on oil.

Physical and Cultural Dimensions of the Offshore Petroleumscape: The Case of Ekofisk

Offshore sites such as Ekofisk represent a new type of "place" in the history of Northern European urbanization. These large, complex structures were sites of intensive work relations, material transformations, and international nodes of cultural and technical exchange. They also outlived their productive stages to mature into places of historical significance. Ekofisk City not only demonstrates the physical extent of the offshore petroleumscape, but also is an example of extended urbanization at sea. Offshore, the petroleumscape acquired the additional social dimensions of living and recreational facilities, producing an everyday culture of the sort identified by Hein as still lacking in existing debates of the petroleumscape.[33] Restricted socially, spatially, and temporally, Ekofisk City was a specific type of urban format that emerged offshore. With two generations of an offshore workforce having been employed there for much of their working lives, it has become a lasting cultural artifact.

The Ekofisk complex had the densest concentration of offshore installations and flow lines in the whole North Sea. Located at the southernmost extremes of the Norwegian continental shelf, with water depths of 70–75 m,[34] the area developed in stages. It comprises the four fields—Ekofisk, Eldfisk, Embla, and Tor—and six further fields (including Cod) that have ceased production, with a total of thirty-two platforms (Figure 6.2). The "center" of Ekofisk was a series of eight platforms and two flare stacks connected by pedestrian

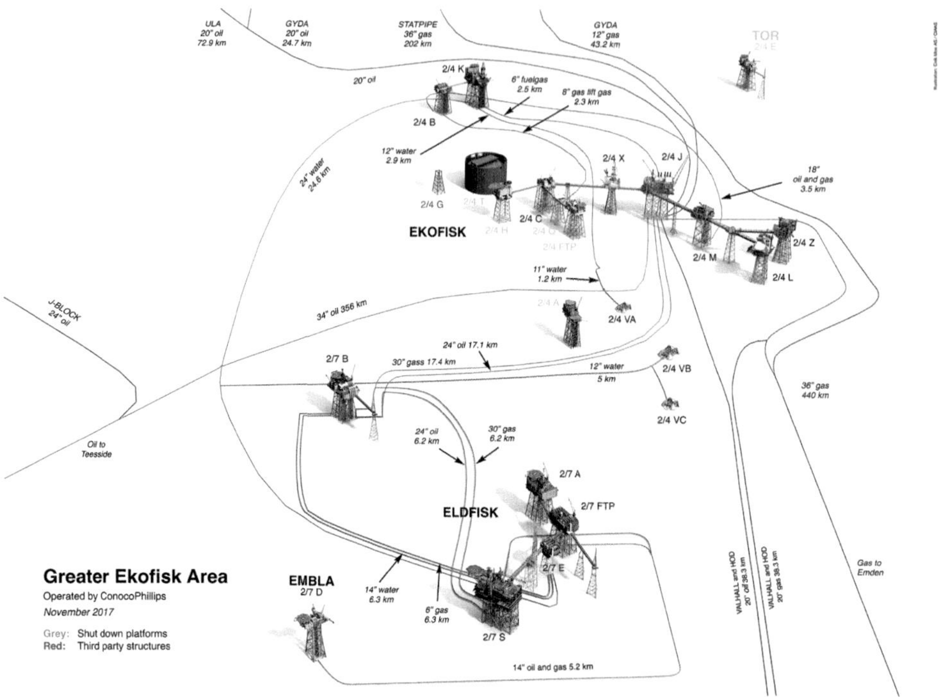

FIGURE 6.2 The Greater Ekofisk Area. *Source*: ConocoPhilipps.

bridges.[35] Ekofisk occupies a central place in Norwegian history, not only because it was the first producing field on the Norwegian continental shelf, but it also presented multiple technical challenges and ultimately established new Norwegian competence in the oil industry. It has been hailed as "a pioneering effort that ushered Norway into the oil age."[36]

Between 1974 and 2004, Ekofisk brought Norway US$190 billion from oil and gas sales, equivalent to 20 percent of Norway's total production during this period.[37] Members of more than fifty different professions worked on the Ekofisk complex, with between 7,000 and 8,000 employees there during the construction phase of 1972, making up a population four times larger than most Norwegian coastal towns.[38] Ekofisk represents a shift in construction, economics, and national policy that transformed Norwegian society.

At Ekofisk City—the name partly reflecting its complexity and partly the shift from temporary installations to more permanent facilities and structures—new fields and discoveries continued to be linked into the complex—it was never clear during the first phase of development how large this city would grow.[39] The Ekofisk site generated continuous and intense circulations of "rig tours,"[40] supply vessels, rigs being hired out for limited contracts and transported to new sites around the North Sea and beyond, and of the flow of oil itself. Parallel to this continuous movement, Ekofisk also received units characterized by new dimensions of limited permanence: between 1972 and 1974, facilities enabling the production of oil around the clock were installed, dedicated accommodation platforms connected with bridges were fixed to the seafloor, and a concrete storage tank was set in place. The concrete foundations for the connecting bridges used a system never before tested offshore. The concrete storage tank—a "cube with dimensions of a standard city block"[41]—had an external

FIGURE 6.3 Ekofisk storage tank. *Source*: Norwegian Petroleum Museum.

diameter of 140 m and a height of 106 m, weighed 215,000 tons, and was capable of storing one million barrels of oil. The transport of this tank out to the field was a historic event, as reported by the Stavanger newspaper: "The monument to a new technological era is today journeying through our district"[42] (Figure 6.3).

At the peak period, in 1981, all platforms were functioning. By 1998, several of the platforms had already shut down production and the question arose as to how the historic nature of the ensemble could be acknowledged as a type of museum to Norway's early oil age. The Ministry of Petroleum and Energy declared it necessary to "ensure that important and historical structures were preserved and made accessible for future generations,"[43] and in accordance with the Cultural Heritage Act, the structures were awarded the status of Industrial Cultural Heritage—"some of the largest and most complex cultural monuments of our time."[44] A process of preservation and documentation was initiated for the first period from 1962 to 1998, with a particular focus on documentation for parts too large to preserve and for less tangible, sociocultural aspects of the offshore working life.

More subtle cultural dimensions of the offshore petroleumscape can be seen in Norwegian practices of naming oil and gas fields. Politicians aimed to link the industrialized sea to Norwegian heritage, in particular to Norse gods, thereby obscuring the processes of modernization and actively promoting oil and gas as part of the national cultural imagination.

In the development of the Norwegian petroleum grid described above, awarded production licenses were named numerically, corresponding to the abstraction of the grid and its unknown potential. Philipps was awarded three licenses, with license 018 covering blocks 1/5, 2/4 (Ekofisk field), 2/7, and 7/11. At the discovery and release of oil or gas from an anonymous block, a new and significant geographic place is established. Ekofisk is an elaborate assemblage of extraction and production infrastructure erected above the choppy surface of the North Sea, beyond the safety margin of the "one-hundred-year wave" (the estimated maximum North Sea wave height that determined the platform level was set at 78 feet/23.77 m at Ekofisk in 1971). But it is more fundamentally an ancient reservoir of hydrocarbons dating from the Paleocene, comprising an oil column of over 300 m in height located at depths between 2,900 and 3,250 below the seafloor—depths which dwarf the spindly twentieth- and twenty-first-century structures perched above the seafloor and piercing its interior. This reservoir is part of the geological system also named after Ekofisk—the Mandal-Ekofisk petroleum system of sedimentary rocks called a pod of active source rock, which encompasses a vast area of around 90 × 280 km below the central North Sea. The Ekofisk reservoir itself is one of many within this formation, its outline measuring around 9 × 12 km. Through technological infrastructure, sites such as Ekofisk connect almost inconceivable scales of time and space, which through production processes have been transformed into nearly equally inconceivable scales of wealth and power.

The naming of the Ekofisk field was random, but vaguely referred to an imaginary North Sea fish:

> Phillips had used letters to identify the various exploration areas on the Norwegian Shelf... It started with A-blocks, B-blocks, etc. The idea was to call the fields by the names of types of fish... in exploration area «C», it was easy to find the name of a fish to correspond—Cod. But what fish had a name that began with E? Eel had already been used on a structure on block 2/7. Earl Walters at Phillips' London office suggested Ekofisk, and no one objected. So Ekofisk it was—even though it should have been spelled Ekkofisk in Norwegian, or Echofish in English.[45]

Subsequent naming exercises on the Norwegian continental shelf were carried out more deliberately as the national oil industry progressed and after both Statoil and the Norwegian Petroleum Directorate were established in 1972–1973. The Ministry of Petroleum and Energy established rules regarding the naming of fields on the Norwegian continental shelf: oil companies could suggest names, which must then be approved by the Norwegian Petroleum Directorate, based on a preestablished list of approved names. According to Norwegian language specialist Botolv Helleland, oil field names fell into five categories: fish and seabirds, mythology, sagas, fairytales, and miscellaneous, with mythological names making up about 45 percent of the total.[46] Philipps named their second field Edda in 1972—a name originating from a group of poems on Norse mythology. The chief god in Norse mythology, Odin, his wife Frigg, his sons Tor, Vale, Balder, and Brage, his brother Vilje, and his spear Gungne all lent their names to Norwegian oil fields.

The Ministry of Petroleum and Energy clearly intended to link the oil fields to symbolic figures and to Norway's cultural history:

> Names are important symbols. This is also the case for petroleum deposits. The names of many fields in Norway are taken from Norse mythology, with strong roots and steeped in national tradition. This is a tradition that should be continued.[47]

Helleland argues that the mythological names reflect the pioneering spirit and sense of adventure that predominated in the initial age of Norwegian oil.[48] While many international companies also operate on the Norwegian continental shelf, national control executed by the Norwegian Petroleum Directorate and the dominance of Statoil resulted in a nationalistic, colonial approach to the naming of fields in 1972–2010, marking Norwegian space at sea and contributing to the forging of oil identities indirectly related to heroic Norse figures.

By 2010, the ministry also stated that the repertoire for these types of names had already been largely exhausted during the first phases of oil production on the Norwegian continental shelf and that names for the new era should reflect the oil industry's importance, "to ensure that they fit into a national context and history."[49] The Norse goddess Frigg presides over the exemplary Norwegian gas field that crosses the international border into UK waters. Both marking the maximum geographical limits to extraction potential on the Norwegian continental shelf and breaking new ground in international law through joint international exploitation, Frigg is an unmistakable nationalist symbol and physical border post in an artificially constructed, mythological sea of oil (Figure 6.4).

FIGURE 6.4 Plaque to the goddess Frigg. *Source*: Nancy Couling.

Through these practices, the construction of the offshore petroleumscape is both linked to something distant and surreal and to heroic acts of mythological proportions which reinforce a sense of national pride and identity. Such narratives correspond closely to the "petroleum imaginaries": as Hein observes, film and literature have "celebrated oil as a heroic partner in creating contemporary society and our identity."[50] Rather than the dissonant reality of concrete and steel installations occupying the sea, the modern industrial narrative is disguised in powerful premodern and superhuman imagery.

The Norwegian petroleumscape did not expand incrementally into the offshore realm; rather, it emerged directly at sea. Due to the location of formations, there was no established landside petroleum extraction industry as there was around the Gulf of Mexico. Norway was already an established maritime nation and it was through expertise in seafaring, maritime trade, and shipbuilding that the Norwegian industries initially adapted to oil.[51] As these traditions were steered into the oil industry, the petroleumscape became synonymous with a new maritime identity of oil. In addition to the official naming practices linked to Norse mythology, workers recruited from traditional maritime industries created an additional stream of Norwegian maritime culture for the offshore petroleumscape.

Fishing had previously provided Norway's primary export earnings. In comparison to long fishing tours of up to seven weeks, followed by one week off, working on a rig at the currently established Norwegian norm of two weeks on and four weeks off is considered an acceptable rhythm. Offshore workers talk about the ways in which maritime culture has been diverted into oil; most coastal Norwegians love the sea, and despite the tough working conditions offshore, workers have a keen sense of adventure and wanted to experience this new challenge.[52] An electrician who worked for thirty-five years at Ekofisk described how regular office work did not suit him and how the large numbers of ex-seamen familiar with maritime life knew how to create a good working atmosphere offshore. Strong bonds were formed with workmates.[53]

Oil was an industry of which the ex-seamen had no previous knowledge, but the oil industry did offer opportunities to apply existing expertise to new endeavors. Along with shipbuilding, Norwegians were also experienced in hydropower, and as the construction challenges in the North Sea emerged in full, it was the technology of concrete construction that Norway adapted and promoted for the offshore context. Concrete Deepwater Structures (Condeep) produced fourteen gravity base structures (GBS) in the North Sea from 1975 to 1995. The last and tallest of the Condeep structures built for the Troll field, located in 300-m-deep waters, was 360 m high, contained 245,000 m^3 of concrete, and 100,000 tons of reinforcing steel. At the time (1995), it was considered the tallest structure ever moved by humans.[54]

In order to acknowledge the human ingredient among the giant offshore machines and the individual contributions to the oil industry, documentation projects, including oral histories, have been carried out in connection with the Norwegian Industrial Heritage project and the British Library and Aberdeen University's "Lives in the Oil Industry" project.[55] Offshore oil quickly established itself and became a newly established "maritime" way of life around the North Sea.

The Post-Oil Petroleumscape

After the 2002 merger of Conoco and Phillips, the company announced the Ekofisk Area Growth project to improve oil and gas recovery at Ekofisk. In 2011, the government approved the development of two additional fields in the region, Ekofisk South and Eldfisk II, hence Ekofisk is expected to continue producing until 2050. The Norwegian petroleumscape is far from retired; in fact, the Ministry of Petroleum and Energy cite their current objectives and challenges as including "improved recovery from fields, development of discoveries and proving undiscovered resources."[56] In the example of Ekofisk, improved knowledge and technology means that the original expected recovery rate of 17 percent has now been improved to around 50 percent.[57] The official Norwegian petroleum strategy is justified by the overriding societal benefits gained through the industry, which has developed into what I would call an overwhelming *ecology of oil* penetrating all levels of Norwegian society. Comparable to a complex system of interdependent parts, like an ecosystem, the logic of oil permeates multiple spatial, organizational, and conceptual levels. This corresponds closely to the concept of feedback loops to maintain dependence characteristic of the "palimpsestic global petroleumscape."[58] In the words of the ministry, "Since the 1970s, the substantial revenues from the activity have helped build the Norwegian welfare society.... The petroleum resources should also contribute to improving the quality of life in Norway in the years to come."[59]

This line of argument has been sharply criticized by Latin American environmentalist Eduardo Gudynas in his discussions of neo-extractivism, where the state has become more involved in resource extraction and invests in social programs to "generate legitimacy" and to pacify local demands, such as those regarding environmental hazards.[60] Norwegians have been largely convinced that their Government Pension Fund is a wise and exemplary investment. In terms of the inherent meaning of sustainability, it is ironic that securing monetary resources for future Norwegian generations has been based on the exploitation of nonrenewable resources. With the current state of global carbon dioxide emissions failing to reach agreed objectives, vast investments will urgently be required in order to repair the global damage caused by petroleum.

Although Norway has not yet turned away from petroleum, there have been moves to prepare for a post-oil future. The offshore terrain established by North Sea petroleum provides the legislative framework, some of the technical expertise, and an established infrastructure for the energyscape of renewables at sea. Wind has emerged as a second-generation energy source in the North Sea and the wind industry is retrofitting the offshore petroleumscape as the oil industry begins the process of decommissioning. Germany and Denmark, both nations with a tradition of innovation in onshore wind energy but without significant offshore hydrocarbons, have been the European pioneers in offshore wind. Legislation moved forward particularly swiftly in Germany, the first European country to establish a Spatial Plan for its Exclusive Economic Zones in the North and Baltic Seas in 2009.[61] The North Sea plan laid out vast priority areas for offshore wind energy development, seen by the German government as an answer to its ambitions regarding the transition from coal and nuclear power to renewables.

The European Environmental Agency has identified the North Sea as an important European location for offshore wind, due to good wind speeds in combination with shallow water (<50 m) and its close proximity to the heavy energy consumers of highly industrialized northern Europe.[62] Offshore sites are released by the state for development, and therefore both land-based issues of ownership and potential NIMBY, an acronym for the phrase "not in my backyard," disputes are avoided. Since the installation of the first offshore wind turbine in 1991 with a height of 52.5 m and a capacity of 0.45 MW, turbine sizes have increased exponentially, exploiting the advantages of the industrial scale offered by the offshore environment. GE Renewable Energy is currently releasing a 12 MW turbine: the Haliade X-12 stands 260 m high (only 60 m less than the Eiffel Tower) and has a rotor diameter of 220 m. Turbines of these dimensions require large downwind spacings; therefore the occupied surface is expanded accordingly to include unprecedented areas of delineated offshore space.

Petroleum ushered in urbanizing instruments which were applied to the North Sea, first producing a gridded territory of resource extraction and then an industrial site of energy production. The establishment of this offshore site has been instrumental in the conception and execution of offshore wind, preparing the way for its development. In their discussion of planetary urbanization, urban theorist Neil Brenner and colleague Nikos Katsikis refer to such previously "natural" areas that have been transformed into instrumentalized production sites across the globe as "operational landscapes."[63] Resonating with this definition, the North Sea has become an operational seascape, which once established becomes a semi-permanent feature. At the termination of a typical twenty-five-year operating license, offshore wind parks are "repowered" with the latest generation of turbines, but due to the difference in scale which determines the foundation layout, existing foundations cannot be used and the site undergoes another round of "creative destruction."[64] Just as the petroleumscape is continuously extending and rebuilding itself, so must offshore wind adapt to technological improvements. Offshore, this means that it is not human habitats that will be displaced, but marine habitats, recently reestablished after the first round of construction, and the marine life they support.

Offshore wind and oil have become close neighbors in the North Sea as all littoral nations expand seawards with renewable energy production. This is not all they have in common; as part of the strategies to "optimize" remaining assets, wind and gas companies discuss possibilities of delivering "clean," cheap energy to unmanned gas rigs to power their production, in particular in the southern part of the North Sea.[65] Further potential synergies exist in the area of logistics—supply, crew, and safety vessels could be shared. The current 1,000 offshore turbines in the North Sea each weigh around 1,000 tons—equivalent to an unmanned oil or gas installation, therefore foundation systems are comparable and certain technological solutions can be exchanged.

In the German North Sea, the manpower required to maintain wind farms has been grossly underestimated. Consequently, accommodation options must be considered and the wind park Dan Tysk (Vattenfall) has constructed a stand-alone fifty-berth accommodation platform 90 km from shore. As the oil industry moves increasingly toward automated operations at the remaining stations, the wind industry is physically expanding offshore, requiring more diversified functions such as convertor platforms, small power stations, and more permanent accommodation.

Companies previously uniquely associated with oil are now moving into wind energy, demonstrating the direct link between the offshore petroleumscape and its successors. In May 2018, Statoil announced it was changing its name to Equinor. Pursuing similar development strategies as many oil companies and state institutions, Statoil has long been expanding into renewables, and recently it boasted the completion of Hywind, the world's first floating wind park off the coast of Scotland. Equinor has secured consent for developing the Creyke Beck B wind park at the Dogger Bank. They claim it is the largest zone available for wind development in UK waters and will accommodate 200 turbines in water depths of 20–33 m. Currently the world's largest offshore wind park is the Walney Extension in the

Irish Sea, with 87 turbines, a total capacity of 659 MW, and covering an area of 145 km^2.[66] The refashioned Equinor presents itself as a socially responsible company moving toward gas, solar, and wind energy, but says less about the opening up of new oil and gas extraction sites in the delicate Barents Sea, one of Europe's last intact marine ecosystems.[67]

The North Sea petroleumscape is proving to be a valuable asset for future energy scenarios. Depleted reservoirs are being used as the sites for carbon capture and storage. For example, the Utsira formation receives carbon dioxide produced by the processing of gas and condensate at the Sleipner platform, operated by Equinor.[68] In the Netherlands, studies are being carried out on the feasibility of converting electricity from renewable sources into hydrogen. Storage of renewable energy is a problem, therefore its transformation into hydrogen enables renewable energy to be more readily stored as well as to be directly used for mobility, transport, and further industrial applications. The existing petroleumscape provides useful spatial resources; possible storage sites include depleted gas deposits on the mainland and the continental shelf as well as large areas containing tanks comparable in size to existing facilities for the storage of oil and other petroleum products at ports such as Rotterdam.[69] Technology enables the physical extremities of the petroleumscape, drilled to depths reaching 3,000 m below the seabed, to be recycled for either the excess "products" of a petroleum-dependent society or for the as-yet uncertain flows of alternative forms of energy.

Conclusion

The North Sea petroleumscape is a sprawling and unfamiliar agglomeration of grids, gods, and giant machines, which challenges our inherited understandings of geographic place. The resulting combination of contradictory elements is characteristic of extended urbanization processes and, in the North Sea, petroleum was a major force in the initial unleashing of these processes. Governments borrowed the classic urban ordering device of the grid to organize petroleum licensing areas, hence essential referential coordinates were established for potential new geographies of oil. These references then developed into clusters of platforms and complexes such as Ekofisk City, which became international centers of work, recreation, and the production of wealth, around which uninterrupted global activity has revolved. While physically remote, dispersed, specialized, and constrained, these sites accommodate critical activities which serve urban centers around the North Sea and beyond.

In the Norwegian North Sea, new offshore places were linked to mythology and Norse culture through strategic naming practices and have since become established virtual places in the national landscape of industrial cultural monuments. The petroleumscape is hence a cultural artifact, entrenched in concepts of national identity. Two generations of workers have constructed and populated it and they are now partly dismantling it; however, its efficiency continues to be optimized by technology and remotely controlled subsea templates represent the state of the art. The petroleumscape was responsible for launching a radical, unprecedented transformation of the North Sea space, and once installed, current developments illustrate its immutability. Together, governments and multinational corporations are retrofitting and reusing it for hydrogen and carbon dioxide storage. The logic of oil continues to permeate strategic narratives of offshore renewables and to perpetuate the vast and complex network of spaces and channels it has carved out of the North Sea.

Notes

1 Stig S. Kvendseth, *Giant Discovery. A History of Ekofisk through the First 20 Years* (Norway: Phillips Petroleum Company, 1988), 64.
2 Henri Lefebvre, *The Production of Space*, trans. Donald Nicholson-Smith (Oxford: Blackwell, 1991).
3 Neil Brenner and Christian Schmid, "Towards a New Epistemology of the Urban?," *City* 19, no. 2–3 (May 4, 2015): 151–82, https://doi.org/10.1080/13604813.2015.1014712.
4 Nancy Couling, "Formats of Extended Urbanisation in Ocean Space," in *Emerging Urban Spaces—A Planetary Perspective*, eds. Philipp Horn, Paola Alfaro d'Alençon, and Ana Claudia Duarte Carduso (Springer International Publishing, 2018), XII, 219.
5 European Committee (UK), "The North Sea under Pressure: Is Regional Marine Co-Operation the Answer?," HL Paper (London: House of Lords, March 17, 2015), https://publications.parliament.uk/pa/ld201415/ldselect/ldeucom/137/137.pdf.
6 Juliette Roding and Lex Heerma van Voss, eds. *The North Sea and Culture (1550–1800): Proceedings of the International Conference Held at Leiden, 21–22 April 1995* (Hilversum: Verloren, 1996).
7 Michael Pye, *The Edge of the World : How the North Sea Made Us Who We Are* (London: Penguin Books, 2015).
8 Hugo Grotius, *Mare Liberum, Sive de Jure Quod Batavis Competit Ad Indicana Commercia Dissertatio* (Dutch Republic: Lodewijk Elzevir, 1609), https://upload.wikimedia.org/wikipedia/commons/7/7b/Grotius_Hugo_The_Freedom_of_the_Sea_(v1.0).pdf.
9 Edward Gordon, "Grotius and the Freedom of the Seas in the Seventeenth Century," *Willamette Journal of International Law and Dispute Resolution* 16, no. 2 (2008): 252–69.
10 Keith Chapman, *North Sea Oil and Gas: A Geographical Perspective* (Newton Abbot: David & Charles, 1976).
11 The first offshore oil wells were drilled off the California coast 6 km east of Santa Barbara in the Summerland oil field in 1896, accessed by wooden piers. The longest of these piers, the Treadwell wharf, was 360 m long.
12 Joseph A. Pratt, Tyler Priest, and Christopher J. Castaneda, *Offshore Pioneers Brown & Root and the History of Offshore Oil and Gas* (Houston: Gulf Pub. Co, 1997).
13 United Nations, "United Nations Convention on the Law of the Sea (A Historical Perspective)," 1998, http://www.un.org/depts/los/convention_agreements/convention_historical_perspective.htm.
14 United Nations, "Convention on the Continental Shelf," April 29, 1958, https://treaties.un.org/doc/Publication/UNTS/Volume%20499/volume-499-I-7302-English.pdf.
15 Alex Kemp, *The Official History of North Sea Oil and Gas; Volume 1: The Growing Dominance of the State*, Whitehall Histories: Government's Official History Series (UK: Routledge, 2014), https://abdn.pure.elsevier.com/en/publications/the-official-history-of-north-sea-oil-and-gas-volume-1-the-growin.
16 European Parliament and the Council of the European Union, "Directive 2014/89/EU Establishing a Framework for Maritime Spatial Planning" (European Union, July 23, 2014), https://eur-lex.europa.eu/legal-content/EN/TXT/PDF/?uri=CELEX:32014L0089&from=EN.
17 Nancy Couling, "Seascapes_ the Planning Challenge of the Century," in *History Urbanism Resilience, Planning & Heritage*, Vol. 4 (The 17th International Planning History Society Conference, Delft: TU Delft Open, 2016), 387–400, http://dx.doi.org/10.7480/iphs.2016.4.1305.
18 David Harvey, "The Geopolitics of Capitalism," in *Social Relations and Spatial Structures*, eds. Derek Gregory and John Urry (London: Macmillan Education UK, 1985), 128–63, https://doi.org/10.1007/978-1-349-27935-7_7.
19 Neil Brenner, *New Urban Spaces: Urban Theory and the Scale Question* (New York: Oxford University Press, 2019), 54
20 Brenner and Schmid, "Epistemology of the Urban?"
21 Henri Lefebvre, *The Urban Revolution* (Minneapolis: University of Minnesota Press, 2003); Neil Brenner, *Implosions/Explosions: Towards a Study of Planetary Urbanization* (Berlin: Jovis, 2014).
22 Nikos Katsikis, "On the Geographical Organization of World Urbanization," *MONU*, Geographical Urbanism, 20 (2014).
23 Chapman, *North Sea Oil and Gas.*
24 Kemp, *Official History.*
25 Ibid.

26 Chapman, *North Sea Oil and Gas*, 35.
27 Oystein Noreng, *The Oil Industry and Government Strategy in the North Sea* (London: Routledge, 2016).
28 Chapman, *North Sea Oil and Gas*.
29 Kvendseth, *Giant Discovery*.
30 Chapman, *North Sea Oil and Gas*.
31 Kvendseth, *Giant Discovery*.
32 Pratt, Priest, and Castaneda, *Offshore Pioneers*.
33 Carola Hein, "Oil Spaces: The Global Petroleumscape in the Rotterdam/The Hague Area," *Journal of Urban History*, February 13, 2018, https://doi.org/10.1177/0096144217752460.
34 Harald Tønnesen and Gunleiv Hadland, *Oil and Gas Fields in Norway: Industrial Heritage Plan* (Stavanger: Norwegian Petroleum Museum, 2011).
35 Tønnesen and Hadland, *Industrial Heritage Plan*.
36 Kvendseth, *Giant Discovery*, 7.
37 Kristin Gjerde Ø, "Ekofisk Industrial Heritage—How to Document and Communicate a Large Technical-Industrial Monument at Sea" (Society for the History of Technology Conference (SHOT), Minneapolis, 2005), https://www.academia.edu/37155653/Ekofisk_Industrial_Heritage3.doc.
38 Kvendseth, *Giant Discovery*.
39 Ibid.
40 Rig tours refers to the part of the working cycle spent on a rig offshore.
41 H. D. Trotter, "The Ekofisk Tank—A Concrete Gravity Structure In The Norwegian North Sea" (Annual Meeting Papers, Division of Production, American Petroleum Institute, 1974), https://www.onepetro.org/conference-paper/API-74-B001.
42 Kvendseth, *Giant Discovery*, 77.
43 Finn Sandberg H and Kristin Gjerde Ø, "Preserving Norway's Oil Heritage," in *Challenging Collections: Approaches to the Heritage of Recent Science and Technology*, eds. Johannes-Geert Hagmann and Alison Boyle, Vol. 11, Artefacts—Studies in the History of Science & Technology (Washington, DC: The Smithsonian Institution, 2019), 150–67, https://doi.org/10.5479/si.9781944466121.
44 "Kulturminne Ekofisk," accessed May 7, 2017, http://www.kulturminne-ekofisk.no/.
45 Kvendseth, *Giant Discovery*, 31.
46 Botolv Helleland, "Place-Names (Geographical Names) on and off the Coast of Western Norway: Names of Oil Fields Compared to Traditional Place-Names," Sixth International Seminar on the Naming of Seas: Special Emphasis Concerning the East Sea (The 29th International Geographical Congress: Special Session II, Geography and Place Names: Political, Seoul: Society for East Sea, 2000), 8–17, http://www.eastsea1994.org/data/bbsData/14630207711.pdf.
47 Ministry of Petroleum and Energy, "Storting White Paper 28 (2010–2011) – Unofficial Translation of Chapter 1. An Industry for the Future – Norway's Petroleum Activities," 2011, 2010, https://www.regjeringen.no/globalassets/upload/oed/petroleumsmeldingen_2011/oversettelse/chapter1_white_paper_28-2010-2011.pdf.
48 Helleland, "Place-Names."
49 Ministry of Petroleum and Energy, "Storting White Paper 28 (2010–2011) – Unofficial Translation of Chapter 1. An Industry for the Future – Norway's Petroleum Activities."
50 Hein, "Oil Spaces."
51 Tønnesen and Hadland, *Industrial Heritage Plan*.
52 Interviews conducted by the author at Ågotnes, near Bergen, Norway in April 2018.
53 Sigmund Trageton, *Oljeliv/Offshore ID* (Norwegian Petroleum Museum, 2018), https://www.norskolje.museum.no/en/how-is-life-with-mum-working-offshore/.
54 Tønnesen and Hadland, *Industrial Heritage Plan*.
55 Terry Brotherstone, "Lives in the Oil Industry. Oral History of the North Sea Oil and Gas Industry," University of Aberdeen, 2010, https://www.abdn.ac.uk/oillives/about/index.shtml.
56 Ministry of Petroleum and Energy, "Storting White Paper 28 (2010–2011) – Unofficial Translation of Chapter 1. An Industry for the Future – Norway's Petroleum Activities."
57 Tønnesen and Hadland, *Industrial Heritage Plan*.
58 Hein, "Oil Spaces."
59 Ministry of Petroleum and Energy, "Storting White Paper 28 (2010–2011) – Unofficial Translation of Chapter 1. An Industry for the Future – Norway's Petroleum Activities."

60 Eduardo Gudynas, "The New Extractivism of the 21st Century: Ten Urgent Theses about Extractivism in Relation to Current South American Progressivism," Americas Programe Report (Washington, DC: Center for International Policy, January 21, 2010), https://www.academia.edu/4242379/The_new_extractivism_of_the_21st_century_ten_urgent_theses_about_extractivism_in_relation_to_current_South_American_Progressivism%20and%20South%20American%20Progressivism.pdf.
61 Bundesregierung Deutschland, "Anlage Zur Verordnung Über Die Raumordnung in Der Deutschen Ausschließlichen Wirtschaftszone in Der Nordsee (AWZ Nordsee-ROV) Vom 21. September 2009" (Bundesanzeiger Verlag GmbH, September 25, 2009), http://www.bgbl.de/xaver/bgbl/start.xav?startbk=Bundesanzeiger_BGBl#__bgbl__%2F%2F*%5B%40attr_id%3D%27anlageband_bgbl109061.pdf%27%5D__1474573555793.
62 "The European Offshore Wind Industry – Key Trends and Statistics 2013" (Brussels: European Wind Energy Association, January 2014), http://www.ewea.org/fileadmin/files/library/publications/statistics/European_offshore_statistics_2013.pdf.
63 Neil Brenner and Nikos Katsikis, "Is the Mediterranean Urban?," in *Implosions/Explosions. Towards a Study of Planetary Urbanization* (Berlin: Jovis, 2014), 248–59.
64 Brenner and Schmid, "Epistemology of the Urban?," 168
65 Rob Hastings, "Optimising Value from the Southern North Sea" (Offshore Energy 2017 Exhibition and Conference, Amsterdam, October 9, 2017).
66 Ørsted, "Walney Extension," https://walneyextension.co.uk/, accessed July 6, 2020.
67 "The Barents Sea Ecoregion: A Biodiversity Assessment" (Oslo: WWF, 2004), http://d2ouvy59p0dg6k.cloudfront.net/downloads/barentsseaecoregionreport.pdf.
68 Tønnesen and Hadland, *Industrial Heritage Plan*.
69 S. F. van Gessel, "Past, Present and Future Perspectives on North Sea Developments. North Sea Heritage and Infrastructure – Geology and Energy Perspective" (Viscous Space – the offshore physicality of the North Sea between solid and liquid, TU Delft, June 21, 2018), www.viscousspace.com.

PART II

Oil, Materiality, and Cultural Practices

7

ARAB OIL TOWNS AS PETRO-HISTORIES

Nelida Fuccaro

The petroleumscape concept draws attention to how oil may underpin and structure everyday life in ways that are taken for granted and go unnoticed.[1] Pipelines may be underground and consumers may have no idea that the products they use are connected to the petroleum industry. Like refineries or oil rigs, oil towns that were built or expanded in the shadow of the oil industry might seem to be an exception to oil's invisibility as they illustrate petroleum's diverse histories in different corners of the globe. They are clearly identified with oil as an historical extractive industry in a way that New York or Oslo are not. As veritable "material" witnesses of this petroleumscape, linking together oil's past and present financial, infrastructural, and political streams, towns and cities closely associated with the industry seemingly defy petroleum's proverbial invisibility as commodity and underground resource.[2] Yet in reality, the history of these urban landscapes and of the experiences and images associated with them can make oil both visible and invisible. In other words, the physical, represented, and lived practices of oil towns are intertwined with petroleum in often subtle and discreet ways.

It is in the literary field—rather than in the historiography of the oil industry—that the black gold has been connected to cities. Contrasting Helon Habila's novel *Oil on Water* (2010) with 'Abd al-Rahman Munif's *Cities of Salt* (1984), Amy Riddle reads Munif's work as portraying oil as physically absent as a substance, but transformative as a market commodity with corrosive social, political, and cultural consequences. In Munif's narrative, oil is a mood of gloom, an atmosphere of loss and oppression, and crucially a dystopic urban landscape of damnation. The novel represents oil as part of the distorted developmental framework of foreign capital, personified by the American multinational corporation that operates in a fictional desert kingdom of the Arabian Peninsula. Yet it also presents its main protagonists—the city of Harran and the people associated with it, both Arab and American—as the almost inevitable physical and social formations of the early modern petroleumscape in the region.[3] In fact, as it unfolds in the novel, the making of Harran represents the prototypical development of the early oil towns of the Arab World and Iran. Built very quickly in a rather inhospitable environment by a foreign company, Harran is

DOI: 10.4324/9780367816049-10

culturally and socially segregated (there is an "American" Harran versus an "Arab" Harran), and brings to the locals a mixture of opportunity and excitement, but also alienation and despair, an ambivalence that was characteristic of early petroleum experiences and anxieties throughout the region.[4] This ambivalence was also part of the experience of Arab labor that did not come from oil-producing regions as evoked by the moving family history of the Khayyat sisters. They tell the story of their Hijazi grandfather, who was employed by the Arab American Oil Company (Aramco) between 1942 and 1975, and of their mother, who had an all-American childhood in one of the company's oil camps. Their upbringing is suggestive of the powerful cross-generational hybrid subjectivities that the family transported from Dhahran to Beirut once their mother relocated to Lebanon after marriage.[5]

Oil Towns as Infrastructural and "Hinge" Urbanism

It is no coincidence that Munif uses the fast developing oil town of Harran as the protagonist and stage set of his petro-story, a corrosive critique of the collusion between Arab governments and Western (most notably American) imperialism in an unspecified age of mounting economic and cultural colonization. In the Arab world, the history of oil urbanization, and of the associated political and social struggles it generated, is usually identified with the spectacular transformation of pre-oil towns into glitzy modern cities: think of Baghdad, Kuwait, and Manama in the 1950s and 1960s, and Riyadh and Dubai in the following decades.[6] Yet the urban domains connected to oil exploitation across the region were diverse, both in terms of the territories they occupied and the infrastructures that supported them. The oil company town was one of the most prominent of these domains, the center stage of petroleum's industrial urbanism, and the brainchild and headquarters of the powerful consortia of foreign companies that set up and controlled the industry throughout the Arab world until the 1970s.[7] Besides embodying the troubled history of oil under colonial and neocolonial conditions, these towns also exemplify the regional history of industrial capitalism: before the early twenty-first century, petroleum was the most important industrial enterprise in the Arab world. This capitalism was both extractive and corporate, and until nationalization it was dominated by a very powerful foreign cartel of multinational companies attached to old colonial powers such as Great Britain and France, and neocolonial newcomers such as the US. In Iraq, Bahrain, Kuwait, Saudi Arabia, and Libya—but also in Iran—oil towns became central to the large industrial conurbations that sustained the infrastructure of oil extraction. Initially built to serve this infrastructure, particularly in the early years of the development of the industry, in later years they functioned as central places of industrial growth and oil-related employment, key nodes in the management of oil operations, from infrastructure to work force. From the time oil was first discovered in Iran in the early twentieth century, oil towns underpinned the new geographies of petroleum refinement and transport that shaped local, regional, and global geopolitics. In underdeveloped and sparsely populated countries such as Saudi Arabia, Bahrain, Kuwait, and Qatar, they also set the pace of urbanization and urban modernization, and their expansion measured the demographic growth of the oil industry.[8]

The first of such towns was Abadan in southwestern Iran built by the Anglo Iranian Oil Company (AIOC) before World War I. Although located in Iran, Abadan was very close to Iraq and to the Arab states of the Persian Gulf. It was connected to both areas by strong cultural and political links and located in a region that had large Arab populations and was historically known as Arabistan.[9] The construction of Abadan was followed by that of Awali in Bahrain (1937), Kirkuk-Arappha in Iraq (1946), Ahmadi in Kuwait (1946), Dhahran, Ras Tannura, and Abqaiq in Saudi Arabia (after 1938), Umm Sa'id in Qatar (1949), and Marsa al-Burayka (also known as Marsa el-Brega) in Libya (1959). Abadan developed around a large oil refinery, a stunning and awesome sight that in the 1950s AIOC liked to portray in propaganda postcards as a holiday resort of sorts. Kirkuk's oil town Arappha (or in popular parlance *Kirkuk al-Jadida*, the New Kirkuk) developed as the terminus of the pipeline that connected Kirkuk's oil fields to the Mediterranean. Similarly, the Libyan town of Marsa al-Burayqa, built by Esso, owed its existence to the oil pipeline that transported crude to the coast. Ras Tannura in Saudi Arabia served the refinery and the port with the same name that was a major point of transport for Saudi crude, and Umm Sa'id in Qatar was developed as a tanker terminal. With the exception of Arappha, which occupied the northwestern corner of Kirkuk's urban area, all these towns were built in sparsely populated and inhospitable areas not too far from major oil fields. Around oil towns, and part of the infrastructure of oil production that gravitated around them, workers' camps mushroomed in the desert. Some were shantytowns that grew spontaneously, others were subsidized by the companies, which provided masonry buildings and basic amenities. These permanent settlements included the largest pumping stations along the Kirkuk-Mediterranean pipeline. Planned as miniature oil towns, they were hailed by the Iraq Petroleum Company (IPC) as harbingers of a new urban future[10] (Figures 7.1 and 7.2).

As the pivots of petroleum infrastructure, these towns supported the new and intersecting regimes of technology, expertise, work, and life created by the petroleum industry. As high-tech enclaves and administrative zones serving oil fields, they constituted nodes of

FIGURE 7.1 Aerial view of Ahmadi, Kuwait, 1956. Kuwait Oil Company Archive.

FIGURE 7.2 Aerial view of Ahmadi, Kuwait, 1956. Kuwait Oil Company Archive.

novel transport and communication links, most notably roads and airfields, but also telephone and telegraph lines. They shaped new technology and energy networks, oil wells, refineries, and pipelines, but also electricity grids. By the early 1950s, they stood at the center of the development of early information and knowledge societies that gravitated around the public relations offices, training centers, and media outlets supported by the companies striving to present the compassionate face of the oil industry to increasingly politicized local publics.[11] The influence of these towns crossed the political and cultural boundaries of the Arab nation-states that emerged or became consolidated after World War I. While physically rooted in national territories like the petroleum they owed their existence to, oil towns emerged as transnational and supranational enclaves, reflecting the cross-regional and global influence of the companies that controlled them.

Visually, experientially, and discursively, these glitzy oil enclaves differed starkly from existing urban and rural settlements, at least at the time of their construction. In some cases, until the 1960s and 1970s, they stood as a neat and orderly radical break with the urban past of the region, a fact that was very much emphasized in the propaganda of oil companies and local governments keen to portray oil as the benevolent modernizer of Arab societies. These towns fostered contrasting experiences as powerful connectors between old and new: between desert or rural hinterlands and industrial sites, between historical centers and oil fields, and between local politics and cultures and the neo-imperial and corporate worlds of British and American oil companies. In many respects, they functioned as urban "hinges"—to use Richard Sennet's metaphor for Istanbul[12]—that brought together disparate peoples, often through spatial and social segregation. They acted as powerful symbols and catalysts of collective memory and experience, and as models of modern urban life and architecture.

Ahmadi greatly influenced the development of Kuwait City in the 1950s and 1960s, in terms of residential and public architecture, family life, and even ideals of female behavior.

To this day, life in Ahmadi in the 1960s and 1970s evokes nostalgia among an older generation of Kuwaitis. As the primary symbol and model of American petro-capitalism in the Middle East, Dhahran and the oil camps built by Aramco in the Eastern Province of Saudi Arabia have produced intimate geographies of family, memory, and empire among the offspring of former Arab residents. These geographies have recently become manifest in recuperated material and mnemonic traces of life in the oil conurbation, traces that include family photographic collections and oil company archival and visual materials. In planning terms, Saudi Arabia's oil conurbation served as the model for the development of the first modern residential districts in Riyadh in the 1950s, particularly Malazz, located in the northeast suburbs of the Saudi capital. In Bahrain, Awali became one of the coveted symbols of the youth consumer cultures that epitomized Bahrain's oil modernization and shaped the country's political mobilization in the 1950s and 1960s.[13]

Oil and Urbanization: Circulation and Velocity/Blockage and Viscosity

As the geographical, social, and political footprint of the fast-developing petroleum industry in the Arab world, oil towns were important sites of mixing, negotiation, and transfer, in spite of the worlds of economic privilege and racial segregation they represented. As the material and discursive traces of early petroleum modernity, they reflected the dialectical powers of oil as a mineral substance, fossil fuel, and industry. To put it differently, and drawing on concepts such as circulation and velocity, these industrial/residential enclaves show how oil in the Arab world had the historical ability to set in motion and incrementally define and redefine the matrix of regional modernization.[14] This process started at different moments in time across the region following the diachronic discovery and exploitation of petroleum resources, first in Iran before World War I, followed by Iraq and Bahrain in the 1920s and 1930s, and Saudi Arabia, Kuwait, Qatar, and Libya after World War II.

The boom of Middle Eastern oil in the world markets after World War II transformed petroleum into the master of the mobilities and materialities that became associated with the process of city-making.[15] The growth of the industry imparted new direction to and intensified the circulation and exchange of people, commodities, and technical and scientific knowledge, playing a key role in the creation of new urban space and societies, inside and outside oil towns. Large waves of human bodies flowed to Iraq, Bahrain, Kuwait, Saudi Arabia, and Qatar from across the region and from Europe, the US, and the Indian subcontinent: skilled and unskilled workers, technicians, oil executives, and people working in key service sectors such as construction, transport, and air travel. In the smaller Arab states of the Gulf and in Saudi Arabia, employment figures for Arabs provided by oil companies until nationalization are generally misleading and do not account for the high mobility of labor across the region. This was the result of the unclear nationality of some of this labor force, but also because companies manipulated labor statistics to meet the terms stipulated by oil concessions regarding the employment of a certain percentage of nationals. After World War II, the industry experienced a labor boom that peaked in the mid-1950s in Saudi Arabia, Iraq, and Bahrain and in the 1960s in Kuwait. Large infrastructural projects such as oil towns, refineries, plants, oil storage tanks, and pipelines brought in considerable numbers of skilled and unskilled personnel from neighboring regions and farther afield. Bahrain relied on a large workforce from Iran for the construction of the refinery at Sitrah between 1936 and 1938. During the early postwar expansion of oil installations in the Eastern province,

Aramco employed large numbers of Italians coming from Eritrea through a recruitment scheme that ended in 1954.[16]

The influx of new objects, particularly cars and consumer items, Western household furniture, and construction materials, also played a considerable part in defining petroleum's new urban worlds. Beginning in the mid-1950s, the petroleumscape of Tehran developed in line with US styles of consumption and car use.[17] Company planners designed and built oil towns as car and consumer-oriented societies following the blueprint of British and American suburbs. These models of urban and suburban living often had a powerful impact on the lives of those who resided outside company towns, oil camps, and industrial areas. As early as 1937 in Manama, the capital city of Bahrain, which experienced precocious oil modernization along the Gulf's Arab coast, local merchants started to import European and American goods for the residents of Awali, Bahrain's oil town. By the early 1950s, the taste of the city's upper classes was radically transformed. The interiors of their houses had tables, chairs, and curtains rather than the traditional carpets and cushions. The supply of electricity encouraged the richest households to buy refrigerators, fans, gramophones, and radio sets. Cars became a coveted status symbol for many households, and their numbers skyrocketed between 1960 and 1967[18] (Figures 7.3 and 7.4).

Finally, the increased circulation of technical reports, glossy magazines, newsletters, and propaganda materials produced in the headquarters of oil companies contributed to creating more politically and socially aware individuals, often in juxtaposition with a vociferous, anti-imperialist, and nationalist press. This new type of literacy encouraged locals on the payroll of companies to acquire new skills and technical expertise in order to escape poverty and an oppressive labor regime that favored white and Indian employees. Company magazines and broadsheets, which started to appear in both Arabic and English in the late 1950s, reached considerable popularity: *al-Najmah al-'Usbu'iyyah* (the *Weekly Star*) in Bahrain, *Iraq Petroleum* and *Ahl al-Naft* (*Oil People*) in Iraq, *Risalah al-Naft* (the *Oil Newsletter*)

FIGURE 7.3 Dhahran, Saudi Arabia, Toiletries store, 1958. Ali Akbar Bushihri Archive.

FIGURE 7.4 Oil Camp, Kuwait. Kiosk and Canada Dry advertising for locals, 1961. Kuwait Oil Company Archive.

and *The Kuwaiti* in Kuwait, and *Aramco World* in Saudi Arabia. Emerging oil departments within local governments also started their own publications. The Directorate of Petroleum and Mineral Affairs in Saudi Arabia is one example: under the leadership of the famous oil technocrat ʻAbdallah al-Tariki, the directorate started publishing a four-page newsletter on oil affairs in the late 1950s.[19]

As suggested by Erik Swyngedouw—a geographer concerned with the relationship between nature, society, and social power—the city and urbanization can be approached as circulatory systems underpinned by but also as generating a series of metabolic relations. Besides mobility and circulation, metabolism and speed are thus key features of urban modernity as a qualitative and quantitative process in a transformation involving both human and material worlds.[20] Petro-urbanism—in the Arab world as in many other corners of the globe—can thus be explained as a process of circulation and recirculation of matter and resources. While the extraction of oil from the ground dislocated and de-territorialized it as a natural resource, the construction of oil towns re-territorialized petroleum as built environment and infrastructure. This process was activated by the wealth generated by oil as a commodity and the technical expertise and infrastructural capital provided by foreign companies.[21] By the 1950s, the metaphor of oil as the most vital part of a human body involved in a series of metabolic reactions had become part of a new vocabulary of resistance to the powerful oil corporations deployed by Arab oil technocrats working for governments and the Arab League. In 1960, Ashraf Lutfi, a Kuwaiti-Palestinian who became secretary general of OPEC in 1965, made an impassioned plea for "Arab Oil for the Arabs":

> Petroleum, then, is to the world as blood is to the human body, and both would expire without them. In this simile, an oilfield might be likened to a human heart. If the

> body is to remain healthy, care must be taken of the heart which pumps blood to the various parts of the body, supplying them with the essential daily nourishment. If the heart should fail, the body would come to a standstill.[22]

The cultural theorist Peter Hitchcock has recently suggested that there is a flipside to the idea of oil's velocity and circulation as the life-enhancing, revolutionary, and creative force of modernity. As a mineral substance, oil sticks and its viscous properties often impair movement, creating inertia.[23] Translated into the urban worlds of petroleum in the age of Western imperialism, this has meant that oil bonanzas often fragmented political and spatial experiences, both individually and collectively. In this respect, the history of early petro-urbanism in the Arab World and Iran can be also read as one of circulation and blockages, of velocity and viscosity, of friction and violence, a history that became especially visible in oil towns as exclusive industrial enclaves.

In these towns, the effects of petroleum flows generated a great deal of instability, both political and physical. As key nodes of industrial conurbations, oil towns represented interconnected points of social and political "vulnerability"—to use Timothy Mitchell's wording—and they became enclaves of order and disorder dominated by the politics, geographies, and disciplines of oil production.[24] As early as the late 1940s, the oil conurbations that developed around Abadan in southwestern Iran, Dhahran in the Eastern Province of Saudi Arabia, and around Kirkuk in northern Iraq, including the pipeline that transported Kirkuk's oil to the Mediterranean, ceased to be frontier regions of exploration and exploitation. They had become frontlines and vanguards of nationalist and left-wing politics, trade unionism, new forms of surveillance, urban planning, and state violence. Before the 1960s, the shantytowns and informal communities that accommodated indigenous and migrant laborers who lived within or near the perimeters of oil towns often had violent histories. In 1942, Iranian workers living in the shantytown of Ahmadabad in Abadan attacked neighboring Indians in what became known as the Bahmashir incident. Infuriated mobs, resentful of the better conditions enjoyed by Indian employees of AIOC, expelled Indians from their houses and chased them.[25] Mobilization against the companies resulted from dissatisfaction with slowly improving living and working conditions, differential treatment meted out to different communities, which was exacerbated by the close (yet separated) proximity in which workers lived in oil towns and camps, and the unfulfilled expectations of modernization that urbanization had created. In 1953–1956 and in 1967, sustained labor unrest beset the Saudi oil conurbation. In 1967, there were indications that in the "native" oil towns of Ras Tanura and Abqaiq, which were inhabited by Saudis, class divisions mobilized youths not exclusively against Aramco but also against Saudis, who enjoyed better living conditions.[26]

A number of interesting questions emerge in relation to how labor unrest and criminal activities triggered changes in urban form. In 1965, the sabotage of oil installations in Marsa al-Burayqa prompted Esso Libya to establish army checkpoints and to erect security barriers around the town, exacerbating the separation of Libyan workers that lived around it in tents and makeshift dwellings. The securitization of oil towns and camps after episodes of unrest and hooliganism was a common practice across the region. In redefining living spaces and experience, labor and political struggles also subverted hierarchies of power. A case in point is the takeover of the large K3 oil station along the Kirkuk-Mediterranean pipeline by labor and Communist activists in 1948. In an act of policing in reverse, strike leaders established a new Communist order inside the station through the organization of committees to manage security, negotiations with IPC, and food provisions.[27] The sites where towns were built could become *lieux de mémoire* linked to past violent events. For instance, Marsa al-Burayqa in Libya stood on the site of a concentration camp for Cyrenaica tribes that had resisted Italian colonial rule.[28]

Another unsettling dimension of the early petroleum age was—to paraphrase Reinhardt Koselleck—the tension between spaces of urban experience and the horizons of expectation.[29] In the fictional world of Munif's *Cities of Salt*, oil permeates the mindset of the characters—in the words of Amy Riddle—as "some kind of wealth or gold to come."[30] In reality, the advent of oil undoubtedly altered expectations of modernity and of the future, but the solidity, security and fulfillment promised by this future could be deceiving, as exemplified by modern architecture and planning. Observers of early Kuwaiti oil urbanism in the late 1940s and 1950s explained the accelerated development of the oil conurbation linking Kuwait City to Ahmadi and to the oil fields in terms of *orderly chaos* and *disconnection*, casting doubts over the benefits accrued by the indigenous population.[31] Seemingly, solid built environments and modern infrastructure—which particularly in remote areas stood as the unmistakable markers of Vitruvian/European modernity—proved to be impermanent. A large number of construction failures—from houses to roads—marked the UAE's late oil boom in the late 1960s because cement and other construction materials were improperly treated with salt water. Eventually, a special UN commission was dispatched to Ras al-Khayma in 1974 to investigate the matter.[32] The tension between real and ideal and expected and experienced was also observed by the British anthropologist Peter Liendhart in his perceptive account of Kuwait in the 1950s, appropriately titled *Disorientations–A Society in Flux*. In the outskirts of Kuwait town, Liendhardt hopes to find a rich site to conduct fieldwork focused on settled Bedouin communities. With some disappointment, he soon realizes that his expectations are not going to be fulfilled, because in Kuwait's new suburbia, Bedouin are being replaced by foreign migrant workers, employed in oil installation and in the thriving construction industry of Kuwait's oil boom.[33]

Twentieth-Century Frontier Urbanism: Imperial, Colonial, and National Modernities

The industrial urbanism of Middle Eastern oil towns flourished in the midst of a generally wild, inhospitable, and rugged frontier, in the desert and mountainous regions of southwestern Iran, northwestern Iraq, Bahrain, Kuwait, Saudi Arabia, and Libya. This frontier was not dissimilar to that which since 1900 had shaped much of Canada's industrial development around staple industries such as cod fishing and mining. As with other commodity frontiers, the development of oil resources, the extension of central government, and the process of nation-building were to different degrees part of a larger narrative of British colonialism and of new trajectories of neo-imperial/colonial expansion ushered in by the advance of American power after World War II.[34] A case in point is Dhahran, the most tangible manifestation of US control over the oil industry in the Arab world, and of the ability of petroleum's powerful corporations to act as state-makers. Dhahran was in fact the instrument through which, what was known historically as, Eastern Arabia was incorporated into the Saudi state, whose modernization and integration into the world economy depended on oil production (Figure 7.5).

FIGURE 7.5 Signpost indicating distances to Kirkuk, Baghdad, Paris, London, and Beirut outside the K3 Pumping Station, Iraq. BP Archive.

Throughout the region, oil towns were unmistakably enmeshed with the corporate values of oil companies that celebrated the maximization of profit and made companies urban entrepreneurs of the early oil age. Yet oil's industrial urbanism reflected diverse ideals of imperial/colonial modernity. The planning of Abadan, Ahmadi, and Kirkuk/Arappha, as convincingly shown by Mark Crinson and Reem Alissa, reflected what Alissa has termed "colonial capitalism." It followed the blueprint of colonial New Delhi as planned by the architect James Mollison Wilson (1887–1965) who worked for the AIOC, IPC, and Kuwait Oil Company (KOC), all of which were under the spell of the British government. While the garden city movement had an important influence on Wilson, the guiding principles of the planning of oil towns under British control were the same ones that underpinned the creation of so-called dual cities around the nonwhite world in the high noon of colonialism: the enforcement of ethnic segregation and the drawing of color lines.[35]

Racial hierarchies also played an important role in the development of Dhahran and Awali, the two oil towns built by American companies Aramco and the Bahrain Petroleum Company (Bapco). In the initial model envisaged by company executives, pay grades rather than color lines were to structure the organization of settled space. As the development of Dhahran and the oil camps that surrounded it unfolded, relations between different resident communities came to resemble that of earlier colonial encounters in Africa and Asia, with the inevitable creation of zones of exclusion, color lines, and the deployment of surveillance techniques to contain social unrest. Developed as an oil camp after 1938, the year oil was discovered in commercial quantities in Saudi Arabia's Eastern Province, Dhahran was built around a fenced enclave, known until the early 1950s as American Camp (later renamed Senior Staff Camp, but popularly referred to as White Men's Camp). Around the American compound, Aramco planned a constellation of separate worker camps populated by Saudis, Pakistanis, Italians, and Indians, many of whom were skilled laborers. In addition to the segregation, there were gross disparities in living quarters, wages, rights, and access to services and amenities. For many locals, who were arguably at the bottom of the socioeconomic order promoted by Aramco, the oil camp complex of Dhahran came to represent the "shock" as well as the "allure" of the modern, acutely felt from their *barasti* dwellings made of palm leaves, which well into the 1950s dominated the landscape of Dhahran's "Arab village."[36]

Central to American oil urbanism was the model of suburbia. Images, spaces, and experiences of this suburbia dominated the life of Dhahran. In 1956, the "White Men's Camp" was described by the journalist Wanda Jablonski as a frontier organization that looked like "a lovely suburb of Houston."[37] Other American observers characterized its villas as Californian-style ranch homes. Until 1971, Awali in Bahrain, an oil town that followed a Caltex standard plan but with a British suburban style, was a key pole in the urban geography of white/European settlement in Bahrain, alongside the British military and political establishments.[38] Oil towns such as Dhahran and Awali were part of broader plans to transplant the American ideology of suburbia in the Arab world after World War II. In other words, they fitted in very well with community-building projects overseas, which construed suburbia as a site of both consumption and capitalist values. This was a reflection of—in the words of Nathan Citino—"the suburban single family home becoming the most sought after commodity in the US alongside cars, both sites and object of consumption." British oil companies continued to see industrial urbanism through the old lens of the "white man's burden" which had shaped British colonialism in the nineteenth century. In contrast, after World War II, most American companies and entrepreneurs saw corporate investment as coterminous with modernization. Central to this investment was the model of suburbia driven by supposedly universal middle-class values.[39]

The irony was that in spite of ubiquitous suburban spaces and facilities, life in oil camps and towns did not appear to perceptive observers as particularly suburban or urban. In the mid-1950s, Roderic Owen, commenting on Umm Sa'id in Qatar and Awali in Bahrain, noted:

> They [the oil camps] are an isolated, unnaturally young, society organized on paramilitary lines where salary scale replaces rank. The result is bound to be a mixture of the holiday camp and the army, with hints of a new satellite town more predominant as the number of wives and families increases ... It is a common criticism to call oil camps suburban.... People in oil camps have no city.[40]

While some European and American residents longed, like Owen, for the excitement of their hometowns and felt deprived of metropolitan life, the locals—particularly those on the company payroll—often saw modern housing and facilities as a symbol of affluence, social advancement, and a better life. Unequal and unfair housing policies and the inability on the part of many locals to gain access to oil towns proved extremely divisive and dangerous. Demands for better housing and access to basic amenities often triggered rioting in the Saudi oil conurbation in the 1950s and 1960s, with workers' demands often framed in the context of labor grievances and radical Arab nationalism.[41] In Iraq's oil townships, inequality eluded the logic of European-versus-indigenous employees of IPC. Iraqi Christians and Armenians, the privileged "minority" groups that occupied higher positions in the industry and lived in modern accommodations, experienced constant threats and intimidation by labor activists. In Arappha/Kirkuk, many Christians complained of insufficient police protection and the insecurity of modern housing, which were more vulnerable to theft and intrusion than traditional dwellings. In American oil towns, company housing policies often created friction among white employees as nuclear families received preferential treatment over single males. The so-called bachelors' quarters in Dhahran and in Awali were not particularly happy places, with residents often complaining of poor facilities and social isolation from other white residents. In the early 1960s, the expanding American female workforce in the Libyan town of Marsa al-Burayqa also lamented their loneliness and discrimination in the workplace.[42]

In the 1950s, companies started to provide home ownership loans as an attempt to integrate the indigenous workforce into capitalist modernity. First introduced in the new industrial towns of North and South America at the beginning of the twentieth century as part of new ideas that underpinned the development of the new industrial town, this social program yielded mixed results in the Middle East.[43] The Aramco home ownership scheme started in 1951 in order to tackle the issue of the mushrooming of shantytowns inhabited by rural migrants employed or searching for employment in the oil fields. The scheme did not prove popular with locals, and by 1960 applications sharply declined. Company surveys found that women in particular did not like living in modern housing: they felt isolated from extended family and abhorred the anonymity of the town. After having considered the political implications of building houses for the Iraqi workforce in Kirkuk's suburban oil enclave of Arappha, IPC decided to initiate a home ownership scheme in the old city center of Kirkuk. This scheme was readily taken up by Turkmen, Arabs, and Kurds, but not by Christians and Armenians, who preferred to live in the secluded comfort of Arappha as company protégées.[44]

After nationalization, company towns became the exclusive property of governments. As white residents were forced to vacate and were replaced by locals, the towns became the architectural and social settings of a new national modernity, primary instruments of state and nation-building. Ahmadi in Kuwait provides an excellent example. The propaganda documentary "Story of a Town," shot in 1978 a few years after nationalization, is a celebration of national achievement with the nationalized company starring as a benevolent state enterprise. It portrays Ahmadi's public and private spaces of sociability—the residential suburb, the modern villa, the garden, the school, the hospital, the social club, the workshop, and the classroom—as the archetypal stage sets of Kuwait as a modern nation.[45]

From the Material to the Representational

Oil companies were the first to promote the use of oil towns for propaganda purposes in the 1950s and 1960s. It is not by coincidence that "Story of a Town" was commissioned and first scripted in 1965, following instructions from the KOC head office in London.[46] Throughout the region, companies' public relations departments packaged alluring images of corporate urbanism in order to publicize locally the benefits of employment in the oil industry, to maximize productivity, and to dissipate discontent and avoid lingering industrial conflict. These new urban imaginaries, carefully constructed to target local publics, focused on particular spaces inside oil towns in order to enforce new modes of urban socialization and to promise local employees happy and fulfilled oil futures. This new corporate culture, which became superimposed on an American model of suburbia, was the result of a new rhetoric of indigenous development adopted by oil companies in the 1950s, when they came under the increasing attack of nationalism and anticolonial and anti-imperialist ideologies, both in the Arab world and internationally.[47]

Canteens were central to this rhetoric. One of the first development enterprises publicized by oil companies was the provision to local employees of subsidized "Oriental" meals—as they were frequently referred to by company managers—in between shifts in oil towns and in camps, in a clean, modern, and organized environment. The leading article of the first issue of *al-Najmah al-Usbu'iyyah*, (the *Weekly Star*), Bapco's Arabic publication, reported in detail the construction of the first modern refectory in Awali, placing emphasis on its state-of-the-art facilities: toilets, neon lighting, and air conditioning. In Iraq, by the early 1950s, all major centers of oil operations, including Kirkuk/Arappha, had refectories that featured prominently in company publicity and brochures. Canteens were not only publicized as an essential entitlement of workers but also as places of leisure and pleasure, part of the training in modern and "civilized" lifestyles offered by the companies. In towns like Ahmadi, which by the late 1950s also accommodated Kuwaiti families, these modern lunchrooms were often shown hosting parties and company and community celebrations.[48]

Yet, particularly in Iraq and Kuwait, by the mid-1950s the suburban villa had become the key setting in the imagery depicting oil urbanism as a new way of corporate life for local company employees. Company magazines started to devote considerable attention to housing developments sponsored by IPC and KOC or subsidized through home ownership schemes. Company housing was not necessarily located in urban areas or in large oil towns. Larger oil camps and stations that mushroomed in the desert were veritable mini towns, with facilities such as canteens, clubs, cinemas, and, in some cases, supermarkets and schools. Feature articles included aerial views, house plans, and images of leisure and social activities taking place.[49] Villas were publicized as urban skylines in their own right, as suggested by a number of striking images of houses built by IPC in the outskirts of Kirkuk. Published in *Iraq Petroleum*, these images show an empty and desolate suburbia which the caption describes as "imaginative town-planning."[50]

Company houses became the accessory that defined the image of the model Iraqi and Kuwaiti employee. Their modern design was often juxtaposed with images of family life that popularized the archetypal oil household as a nuclear family. Publicizing "moving in" days often served this purpose. An article published in *The Kuwaiti* in 1954, the English language magazine published by KOC, shows the first Kuwaiti family to take possession of a company villa in South Ahmadi. Four years later, IPC celebrated the Basra Home Ownership Scheme with the publication of a similar feature article in *Iraq Petroleum*.[51] This type of advertising was also used to maximum effect in articles that publicized model villas and the organization of viewings for prospective buyers eligible for home ownership schemes. These articles are highly suggestive of how the company instructed its workers in the art of modern urban living. They included pictures of fully furnished interiors, modern kitchens and dining rooms, linoleum floors and built-in cupboards.[52] By the early 1960s, company publications in both Arabic and English were also instrumental in linking the space and fittings of the modern oil villa to new models of female domesticity. These articles created the image of the oil town housewife, an image that in Kuwait had a profound influence on the lifestyle of the young female population in the 1960s and 1970s.[53]

Conclusion

Before nationalization, oil towns were central to the development of the petroleum industry in the Arab world. Looking at these towns as spatial, political, social, and discursive formations of foreign capitalism and imperialism—but also as the harbingers of indigenous modernities—offers insight into a complex set of oil relationships: between oil as a market commodity, mineral substance, and industrial/imperialist enterprise on the one hand, and as the material and ideological building block of modern societies on the other. These urban petro-histories objectified as part of the petroleumscape are also suggestive of a set of highly mobile connectivities, which accelerated as the development of the industry gathered momentum after World War II, when most of oil towns were built. These connectivities linked foreign companies, local governments, and indigenous populations in concentric circles, binding them at local, national, regional, and international/global scales. To go back to Ashraf Lutfi's allegory of petroleum as the blood in a human body, both the oil field and the oil town can be considered the pulsating heart of the early Arab oil industry.

Notes

1 Carola Hein, "Oil Spaces: The Global Petroleumscape in the Rotterdam/The Hague Area," *Journal of Urban History* 44, no. 5 (2018): 887–929, https://doi.org/10.1177/0096144217752460.
2 Ernst Logar, *Invisible Oil* (Vienna: Ambra Verlag, 2011).
3 Amy Riddle, "Petrofiction and Political Economy in the Age of Late Fossil Capital," *Meditations: Journal of the Marxist Literary Group* 31, no. 2 (2018), 55–74.
4 Abdelrahman Munif, *Cities of Salt* (New York: Vintage, 1987). On the building of American and Arab Harran, see pp. 198–239. For this ambivalence in the context of Bahrain see Nelida Fuccaro, "Shaping the Urban Life of Oil in Bahrain: Consumerism, Leisure, and Public Communication in Manama and in the Oil Camps, 1932–1960s," *Comparative Studies in South Asia, Africa and the Middle East* 33, no. 1 (2013): 60–67.
5 Yasmin Munira and Rola Khayyat, "Pieces of Us. The Intimate Imperial Archive," *Journal of Middle East Women's Studies* 14, no. 3 (2018): 268–90.
6 See, e.g., Farah al-Nakib, *Kuwait Transformed: A History of Oil and Urban Life* (Stanford: Stanford University Press, 2016); Nelida Fuccaro, *Histories of City and State in the Persian Gulf: Manama since 1900* (Cambridge: Cambridge University Press, 2009); Ahmad Kanna, *Dubai: The City as Corporation* (Minneapolis: Minnesota University Press, 2011); Caecilia Pieri, *Bagdad: La Construction d'une Capitale Moderne (1914–1960)* (Beirut: IFPO, 2015).
7 On company towns in Iran and the Arab world, see Ian J. Seccombe and Richard Lawless, *Work Camps and Company Towns: Settlement Patterns and the Gulf Oil Industry* (Durham: University of Durham, Centre for Middle Eastern and Islamic Studies, 1987); Nelida Fuccaro, "Histories of Oil and Urban Modernity in the Middle East," *Comparative Studies in South Asia, Africa and the Middle East* 33, no. 1 (2013): 1–6; Mark Crinson, "Abadan: Planning and Architecture under the Anglo-Iranian Oil Company," *Planning Perspectives* 12, no. 3 (1997): 341–59; Kaveh Ehsani, "The Social History of Labor in the Iranian Oil Industry: The Built Environment and the Making of the Industrial Working Class (1908–1941)" (PhD diss., University of Leiden, 2014); Reem Alissa, "Building for Oil: Corporate Colonialism, Nationalism and Urban Modernity in Ahmadi, 1946–1992" (PhD diss., University of California, Berkeley, 2012); Elisabetta Bini, "Building an Oil Empire: Labor and Gender Relations in American Company Towns in Libya, 1950s–1970s," in *Working for Oil: Comparative Social Histories of Labor in the Global Oil Industry*, eds. Touraj Atabaki, Elisabetta Bini, and Kaveh Ehsani (Cham: Palgrave Macmillan US, 2017), 313–36.
8 Seccombe and Lawless, *Work Camps and Company Towns*, 29–58.
9 For a discussion of the Arab workforce in Abadan, see Rasmus Christian Elling, "War of Clubs: Struggle for Space in Abadan and the 1946 Strike," in *Violence and the City in the Modern Middle East*, ed. Nelida Fuccaro (Stanford: Stanford University Press, 2016), 189–210.

10 Nelida Fuccaro, "Reading Oil as Urban Violence: Kirkuk and Its Oil Conurbation, 1927–58," in *Urban Violence in the Middle East: Changing Cityscapes in the Transition Between Empire and Nation State*, eds. Ulrike Freitag, Nelida Fuccaro, Claudia Ghrawi, and Nora Lafi (New York: Berghahn, 2016), 228.
11 Nelida Fuccaro, "Telling the Human Story of Oil: Oil Propaganda and Corporate Representations of Urbanism in Iraq, Kuwait and Bahrain," paper presented at the conference "The Comparative Social Histories of Labour in the Oil Industry," International Institute of Social History, University of Amsterdam, 14–15 June 2013; David H. Finnie, *Desert Enterprise: The Middle East Oil Industry in Its Local Environment* (Cambridge, MA: Harvard University Press, 1958), 90–141.
12 Richard Sennet, "The Hinge City," in *Living in the Endless City: The Urban Age Project by the London School of Economics and Deutsche Bank's Alfred Herrhausen Society*, eds. Ricky Burdett and Dejan Sudjic (London: Phaidon Press, 2011), 218–23.
13 Reem Alissa, "The Oil Town of Ahmadi Since 1946: From Colonial Town to Nostalgic City," *Comparative Studies of South Asia, Africa and the Middle East* 33, no. 1 (2013): 54–58; Fuccaro, "Urban Life of Oil", 68–69; Munira and Khayyat, "Pieces of Us"; Pascal Ménoret, *Joyriding in Riyadh: Oil, Urbanism and Road Revolt* (Cambridge: Cambridge University Press, 2014), 80–81.
14 Peter Hitchcock, "Velocity and Viscosity," in *Subterranean Estates: Life Worlds of Oil and Gas*, eds. Hannah Appel, Arthur Mason, and Michael Watts (Ithaca, NY: Cornell University Press, 2015), 46.
15 For an instructive comparative discussion on how intensified travel shaped debates on modernization in the Arab world in the Cold War era, see Nathan J. Citino, *Envisioning the Arab Future: Modernization in U.S.-Arab Relations, 1945–1967* (Cambridge: Cambridge University Press, 2017), 97–141.
16 Seccombe and Lawless, *Work Camps and Company Towns*, 8–27. On the very poor living conditions of the skilled and semi-skilled Italian workers employed by Aramco, see Seccombe and Lawless, 72–74.
17 Carola Hein and Mohammad Sedighi, "Iran's Global Petroleumscape: The Role of Oil in Shaping Khuzestan and Tehran," *Architectural Theory Review* 21, no. 3 (2017): 349–74.
18 Fuccaro, *Histories of City and State*, 197, 200; Fuccaro, "Urban Life of Oil," 63–67.
19 Stephen R. Duguid, "Technocrats, Politics and Planning: The Formulation of Arab Oil Policy, 1957–1967" (PhD diss., Simon Fraser University, 1976), 75–76. For a discussion of the making of new oil publics after World War II, see Nelida Fuccaro, "Oilmen, Petroleum Arabism and OPEC: New Political and Public Cultures of Oil in the Arab World, 1959–1964," in *Handbook of OPEC and the Global Energy Order: Past, Present and Future Challenges*, eds Giuliano Garavini and Dag Harald Claes (London: Routledge, 2020), 15–30.
20 Erik Swyngedouw, "Circulations and Metabolisms: (Hybrid) Natures and (Cyborg) Cities," *Science and Culture* 15, no. 2 (2006): 105–21.
21 For a discussion of cityscapes as part of the territorial infrastructure and geographies of fossil fuels, see Rania Gosh and El Hadi Jazairy, "Geography and Oil: The Territory of Externalities," in *Infrastructure Space*, eds. Ilka Ruby and Andreas Ruby (Berlin: Ruby Press, 2017), 361–76.
22 Ashraf Lutfi, *Arab Oil: A Plan for the Future* (Beirut: Middle East Research and Publishing Centre, 1960), 1.
23 Hitchcock, "Velocity and Viscosity," 47–60.
24 Timothy Mitchell, *Carbon Democracy: Political Power in the Age of Oil* (London: Verso, 2011), 103. For an elaboration of the concept in relation to Kirkuk's oil conurbation see Fuccaro, "Reading Oil," 231–37.
25 Rasmus Christian Elling, "On Lines and Fences: Oil, Community and Violence in an Oil City," in Freitag, Fuccaro, Ghrawi, and Lafi, *Urban Violence in the Middle East*, 197–221.
26 Claudia Ghrawi, "Structural and Physical Violence in Saudi Arabian Oil Towns," in Freitag, Fuccaro, Ghrawi, and Lafi, *Urban Violence in the Middle East*, 243–64; Claudia Ghrawi, "A Tamed Urban Revolution: Saudi Arabia's Oil Conurbation and the 1967 Riots," in Fuccaro, *Violence and the City*, 169–88.
27 Fuccaro, "Reading Oil," 235–36.
28 Bini, "Building an Oil Empire," 321–22.
29 Reinhart Kosellek, *Future Pasts: The Semantics of Historical Time* (New York: Columbia University Press, 2004).

30 Riddle, "Petrofiction," 65.
31 National Archives United Kingdom: LAB 13/483, Report: "Kuweit –Dates of Visit 3.2.49 to 10.2.49" by M. T. Audsley, 2 April 1949.
32 Matthew MacLean, "Building the Nation: A History of Rock, Cement, and Concrete in the United Arab Emirates," paper presented at the conference "Mobilities and Materialities of the Gulf and Arabian Peninsula," American University of Kuwait, 17–19 April 2017.
33 Peter Lienhardt, *Disorientations: A Society in Flux, Kuwait in the 1950s*, ed. Ahmad al Shahi (Reading, Ithaca Press, 1993), 29.
34 For an understanding of the making of the Middle East and Kuwait as a British imperial resource frontier driven by oil extraction, see William Beinardt and Lotte Hughes, "Oil Extraction in the Middle East: the Kuwait Experience" in *Environment and Empire*, Oxford History of the British Empire, Vol. 6 (Oxford: Oxford University Press, 2007), 251–68.
35 Crinson, "Abadan"; Alissa, "Building for Oil." On the garden city, see Ebenezer Howard, *Garden Cities of Tomorrow* (London: S. Sonnenschein, 1902).
36 Robert Vitalis, "Aramco World: Business and Culture on the Arabian Oil Frontier," in *Counter-Narratives: History, Society and Politics in Saudi Arabia and Yemen*, eds. Madawi Al-Rasheed and Robert Vitalis (New York: Palgrave, 2004), 160–66.
37 Anna Rubino, *Queen of the Oil Club* (Boston, MA: Beacon Press, 2008), 109.
38 Roderic Owen, *The Golden Bubble: Arabian Gulf Documentary* (London: Collins, 1957), 60.
39 Nathan J. Citino, "Suburbia and Modernization: Community Building and America's Post-World War II Encounter with the Arab Middle East," *The Arab Studies Journal* 13–14, nos. 1–2 (2005–2006): 42–43.
40 Owen, *Golden Bubble*, 134.
41 Ghrawi, "Structural and Physical Violence" and "A Tamed Urban Revolution"; Citino, "Suburbia and Modernization," 44.
42 Citino, "Suburbia and Modernization," 45; Fuccaro, "Reading Oil," 228–29; Fuccaro, "Urban Life of Oil," 68.
43 Oliver J. Dinius and Angela Vergara, eds., *Company Towns in the Americas: Landscape, Power, and Working Class Communities* (Athens: University of Georgia Press, 2011), 2–3.
44 Citino, "Suburbia and Modernisation," 46–49; Fuccaro, "Reading Oil," 227; Bet Shlimon, "The Politics and Ideology of Urban Development in Iraq's Oil City, 1946–1958," *Comparative Studies in South Asia, Africa and the Middle East* 33, no. 1 (2013): 30–32.
45 "Story of a Town" (1978), English and Arabic, 21 minutes, producer and director Michael Furlong, CD, KOC Public Relations and Information Group.
46 British Petroleum Archive (hereafter BP) 106660/009, Master Scene Script for "The Story of a Town," 23 September 1965, by Michael Forlong, Film Producers' Guild.
47 Nelida Fuccaro, "Telling the Human Story of Oil: Propaganda and Corporate Representations of Urbanism in Iraq, Kuwait and Bahrain," paper presented at the conference "The Comparative Social Histories of Labour in the Oil Industry," International Institute of Social History, University of Amsterdam, 14–15 June 2013.
48 "Insha' ma'tam jadid lil muazzafin fil Awali" *al-Najmah al-Usbu'iyyah*, no. 1, 9 January 1957; "New Canteen Opened at Kirkuk Fields" *Iraq Petroleum*, n. 12, July 1952, Vol. 1. BP 65577: In 1951 the company claimed that it had subsidized more the 2 million meals. "Iraq Oil in 1951" p. 1.
49 "Umm al-'Aish" *al- Kuwayti*, 2 December 1961; "Life in a Pipeline Station–Part I The General Scene/ Part II The Industrial Scene/ Part III The Social Scene," *Iraq Petroleum* 1, nos. 6–8 (January/February/March 1952).
50 "Skyline–Kirkuk City," *Iraq Petroleum* 7, no. 12, July 1958.
51 "First Day Payroll Families Move into Company Accommodation" *The Kuwaiti*, 6 May 1954, n. 318. "Home Ownership in Basrah" *Iraq Petroleum*, n. 9, April 1958, Vol. 7.
52 "The 'Show House' of the Basra Petroleum Company's Home Ownership Scheme," *Iraq Petroleum* 8/9 (November/December 1959).
53 Alissa, "Oil Town of Ahmadi," 52; Fuccaro "Urban Life of Oil," 70–71.

8

BUILDING BRAZIL'S PETROLEUMSCAPE ON LAND AND SEA

Infrastructure, Expertise, and Technology

Drielli Peyerl

For centuries, Brazil served colonial actors as a commodity exporter—mainly of minerals and sugarcane. In the case of petroleum, the country has taken a different path. Early on, the government considered petroleum its best chance for achieving energy self-sufficiency. Its plans, however, were not quickly realized: it took a long time before petroleum was discovered in the territory and additional delays resulted from dependence on imported technologies, limited knowledge of the territory's geology, and a lack of qualified Brazilian technicians. To achieve self-sufficiency in oil production, the Brazilian government cultivated its know-how and trained its own workforce. Brazil's oil history demonstrates how flows of petroleum and the finances related to it can shape a landscape and a society. It also shows that the petroleumscape, once established, involves a feedback loop that brings corporate and public actors together to promote additional petroleum extraction, refining, and use.[1] Once oil was discovered in Brazil, the oil industry developed quickly, which led to the construction of roads, housing, an education system, and many changes to the landscape and local infrastructure that were widely perceived as improvements.

The emergence of the petroleumscape in South America reflected both national interest and the spread of the first petroleumscape that expanded from the US throughout the American continent (see Hein and Lessoff, this volume). Besides that, nationalist movements and the development of the oil industry in other South American countries accelerated the formation of Brazilian national policies that pursued domestic oil extraction and refining. In the case of Argentina, Yacimientos Petrolíferos Fiscales was founded in 1922, and in Bolivia, the state founded the company Yacimientos Petrolíferos Fiscales Bolivianos at the end of 1936. The Bolivian government expropriated the activities related to oil exploration and refining that were being conducted by the US company Standard Oil.[2] In Brazil, the federal government commanded by Presidente Getúlio Vargas created the National Petroleum Council (Conselho Nacional do Petróleo or CNP) in 1938, nationalizing the oil industry even before oil's discovery—a unique case in the history of the continent. In 1939, the first oil well was discovered in Brazil, in the Lobato neighborhood of Salvador City, in the state of Bahia. The discovery dramatically changed Salvador, leading to rapid population growth,

DOI: 10.4324/9780367816049-11

mainly from people arriving in search of work. On a national level, the discovery encouraged oil research in other regions and the establishment of refineries. Until 1953, the CNP focused on achieving industrialization, conducting oil research, discovering new sources of oil, and eliminating foreign capital pressures from the oil sector.[3]

The Brazilian petroleumscape is unusual for at least two reasons. It was built as part of a national project over a century and it encompasses all aspects of spatial and social development, including education. This comprehensive approach became possible when national control was established over petroleum. In 1953, the federal government created the company Petróleo Brasileira S.A., abbreviated as Petrobras, and charged it with oil research as well as the mining, refining, trading, and transporting of oil from wells or shale from its derivates.[4] Petrobras became a state petroleum monopoly and started to invest in an extensive exploration program in order to find large quantities of oil in the territory.[5] As a national entity controlling both oil and urban planning, Petrobras also served as a tool for industrial development. The growing petroleumscape included not only refineries and office buildings, but hospitals, grocery stores, and schools. Many jobs were created at different levels of operation, including those for drillers, blade operators, heavy truck drivers, dynamite load handlers, aerial photo interpreters, designers, mechanics, seismograph operators, geologists, and paleontologists.[6] As Petrobras transformed from an integrated corporation to a major offshore operator, its growth brought benefits to Brazilians who came to support the further development of the industry. The scale of its operations and the intersection of different interests also meant that Petrobras become one of the world's biggest exporters, arousing the interest of investors in other countries.[7]

This chapter explores three stages in the development of the Brazilian petroleumscape (Figure 8.1) in three distinct regions. It first considers Petrobras' activities in the Recôncavo basin, especially in Salvador, Bahia, where infrastructure construction laid the foundation for spatial development and served as an incentive for domestic research and education in the field of oil exploration. Then, it considers the petroleumscape's extension to the hinterland, the Amazon basin, Nova Olinda do Norte, in the state of Amazonas, a region considered suitable for oil exploration due to the presence of massive asphalt deposits in the neighboring country of Venezuela. Finally, this chapter considers the coast, where pre-salt layers (geological formations on the continental shelves) were discovered. Each of these cases involves onshore and offshore exploration and transnational exchanges between international companies, foreign and Brazilian professionals, and CNP/Petrobras. Each case reveals different patterns of investment; however, the national aim in each was to develop the country by means of petroleum.

Building the Brazilian Petroleumscape as a Political and National Project

As consumption of fuel for lighting, industry, and mobility increased in the second half of the nineteenth and the beginning of the twentieth century, countries around the world started to look for petroleum sources. Brazil's scale (8.516 million km^2) and the exceptional economic, geographical, and environmental diversity of its twenty-six states were not initially conducive to oil exploration. The search for oil in the country started at the end of the nineteenth century. The first deep drilling for oil was reported in 1897 in the municipality of Bofete, in the state of São Paulo; however, sulfurous water was found. The historical lack

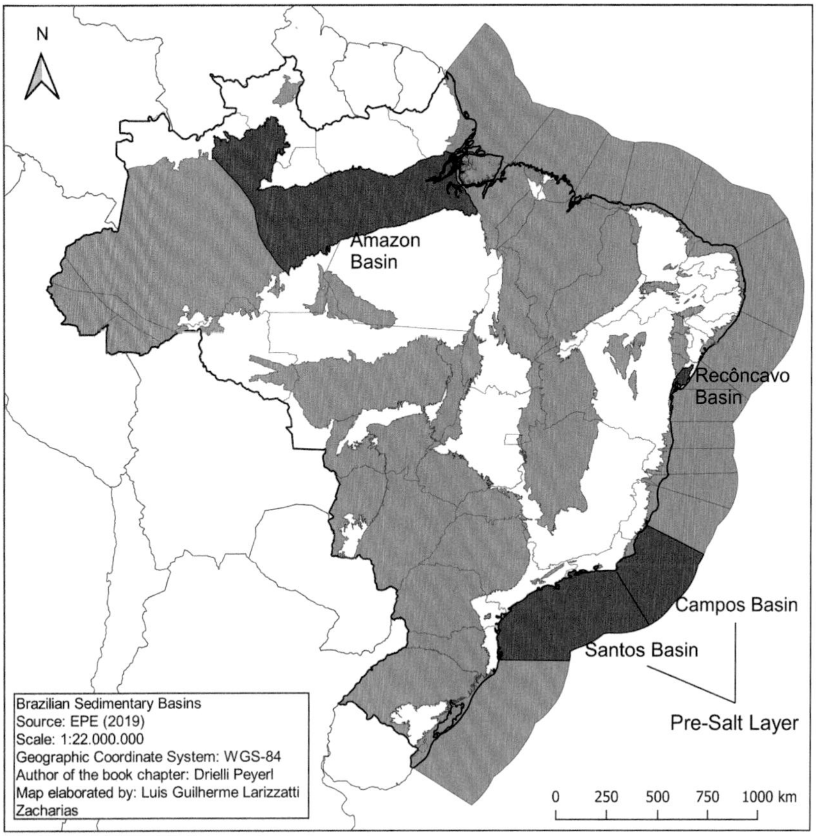

FIGURE 8.1 Map of the Brazilian sedimentary basin. *Source*: Drielli Peyerl; map drawn by Luis Guilherme Larizzatti Zacharias.

of geological knowledge of the territory, limited technology, and a shortage of professionals further impeded petroleum development.[8]

In the beginning of the twentieth century, instead of focusing on petroleum exploration, the Brazilian government invested mainly in the search for coal in its territory.[9] In 1904, the Commission of Studies on the Stone Coal Mines of Brazil (Comissão de Estudos das Minas de Carvão de Pedra do Brazil) was created and supported by the Brazilian government. This commission, led by the North American geologist Israel Charles White (1848–1927), carried out the first coal surveys in southern Brazil. In 1908, two years after the end of the commission, White published a report that suggested the use of new techniques such as briquette for the use of domestic coal.[10] He also wrote two pages about the petroleum potential of the territory in response to countless questions from the government about the probability of finding oil there. He judged the possibilities low of discovering oil in commercial quantity in any part of southern Brazil, except in lands of the so-called Devonian era of the Amazon.[11] At that point, the dream to find oil in Brazil became more distant. In fact, the repercussions of White's report discouraged national and international investments in the oil sector for many years.

In the first decades of the twentieth century, other countries in South America also explored their territories for oil. By 1915, Venezuela had shown great promise as a major oil producer and had reaped the fruits of its early research. Its success yielded significant benefits for the careers of many geologists as well as funds that were used to rebuild the capital, Caracas.[12] The discovery of oil in Venezuela raised the probability of oil in the Amazon basin in Brazil. An impressive number of oil drillings followed, beginning in 1919 with the Geological and Mineralogical Survey of Brazil; but they were also unsuccessful.[13] Beginning in 1930, Brazil began transitioning from an agricultural economy to an industrial one, increasing the need for petroleum.[14] Initially, the largest investments in the building of roads occurred in the south and southeast regions and in big urban centers such as Rio de Janeiro.[15] The government pursued a more aggressive policy in the transport sector, gradually replacing the use of coal with gasoline and diesel. However, the lack of supply resulted in dependence on oil imports, mainly from the US.

The early growth of the petroleum industry in the US made US companies key players in developments in other parts of the world (see Hein, this volume). In the 1930s, the development of geophysics in US oil exploration revolutionized the search for oil in Brazil. Geophysical works executed by a North American company, United Geophysical, contributed to the first discovery of oil fields in the country, in the Recôncavo basin in 1939.[16] The North American companies provided technology and professionals to explore the territory. They had prior experience in oil exploration and specifically in using geophysical surface methods, such as seismic and electrical methods.

The presence of foreign companies in Brazil was one of the reasons that in 1938, the Brazilian government created the CNP. After that, Brazil pursued a very strong national policy and the CNP decided to invest in developing the country's own expertise. The exploration and refining of petroleum resulted in disputes between nationalists and *entreguistas* (those who supported opening the sector to external capital).[17] This conflict culminated in a campaign promoted by nationalists with the slogan "The petroleum is ours" (*O petróleo é nosso*), which advocated for national control over oil.[18] All these factors led to the creation of one of the largest oil companies in the world, Petrobras, in the 1950s.

Changing the Landscape with the Search for and the Discovery of Oil in Brazil

The search for oil—even when unsuccessful—drove the development of the country, of both its institutions and its infrastructure. In January 1939, with the discovery of the first oil well in Salvador, all the petroleum deposits in the region "within a radius of 60 kilometers of Well 163 (the discovery well), were made a national reserve; exploration of the entire area was reserved to the CNP."[19] Politicians used CNP's discovery of oil in Lobato as a symbol of wealth and self-sufficiency about to be achieved (Figure 8.2). Tremendous change took place in regions where the possibility of finding oil was favorable.

After four years of active exploration in the Lobato region, seventeen oil wells were explored (seven with petroleum, ten dry). All the oil wells were subcommercial, meaning that they did not produce oil in the quantity necessary for it to be considered economically viable.[20] In 1941, Brazil's first commercially viable oil field was discovered, in the municipality of Candeias, also in Bahia. After this discovery, CNP focused on developing and applying geophysics methods in the attempt to find other oil fields in Bahia, specifically in

FIGURE 8.2 Visit of President Getúlio Vargas to the exploration fields on the Recôncavo Basin, Bahia, in October 1940. *Source*: "A excursão presidencial ao Norte – Bahia", December 1940, p. 31. National Library of Brazil.

the Recôncavo basin.[21] In the 1940s, directly related to local oil exploration, a process of transformation started in the city when banks and import and export offices were established in Salvador.

In some cities where basic infrastructure was lacking—particularly in the Brazilian northeast, where CNP was searching for oil, and in the main economic centers in the states of São Paulo and Rio de Janeiro—CNP also invested in road building, urban construction, and the establishment of refineries.[22] The federal government's support of CNP's exploratory activities triggered migration to the region, which contributed to urbanization, because it required a workforce from several sectors. Initially, conditions were considered inhospitable by CNP technicians, including those related to the health sector and urban infrastructure. Therefore, from 1940 to 1942, the work of CNP was restricted mainly to the area surrounding the discovery of the first oil well, known as the Recôncavo basin. The deficiency of the preliminary geological studies carried out by the CNP led to the hiring of foreign geologists and international companies.

Petroleum exploration created infrastructures that would shape the development of entire regions, even where oil was not found. In 1943, in Brazil's northeast, the work expanded to the Tucano basin (near the Recôncavo basin) and the drilling at the site resulted in highly sulfuric thermal water rather than oil. Later, this discovery led to the construction of hotels and attracted countless visitors, who believed the thermal water had health benefits. In the same period, when the CNP was active in the state of Maranhão, oil drilling led to the construction of roads, allowing access to these regions. Despite the results of surveys indicating that the wells were dry, the construction of a highway of 250 km was added to the state's transport system. In the case of the state of Rio Grande do Norte, some attempts to find oil resulted in the discovery of potable water wells in a region considered arid.[23]

The 1940s saw global and local problems for the development of the Brazilian oil industry mostly due to difficulties in importing petroleum. During World War II, the CNP found it difficult to import equipment from the US because of a lack of ships to carry it.[24] Also, Brazil was getting most of the petroleum it consumed from the Caribbean island of Aruba through the Lago refinery.[25] As tankers became unavailable as a result of the war, shortages ensued.[26] Brazil started to ration the use of oil, mainly in Bahia and the southern part of the country, anticipating a shortage due to the war. At the end of August 1941, "some interior cities were partially blacked out, owing to fuel shortages; industry suffered."[27] Even with the problems brought about by the war, the construction of the hinterland infrastructure continued along with accelerated urbanization. In the process of industrialization, the southeast (the states of Espírito Santo, Minas Gerais, Rio de Janeiro, and São Paulo) experienced accelerated industrial development as a result of having borders on the ocean with ports that allowed access to international markets.

The growth of the refining industry and the search for new locations for oil exploration demanded more and more professionals and technicians and ultimately fueled domestic education and research. Throughout the 1950s, Petrobras collected geological data on Brazilian sedimentary basins to determine the possibility of finding oil. In most cases in Brazil, oil exploration resulted in dry wells, but it still contributed to the local development of cities due to the presence of professionals who required public services. A variety of people migrated to these places in search of income. In the case of the Recôncavo basin region, an oil industry was beginning to develop around the newly discovered fields. New cities formed in the state of Bahia that were economically based on oil exploration. In 1952, CNP invested in the creation of a sector that could train its own professionals, thus creating the Supervision Sector for Technical Improvement (SSAT), with the objective of generating its own technical and specialized labor. The Brazilian government had realized the importance of growing its own expertise. However, the lack of trained professionals in the area of geosciences with knowledge of sedimentary basins remained a problem for the CNP, which led to the hiring of foreign labor, despite being contrary to the nationalist project.

Expanding the Petroleumscape into the Amazonas Region

The initial success in developing spaces for and through petroleum exploration spurred attention to areas that demonstrated possibilities of finding oil, such as the Amazonian hinterland. In 1950, Brazil was an essentially rural country of the approximately 50 million inhabitants; almost 65 percent of the population lived in the countryside.[28] The creation of a complex oil industry with all the necessary infrastructure would do a lot to change these numbers.[29] In 1953, the federal government created Petrobras through Law No. 2.004, and the company gradually absorbed the activities of the CNP. One of Petrobras' goals was to organize a Department of Exploration, modeled on the most successful international companies. Its purpose was to find domestic sources of oil in great quantity. For this purpose, in 1954, Petrobras hired the North American geologist Walter Karl Link as chief superintendent of the Department of Exploration of Petrobras, with the responsibility of finding oil and ultimately of achieving Brazil's self-sufficiency.

According to Link, the conditions of the Brazilian oil industry were precarious at the time: there was a lack of equipment and qualified professionals and insufficient knowledge of Brazilian geology. These problems had existed since the first decades of the twentieth century. Petrobras invited Link to form the best exploration team money could buy.[30] Link focused the activities of the Department of Exploration mainly on two initiatives:

> a) development of the exploration program in Bahia, and b) investigation of the two biggest continental sedimentary basins in Brazil (the Solimões basin in the state of

Amazonas due to its vast size, and the Paraná basin due to its geographical and economic importance).[31]

The state of Bahia was the scene of numerous investments by Petrobras and, prior to Petrobras' existence, the CNP. All this investment led to the economic development of the area. Before the arrival of Petrobras, Salvador and its region had a reputation as impoverished and infested with parasites and disease.[32] The discovery of oil helped change this reputation. Petrobras' presence in Bahia contributed to urban development and transformed the region from agro-industrial to thoroughly industrial. The country's first government-owned refinery, the Landulpho Alves–Mataripe refinery in São Francisco do Conde, began operating in 1950. To serve the new industry, the company first sought to employ local fishermen and farmers to work both on the building of houses and on performing basic tasks at the refinery such as driving trucks.

The CNP and Petrobras then invested in training its employees to avoid future shortages.[33] Workers involved in the construction of the Landulpho Alves–Mataripe refinery were housed in improvised shelters, initially built to hold 2,000 workers, and generally, these were temporary jobs. Some of the hired professionals, such as electricians, plumbers, and welders, sought accommodation in a nearby city, using company trucks to get back and forth. In order to improve the situation, Petrobras opted for the construction of prefabricated houses. Once the accommodations were completed, professionals started to bring their families to come and live in the city, depending on generators to power residential lighting. Leisure venues were also built, such as a club where it was possible to watch movies.[34] Ultimately, such construction resulted in the growth of cities that formed a nucleus for the development of the oil industry in a national context (Figure 8.3).

The impact of oil exploration was not limited to land development. Brazil still depended substantially on oil imports. The construction of refineries benefited the national economy by importing crude oil, refining and transforming the product into derivatives such as gasoline and diesel.[35] In 1950, CNP invested in the creation of the National Tanker Fleet (FRONAPE), which was intended to fulfill the country's transportation and storage needs pertaining to petroleum and related products, increasing the number of jobs in the city.[36] FRONAPE maintained an office in Salvador, where it managed an active operation. On average, twelve domestic and foreign oil tankers arrived monthly, boosting the region's economy. FRONAPE was the first major initiative to solve the problem of shipping oil products in the country during this period. The investment in sectors of oil outside that of exploration also developed layers of the petroleumscape.[37]

From its inception, Petrobras was interested in exploring and developing the Amazon basin, beyond Bahia. Obstacles included transportation (vast rivers and wetlands), disease, geological problems imposed by intrusive rocks such as basalt flows and diabase, and inadequate logistical support. At the end of the 1950s, 60 percent of the exploration department's budget was invested in the Amazon basin. One of the areas it was exploring in the Amazon, specifically in Nova Olinda do Norte, was described as "a small river ravine with one or two huts."[38] In May of 1955, before oil was found in this location, the presence of drilling rigs and oil research led to the growth of a population nucleus around the exploration area.[39]

The research in Nova Olinda do Norte resulted only in a small amount of oil and the wells were abandoned (Figure 8.4). Petrobras continued with geological and geophysical investigations in the Amazon basin to find places where stratigraphic and structural conditions

FIGURE 8.3 Oil exploration in Bahia in the early 1960s. *Source*: Frederico Waldemar Lange Archive (1911–1988), University of Ponta Grossa, Box 115.

FIGURE 8.4 Nova Olinda do Norte, around the industrial facilities of Petrobras. *Source*: "Petrobras", 1961, p. 5. Frederico Waldemar Lange Archive (1911–1988), University of Ponta Grossa, Box 32.

had a theoretically favorable association with oil accumulation. By the 1960s, the investments shifted away from exploratory research in the state of Amazonas. Reasons included the geographical situation, lack of storage, and problems involved in transporting oil to refineries.

An analysis of crude oil production between 1955 and 1960 in Brazil, comparing the US and Latin American countries (see Table 8.1), shows that the Brazilian oil industry was still on the rise at the time and Petrobras was still searching for oil wells. Even though they had not so far achieved great success, Petrobras continued to pursue national self-sufficiency and continued to invest in technology that could be used for oil exploration.

In addition to investing in that technology, starting in 1955, Petrobras created the Petroleum Improvement Center (CENAP), which operated a comprehensive program to prepare a specialized labor force, mainly focused on the exploration and industrialization of oil reserves.[40] In 1957, Petrobras also approved guidelines for its Training and Improvement of Personnel Plan, an attempt to make up for the deficiencies of the Brazilian education system

TABLE 8.1 World Crude Oil Production 1955/1959

Average Daily Production (Thousands of Barrels)

Countries/Year	1955	1956	1957	1958	1959
US	6.807	7.151	7.170	6.709	7.042
Venezuela	2.157	2.457	2.779	2.606	2.768
Colombia	111	121	125	128	145
Argentina	84	87	93	98	120
Brazil	6	11	28	52	65
Peru	47	50	53	51	48
Chile	7.0	9.7	12	15.3	17.6
Bolivia	7.4	8.7	9.8	9.4	8.8
Ecuador	9.7	9.3	8.7	8.5	7.9

Source: "Petrobras", 1961, p. 02. Frederico Waldemar Lange Archive (1911-1988), University of Ponta Grossa, Box 32.

and linked to professional training for the petroleum industry.[41] Through CENAP, courses were offered on such topics as refining, drilling for oil, petroleum geology, maintaining petroleum equipment, and petroleum engineering.[42] In 1963, Petrobras created the Leopoldo Américo Miguez de Mello Research and Development Center (CENPES), which started operating in 1966. Taking a different approach than CENAP, CENPES focused its activities on a postgraduate program devoted primarily to the field of oil exploration and production.[43]

Petrobras longed to achieve national self-sufficiency, but while that goal remained elusive, the oil company invested in constructing refineries and many began operations. These included the Presidente Bernardes refinery in Cubatão, São Paulo, in 1955; the Duque de Caxias refinery in the state of Rio de Janeiro in 1961; the Gabriel Passos refinery in Betim, Minas Gerais, in 1968; the Alberto Pasqualini refinery in Canoas City, Rio Grande do Sul, also in 1968; the Paulina refinery in Paulínia, São Paulo, in 1972; the Auracária refinery in Auracária, Paraná, in 1977; and the Henrique Lage refinery in São José dos Campos, São Paulo, in 1980.[44] The creation of refineries in different locations aided the Brazilian oil industry by reducing the cost of transporting oil products over long distances. It also contributed to local development in various regions of the country.

The continuous search for oil wells in Brazil, and investments, mainly in the refineries, contributed to modifications of the landscape throughout Brazil. For instance, Petrobras stoked national and international attention when it displayed its logo in competitions called Thousand Brazilian Miles. In São Paulo at the end of 1960, Petrobras participated in the event "V Thousand Brazilian Miles" (Figure 8.5), a competition intended to showcase the national vehicle and auto parts industry, but also the fuel produced by Petrobras refineries, which was used by participants at the event. This type of event showed the participation of Petrobras in social and cultural life.

Petrobras' exploration and construction of oil structures in the Amazon and its intervention in the landscape through huge investment and construction projects in other states, including petrochemical complexes such as Presidente Bernardes refinery in Cubatão, São Paulo, was the source of air pollution and other environmental problems. For many years, the World Health Organization considered Cubatão one of the most polluted cities in the world. In the 1980s, Cubatão became the largest petrochemical center in Latin America, and was referred to as "the Valley of Death" by news media and academic researchers

FIGURE 8.5 "V Thousand Brazilian Miles," with Petrobras gasoline. *Source*: Petrobras, 1961, p. 12. Frederico Waldemar Lange Archive (1911–1988), University of Ponta Grossa, Box 32.

because of its extreme levels of industrial pollution.[45] Other sites had similar problems with environmental pollution stemming from the petrochemical industry, including the Camaçari refinery in Salvador.[46] From 1980 to the present, Petrobras has been investing in programs to reduce pollution from industrial processes and to improve wastewater, air emission, and solid waste treatment systems.[47]

In the Amazon basin, in the 1980s, oil and natural gas exploration activities resumed due to the discovery of the largest proven onshore oil and natural gas reserve in Coari, 650 km from Manaus, capital of the state of Amazonas. The local economy began growing again and the liquefied petroleum gas produced in the region has since supplied the entire state and parts of other states. Petrobras has committed to limiting the negative environmental impact on the region, for example, by avoiding the creation of urban centers close to the exploration sites and trying to preserve the forest as much as possible.[48]

Research and Development as Drivers of Offshore Exploration

Brazil ultimately achieved oil self-sufficiency not from its efforts onshore, but through offshore exploitation. The experiences of the first two stages of petroleum exploration in Brazil had prepared the country for this endeavor. From 1968 on, Petrobras directed most of its work to petroleum research in the sea. The earlier expertise in infrastructural construction and research helped it take on new challenges involving the marine environment, including rocks below the ocean floor; the great distances between platforms and wells on the ocean floor and between platforms and the continent; and the invisibility of operations underwater.[49] Offshore oil brought the royalties needed for investments in education, social and economic infrastructure, health, security, culture, sport, research, science and technology, civil defense, environment, and programs aimed at mitigating and adapting to climate change, for the treatment and social reintegration of drug addicts and programs to eradicate poverty.[50]

In 1968, the first offshore discovery was registered in a well drilled in shallow waters in the field of Guaricema in the Sergipe-Alagoas basin. Also in 1968, the first offshore drilling took place in the Campos basin, in the field of Garoupa, in the state of Rio de Janeiro. The following year saw discoveries at Campo de São Mateus and at Ubarana, both in the

Potiguar basin in Espírito Santo.[51] Meanwhile, international oil prices had begun to escalate, leading to the shocks of the 1970s and the opening of new exploratory frontiers worldwide, including the North Sea.[52] At the end of the 1970s, in the words of historian Tyler Priest, "Petrobras forged even more boldly ahead with subsea technology in its Campos basin development, eventually becoming the leading innovator in this area."[53]

Moreover, by the 1970s, Petrobras had well-established operational bases, including an ancillary petroleumscape along the Brazilian coastline with offices and a complex and coordinated logistics system involving warehouses, workshops, ports, and airports in several capitals, known as "districts."[54] All these districts were controlled by the Petrobras headquarters in Rio de Janeiro, a place chosen for a number of reasons, including the presence of academic institutions such as the University of Rio de Janeiro and convenient transportation as well as a reputation for modernization and development. In addition, including investments in offshore activities, Petrobras started investing in other areas, such as with its creation of BR Distributing, a consumer sales division established in 1971; BRASPETRO, a division devoted to finding sources of oil abroad, established in 1972; Petrofertil, a fertilizer company established in 1973; and Interbras, devoted to international trade and services, established in 1976.[55]

The construction of the spatial petroleumscape was closely related to advances in geological technology. Discoveries took yet another step, when Petrobras went from a period of shallow water exploration (1965–1984) to a deep-water phase (1984–1997) taking advantage of new deep water technology.[56] Since 1984, the oil and gas industry has applied the new technology notably in the Campos basin, including new technologies such as 3D seismic exploration, which optimized time and costs from discovery to field development.[57] In 1986, Petrobras created the Program of Technological Training to explore offshore areas. Because of its numerous discoveries over the years, Petrobras became the international leader in deep-sea exploration technology.[58]

In 1997, after decades of Petrobras holding a petroleum monopoly, the New Petroleum Law was enacted (9478/97), regulating the oil sector in the country and creating the National Agency for Petroleum, Natural Gas and Biofuels (ANP). Both decisions started a new era for the Brazilian oil industry contributing to a more competitive oil sector market and openness for investment by international oil companies. Petrobras still constitutes an industry with a strong presence in some coastal cities, directly influencing the process of urbanization, the real estate market, and infrastructure due to the receipt of royalties.

By the mid-2000s, Brazil started to benefit from the scientific and technological infrastructure it had built for several decades. In 2006, the discovery of the pre-salt oil region (extending from the coast of Espírito Santo to Santa Catarina) made Brazil a net exporter of oil.[59] Thus, a new pattern of interaction between Petrobras and its large foreign suppliers such as Shell had started, in relation to the development of technologies applicable to oil exploration in offshore area.[60]

Three Regions, Three Time Periods, Three Petroleumscape Transformations

Since the last century, the Brazilian government and society developed a network of integrated space connected to petroleum. These spaces involved, as Carola Hein has theorized, not just a physical landscape but a social imaginary and forms of citizen behavior.[61] CNP and Petrobras became a collective representation of Brazilian society associated with economic development and progress. Initially, oil was responsible for population growth, housing construction, and the development of basic services in Salvador. The construction of the refinery and universities changed the landscape of that city, bringing rapid economic development, including migration from rural to urban areas. Brazil's petroleumscape continued to be constructed by the state through education, the consolidation of a national oil industry, and the formation of its own skilled labor force. With the discovery of oil in pre-salt layers, self-sufficiency was achieved and even greater wealth was produced. Oil wealth—associated with ease, comfort, and freedom—brought problems as well as solutions.

The creation of Petrobras, a huge company intimately connected with the government and its financial and planning powers, established conditions not just for an extensive petroleumscape, but also for large-scale corruption. In 2014, after a long investigation by public prosecutors and the Federal Police of Brazil, Petrobras was found to have enabled "billions of dollars in kickbacks from large contracts paid by suppliers to executives of the oil company and politicians."[62] But even when associated with such illegality, it has managed to recover. Although the corruption charges weakened the confidence of many foreign investors,[63] they did little to damage Brazilians' pride in the company. There is a tendency to see the problem as a matter of the inevitable dishonesty of politicians and of big business. Petrobras had and still has a role in various sectors in addition to petroleum and remains active in education, for example, through Petrobras University. Demonstrating concern regarding environmental and sustainability issues, Petrobras has invested in research on renewable energies and biofuels and CO_2 reduction in upstream activities, including exploration, drilling, and extraction.[64]

Positive national sentiment has remained strong toward CNP and Petrobras also because of the countless attempts over the years to find onshore oil. Even before finding domestic sources of oil in sufficient quantity, the company worked to maintain the goodwill of consumers and citizens, spreading commerce around the country through gas stations and petroleum products, employing workers, and becoming an accepted part of everyday life. As a company, it has served as a model for other countries and, in Brazil and many other places, Petrobras and petroleum in general are associated with progress and development driven by national political decisions.

Acknowledgments

Drielli Peyerl is especially grateful for the financial support of grant process 2017/18208-8 and 2018/26388-9 from the São Paulo Research Foundation (Fundação de Amparo à Pesquisa do Estado de São Paulo [FAPESP]). The author gratefully acknowledges support from SHELL Brazil and FAPESP through the "Research Centre for Gas Innovation" (RCGI) (Fapesp Proc. 2014/50279-4), hosted by the University of São Paulo, and the strategic importance of the support given by ANP (Brazil's National Oil, Natural Gas and Biofuels Agency) through R&D levy regulation.

Notes

1 Carola Hein, "Oil Spaces: The Global Petroleumscape in the Rotterdam/The Hague Area," *Journal of Urban History* 44, no. 5 (2018): 887–929.

2 Luciano Martins, *Pouvoir et développement économique – formation et Evolution des structures politiques au Brésil* (Paris: èditions Anthropos, 1976).

3 Drielli Peyerl, *The Oil in Brazil* (Cham, Switzerland: Springer Nature Switzerland AG, 2019).
4 Brazil. Law n. 2.004, 3 October 1952, www.planalto.gov.br/ccivil_03/L2004.htm, accessed January 2013.
5 José Luciano de Mattos and Maria Ana Quaglino, *A questão do petróleo no Brasil: uma história da Petrobrás* (Rio de Janeiro: Fundação Getúlio Vargas, 1993).
6 Pedro de Moura and Felisberto Carneiro, *Em busca do petróleo brasileiro* (Ouro Preto: Fundação Gorceix, 1976).
7 Edmilson Moutinho dos Santos and Drielli Peyerl, "The Incredible Transforming History of a Former Oil Refiner into a Major Deepwater Offshore Operator: Blending Audacity, Technology, Policy and luck, from the 1970s Oil Crises up to the 2000s Pre-Salt Discoveries," in *History, Exploration and Explotation of Oil and Gas*, eds. Silvia F. de M. Figueirôa, Gregory Good, and Drielli Peyerl (Cham, Switzerland: Springer Nature Switzerland AG, 2019), 87–100.
8 Juarez Távora, *Petróleo para o Brasil* (Rio de Janeiro: José Olympio, 1955)
9 Peyerl, Figueirôa, Silvia, Moretto, Mouette, dos Santos, and Moutinho dos Santos, "The Problem of Domestic Supply of Fossil Fuels in Brazil," *Oil-Industry History* 18 (2018): 97–106.
10 Israel Charles White, *Relatório Final: Comissão de Estudos das Minas de Carvão de Pedra do Brazil* (Rio de Janeiro: Imprensa Nacional, 1908).
11 Ilmar Penna Marinho Júnior, *Petróleo: política e poder – um novo choque do petróleo?* (Rio de Janeiro: José Olympio, 1989).
12 Edgar Wesley Owen, *Trek of the Oil Finders: A History of Exploration for Petroleum* (Tulsa, OK: The American Association of Petroleum Geologists, 1975); Carola Hein, "Maurice Rotival: French Planning on a World-Scale (Part I)," *Planning Perspectives* 17, no. 3 (2002): 247–65; Carola Hein, "Maurice Rotival: French Planning on a World-Scale (Part II)," *Planning Perspectives* 17, no. 4 (2002): 325–44.
13 Peyerl, *Oil in Brazil.*
14 Sônia R. de Mendonça, "As bases do desenvolvimento capitalista dependente: da industrialização restringida à internacionalização," in *História Geral do Brasil,* ed. Maria Y. Linhares (Rio de Janeiro: Elsevier, 1990): 316–26.
15 Hamilton de M. Monteiro, "Da República Velha ao Estado Novo," in *História Geral do Brasil*, ed. Maria Y. Linhares (Rio de Janeiro: Elsevier, 1990).
16 Drielli Peyerl and Silvia Fernanda de Mendonça Figueirôa, "Applied Geophysics in Brazil and the Development of a National Oil Industry (1930–1960)," *History and Technology* 36 (2020): 1–22.
17 Gabriel Cohn, *Petróleo e Nacionalismo* (São Paulo: Difusão Européia do Livro, 1968).
18 Peyerl, *Oil in Brazil.*
19 Peter S. Smith, *Oil and Politics in Modern Brazil* (Canada: Macmillan Company of Canada Limited, 1976), 37.
20 M. da S. Pinto, *Mário da Silva Pinto*: depoimento [1987]; Interview (Rio de Janeiro: CPDOC/FGV – SERCOM/Petrobrás, 1988).
21 Peyerl and Fernanda de Mendonça Figueirôa, "Applied Geophysics."
22 Peyerl, *Oil in Brazil.*
23 Petrobras, *A Petrobrás prepara o seu pessoal técnico: manual* (Rio de Janeiro: Centro de Aperfeiçoamento e Pesquisa de Petróleo, 1957). Frederico Waldemar Lange Archive (1911–1988), Box 114.
24 Smith, *Oil and Politics*, 42.
25 Smith, *Oil and Politics.*
26 Ibid.
27 Ibid., 43.
28 de M. Monteiro, "Da República Velha ao Estado Novo."
29 André Ribeiro, *Leopoldo – Os caminhos de Leopoldo Américo Miguez de Mello* (Rio de Janeiro: Petrobras/CENPES, 2010).
30 Drielli Peyerl, Silvia F. M. Figueirôa, and Elvio Pinto Bosetti, "The North American Geologist Walter Karl Link and Oil Exploratory Research at Petrobras (1954–1960)," *Earth Sciences History,* 35 (2016): 387–98.
31 Ibid., 394.
32 M. da S. Pinto, *Mário da Silva Pinto*: depoimento [1987]. Interview (Rio de Janeiro: CPDOC/FGV – SERCOM/Petrobrás, 1988).
33 *Exploração na Bahia*, report, Frederico Waldemar Lange Archive (1911–1988); Caixa 115, 1958.
34 Carlos E. P. Barreto, *A saga do petróleo brasileiro: "a farra do boi"* (NBL Editora, 2001).

35 Drault Ernanny de Mello e Silva, *A questão do petróleo* (Rio de Janeiro: [not known], 1952).
36 Peyerl, *Oil in Brazil.*
37 Carola Hein, "Oil Spaces: The Global Petroleumscape in the Rotterdam/The Hague Area," *Journal of Urban History* 44 (5) (2018): 887–929.
38 Petrobras, Rio de Janeiro: VII, no. 179 (January 1, 1961): 4.
39 Ibid.
40 Alceu Pinheiro Fortes, *CENAP – Petrobras: uma breve memória 1954–1964* (Rio de Janeiro: Petrobras, 2003).
41 Petrobras, *A PETROBRÁS prepara o seu pessoal técnico: manual. Rio de Janeiro: Centro de Aperfeiçoamento e Pesquisa de Petróleo* (1957), Frederico Waldemar Lange Archive (1911–1988). Box 114.
42 Antonio Seabra. Moggi, "Pessoal para o avanço Tecnológico a experiência da Petrobrás," *Diário de Notícias Newspaper* (Rio de Janeiro, Ed. 13840, December, 1967).
43 Peyerl, *Oil in Brazil.*
44 "Petrobras," www.petrobras.com.br/en/about-us/our-history/, accessed April 2018.
45 Daniel J. Hogan, "Population, Poverty, and Pollution in Cubatão, São Paulo," *Geographia Polonica*, no. 64 (1995): 201–24.
46 Tânia Franco, ed. *Trabalho, Riscos Industriais e Meio ambiente: rumo ao desenvolvimento sustentável?* (Salvador: EDUFBA, 1997).
47 "Petrobras," www.petrobras.com.br/en/about-us/our-history/, accessed April 2018.
48 P. V. Santana, "Dimensões espaciais de cidades amazonenses: do dinheiro do petróleo aos serviços públicos de educação," in *Desenvolvimento regional no Brasil: políticas, estratégias e perspectiva*, ed. A. M. Neto, C. N. Castro, and C. A. Brandão (Rio de Janeiro: Instituto de Pesquisa Econômica Aplicada – IPEA, 2017): 157–96.
49 José Mauro de Morais, *Petróleo em águas profundas: uma história tecnológica da Petrobras na exploração e produção offshore* (Brasília: Ipea and Petrobras, 2013).
50 Brazil. Law n. 12.734, November, 2012, http://www4.planalto.gov.br/legislacao/, accessed June 2020.
51 J. B. O. Neto and A. J. D. Costa, "A Petrobrás e a exploração de Petróleo offshore no Brasil: um approach evolucionário," *Revista Brasileira de Economia, Rio de Janeiro* 61, no. 1 (January/March, 2007): 95–109, 100.
52 Tyler Priest, *Offshore Imperative – Shell Oil's Search for Petroleum in PostWar America* (College Station: Texas A&M University Press).
53 Ibid., 187.
54 Marco A. Pinheiro Machado, *Pré-sal: a saga – A história de uma das maiores descobertas mundiais de petróleo* (Porto Alegre: L&M, 2018).
55 Cristovão Brito, *A PETROBRAS e a gestão do território no Recôncavo Baiano* (Salvador: EDUFBA, 2008).
56 C. H. L. Bruhn, A. C. C. Pinto, P. R. S. Johann, M. C. Salomão, and E. B. Freire, "Campos and Santos Basins: 40 Uears of Reservoir Characterization and Management of Shallow to Ultra-deep Water, Post- and Pre-Salt Reservoirs – Historical Overview and Future Challenges," *Offshore Technology Conference* (24–26 October, Rio de Janeiro, Brazil, 2017).
57 *Technology Petrobras* (Petrobras, 2013), and R. Fainstein, A. Krueger, and W. U. Mohriak, "Ultra-Deepwater Seismic Plays Offshore Brazil – Future Drilling off Santos and Campos Basins," *Interpretation* (2019): 1–34.
58 J. B. O. Neto and A. J. D. Costa, "A Petrobrás e a exploração de Petróleo offshore no Brasil: um approach evolucionário," *Revista Brasileira de Economia* 61, no. 1 (January/March, 2007): 95–109, 100.
59 A. S. Magalhães and E. P. Domingues, "Blessing or Curse: Impacts of the Brazilian Pre-Salt Oil Exploration," *EconomiA* 15 (2014): 343–62.
60 Andre Tosi Furtado, "Pré-sal, Desenvolvimento Industrial e Inovação," *Revista Paranaense de Desenvolvimento* 34, no. 125 (July/December 2013): 79–100, 92.
61 Hein, "Oil Spaces."
62 Renato Lima-de-Oliveira, "Corruption and Local Content Development: Assessing the Impact of the Petrobras' Scandal on Recent Policy Changes in Brazil," *Extractive Industries and Society* 2, no. 7 (2019): 274–82, 274.
63 Amelie Pedersen, "The Impact of Corruption on the Performance of State-owned Companies – Case of Petrobras Brasil" (MA thesis, Fundação Getulio Vargas Escola de Administração de Empresas De São Paulo, 2016).
64 Petrobras, http://www.petrobras.com.br/en/about-us/our-history/, accessed July 2020.

9

PRECIOUS PROPERTY

Water and Oil in Twentieth-Century Kuwait

Laura Hindelang

In early 1962, Kuwait's first substantial subterranean water reservoir was discovered at Raudhatain in northern Kuwait.[1] The Ralph M. Parsons company, which was conducting hydrogeological surveys on behalf of the government of Kuwait, was a US firm formerly active in building oil refineries. Geologists described how "the fresh water [had] gathered in a geological basin one side of which is an anticline of the structure forming the Raudhatain oil-field."[2] Raudhatain's water (most of it fossil) and petroleum effectively sprang from the same geological formation and were accessed by similar technologies. This multifaceted historical relationship between Kuwait's water(scape) and petroleumscape, with its spatial and architectural, social, and political as well as symbolic and representational layers, is the topic of this chapter.

One can trace petroleum's impact on twentieth-century Kuwait in many ways: airplane and automobile culture, gas stations, air-conditioning, and the proliferation of plastics—all depend on petroleum. In Kuwait, the oil industry but also the oil revenue-financed government transformed the city-state's urban and desert landscapes, its architectural forms, and the built environment. However, despite the growing omnipresence of petroleum-derived products and lifestyles, petroleum as a raw material, as an unprocessed liquid, has usually remained invisible in urban space. Chemically, oil and water do not mix, but in Kuwait, as the brief example of Raudhatain illustrates, the history of oil and the history of water flow together. Yet, the visual-spatial absence of oil has obscured the two fluids' interdependent conditions of existence. Water has been given a direct material and spatial presence in a way that oil has not, whereby potable water, whose production, transport, and distribution has relied substantially on petroleum and the petroleumscape in one way or another, became the representative liquid of Kuwait's oil-based modernization.

Water has been celebrated with direct contact in ways petroleum was not because of oil's physical characteristics (toxic, inflammable, smelly). At the official inauguration of the opening of the Raudhatain water reservoirs, Kuwait's deputy prime minister drank from the powerful stream of clear water shooting down from a massive pipe connected

DOI: 10.4324/9780367816049-12

FIGURE 9.1 Photograph taken at the official inauguration of the Raudhatain Water Resources Project, where Sheikh Sabah al-Salem al-Sabah, deputy prime minister and minister of foreign affairs, sampled Raudhatain water. *Source*: Unknown photographer, September 19, 1962. Image from the BP Archive, University of Warwick, ARC233262_061.

to the new water scheme (Figure 9.1). The British ambassador to Kuwait reported from the event that:

> it was interesting to watch the surge of delighted people, led by the Deputy Prime Minister, to the flowing water immediately after the opening ceremony, and their eagerness to touch and taste this rare commodity which now made its first appearance here in its natural state.[3]

In the course of Kuwait's modern history, both potable water and petroleum have acted as pivotal commodities that shaped the modernization of the country and the making of modern Kuwait City.[4] They are Kuwait's precious properties.

This chapter argues that petroleum has been written into Kuwait City's urban space, social practices, and also into its urban visual culture, all of which forms the petroleumscape,[5] through a form of *waterscape*, the spatial, material, social, and symbolic production of water. Integrating perspectives from the architectural history and the environmental/energy history of the Middle East, this chapter examines key historical moments where layers of the petroleum(scape) and water(scape) intersect and form a palimpsestic relationship. We can trace these intersections in the form of institutions, companies, governments, images, built form, and spatial configurations. It is a transnational and quasi-colonial relation due to Kuwait's status as a British quasi-protectorate until 1961 and the European-American monopolization of much of the oil industry that operates globally. Within this relationship, images—visual-symbolic representations—and infrastructures do not just reflect a status quo, but rather they create narratives and sites in a—often diachronic—feedback loop.

Arguing that the commodification of water started early on, preparing the ground for petroleum, this chapter starts with Kuwait's river water imports in the early twentieth century. It then focuses on early attempts at prospecting for water, which the British turned into a search for oil. Mid-century water desalination, which only became feasible with the arrival of cheap fossil fuels and oil revenues, provided Kuwait with sufficient, locally produced potable water for the first time. In the second half of the century, large-scale water production and distribution systems with monumental sculptural architecture appeared across the country, signaling the oil state's capacity to provide for its citizens. Recently, water coolers in various shapes have come to adorn the sidewalks of Kuwait City, demonstrating citizens' will and power to give water for free—although, one should caution, only as long as oil lasts. Overall, Kuwait's waterscape offers an analytical lens for investigating petroleum's spatialization and social and symbolic representation beyond the oil fields. Moreover, this chapter demonstrates that in the Middle East, petroleum effectively "reached a new group of people eager to believe in new lifestyles and imaginaries" and heavily impacted its urbanism/spatial configurations by the mid-twentieth century and not—as is commonly assumed and most studies suggest—as late as the 1970s (due to the oil price boom/oil crisis).[6]

The Demand for Potable Water Gives Way to the Search for Petroleum

The development of life-sustaining natural resources, in particular a water supply, preceded and made possible oil's role in Kuwaiti society, as British archival sources and Kuwaiti accounts reveal. In the early twentieth century, when growing water scarcity made it urgent to find alternative sources, Kuwaitis began importing water by boat. Water had become scarce following the dry winter of 1907–1908 and, under Sheikh Mubarak's reign (1896–1915), the population had been growing.[7] During the summer of 1908, shipowner Mohammed Al-Yacoub first brought river water from the Shatt al Arab to sell in Kuwait. A much frequented intake location along the Shatt al Arab became a spot "just above Khorramshahr"—the port city that was largely developed by the Anglo-Persian Oil Company (APOC) as part of their oil industry infrastructure in Khuzestan in Iran.[8] Prior to these water imports, town dwellers and Bedouins had depended on rainwater harvesting and artesian wells inside and outside the town. Both the inner town wells and the wells outside the city walls contained brackish water, often of poor quality.[9] Still, the numerous Al-Shamiya wells, for example, had apparently supplied "practically the entire population of Kuwait with water."[10] It is noteworthy that the newly imported river water was not only of a better quality but also cheaper than well water, which was transported by donkey and camel. Attesting to the new water's popularity, by the 1930s, already forty-nine boats of a type called *būm* were exclusively used for water transport, of which "an average of 6 arrive daily, each with about 8,000 gallons of water worth roughly Rs. 100/-."[11] In 1939, local merchants established the Kuwait Water Company to organize the water transport and the company built three reservoirs for the storage of their product at the harbor.[12]

The scarcity of water led to a variety of techniques, infrastructure, and jobs involved in the transport and distribution of water in the town, early on turning the fluid into a priceable product. The commodification of water unfolded in layers of physical spaces and social practices.[13] When returning from the Shatt al Arab to Kuwait, each *būm* was loaded with big wooden barrels containing river water. On arrival, the water was either stored in the reservoirs or was poured into goatskins, strapped on a donkey's back, and sold to private households. In addition, through the streets of Kuwait water porters (*al-kandarī*) carried metal containers attached to either end of a wooden stick. The containers had originally been used for kerosene and were now filled with water for sale. These kerosene tins functioned as "the universal water measure."[14] Already, the world of petroleum—in the form of kerosene tins—facilitated the practices and infrastructures of water supply in Kuwait and oil manifested as a culture with a related set of social and aesthetic practices, even though the Kuwaiti oil industry would not begin operating until 1946.

The search for domestic sources of water facilitated the development of the petroleum-scape. In 1912, Shaikh Mubarak of Kuwait asked the British Government of India, based on the British-Kuwaiti Protection Agreement of 1899, for assistance in developing "an adequate water supply."[15] British officials saw an opening for petroleum prospecting, disguised as hydrogeological survey, and seized the moment. Sir Percy Cox, who served as political resident in the Persian Gulf at Bushire, stressed that "the present opportunity for effecting a careful examination of the Oil Fields of Kuwait territory by an expert geologist, whilst nominally examining the potentialities of artesian wells in this region, is a most favourable one and not likely to recur."[16] And with a distinctively imperialist outlook, he highlighted "the importance, on Imperial grounds, of the retention of the Oil Fields in the Persian Gulf region under exclusively British control," along similar lines as William Knox D'Arcy, director of the APOC, who argued for British control of Iranian oil production.[17] Evidently, British officials considered petroleum to merit large-scale surveys in a way that water did not, given that—at the time—British livelihood did not depend on securing water in Kuwait.

It is somewhat ironic that in 1924, Swiss geologist Arnold Heim, one of the early experts to assess the availability of exploitable resources in Eastern Arabia, concluded that there was not enough geological evidence to opt for oil prospecting, but there was for water.[18] In fact, surveying for underground geological water deposits shared many common features with prospecting for hidden petroleum.[19] Although Heim was eventually proven mistaken, it is telling that his contractor, the London-based Eastern and General Syndicate Limited, was not interested in drilling for water and British and American efforts to search for petroleum subsequently increased. As a result, in 1934, British Petroleum (formerly APOC) and Gulf Oil, an American firm, formed the Kuwait Oil Company (KOC) to explore Kuwait's oil deposits. In 1938, the KOC struck crude in the Burgan Oil Field. The Kuwaiti quest to find water became key to foreign prospecting for oil, disclosing not only the seemingly smooth shift from quasi-colonial to increasingly corporate British involvement, but also the inextricability of waterscape and petroleumscape in Kuwait.

Desalinating Water in Kuwait: A Technology Transfer from the Kuwait Oil Company

Given that foreign interest in oil did not solve Kuwait's drinking water problem, the Kuwaiti leadership tried to explore other options. Eventually, sea water desalination emerged as a promising path, but not until the necessary technical, financial, and energy resources were made available through the emerging oil industry in the mid-twentieth century. However, the idea of desalinating the Persian Gulf's salt water was not entirely new.[20] Already Sheikh Mubarak, under whose reign the water imports had started, had favored the construction of a distillation facility after he had reportedly witnessed one running in Aden.[21] A first attempt at operating a desalination plant installed by the British company of Messrs. Strick Scott in 1918–1919 failed and Kuwait remained with brackish well water and imported river water.[22]

Following the initial oil shipment from Kuwait to Britain in 1946, then-ruler Sheikh Ahmad al-Jaber al-Sabah (r. 1921–1950) reopened the question with the British Political

Agency in Kuwait, expressing an interest in producing additional drinking water through seawater desalination, conscious that the KOC was in the process of building a distillation plant at Ahmadi.[23] In the context of the expanding oil industry and the city-state's modernization project, Kuwait City experienced a startling demographic and economic surge, a rapid shift toward petroleum-based lifestyles, and consequently a new level of water demand. Around that time, according to one estimate, imported water had increased to a daily supply of about 100,000 gallons of river water delivered by twenty-two boats to accommodate Kuwait's water consumption.[24]

Eventually, the Kuwaiti government commissioned Ewbank & Partners as consulting engineers for a new and much larger power and desalination plant. It does not come as a surprise that Ewbank & Partners, a company with strong ties to the petroleum sector and substantial experience with constructing oil refineries in Iran and Britain, had first been hired by the KOC for their desalination facility. Put into operation in early 1950, the KOC plant was not only able to satisfy the KOC's demands but was capable of supplying Kuwait City, too.[25] Initially, the KOC provided the town with 120,000 gallons a day, rising to 200,000 from August 1950 onward,[26] by running a 150 mm steel pipeline from the oil harbor Mina al-Ahmadi, where the desalination plant was located, to the capital.[27] Evidently, the oil industry was integral to the industrial production of desalinated, potable water in Kuwait, thus anchoring the petroleumscape in the local water needs and the projects designed to overcome environmental limitations through engineering. Subsequently, the desalination plants established a new relationship between Kuwait and the sea. With the decline of seafaring and maritime trade, the sea gradually became a commodity, first as natural resource for water production and subsequently as a view (see Couling, this volume).

The First Power and Desalination Plant in Kuwait City

Large-scale desalination as undertaken at the Mina al-Ahmadi plant was only feasible if fossil fuels were cheaply available. Sheikh Ahmad had wanted "to utilise as fuel the gas now being burnt off at the fields by the Oil Company."[28] In fact, the wasteful flaring of natural gas in the oil fields around Ahmadi offered one of the few opportunities to visually experience petroleum (or rather, one of its by-products). The political authorities and the KOC tried to stop images of the flaring from circulating as they were considered negative publicity, indicative of wastefulness. Putting the gas to good use in desalination efforts was seen as good for public relations. Such public relations narratives as well as the official opening of the city's plant created important representational layers of the emerging petroleum/waterscape.

On March 29, 1953, both Kuwaiti and British officials inaugurated the massive power and desalination plant in Shuwaikh, a new industrial area southwest of Kuwait Town. The government secretary Abdulla Mulla Saleh praised the plant as "one of the most vital projects for Kuwait" that finally fulfilled "a dream, that has for a long time been in our minds"[29] (Figures 9.2 and 9.3). Although petroleum was not directly mentioned, it had clearly become the way to make dreams come true. Moreover, the plant not only impressed in numbers and cut a fine figure photographically, but it also materialized the conviction of social development through technical and economic progress toward a Western (democratic) model.[30] Development and modernization theorists worldwide considered petroleum the perfect fuel and fortune generator to forge this path of progress. The plant fit this logic of how oil revenues could be used effectively. Kuwait's first comprehensive scheme to generate

FIGURE 9.2 Administration building with water intake (left) and "Power Station A," as viewed from the water intake jetty. *Source*: Unknown photographer, c. 1956, RIBA Collections, RIBA112508.

electricity and to desalinate and distribute water was financed by oil revenues and fueled by petroleum, thereby making crude indispensable to Kuwait's modernization on multiple levels. It is noteworthy that the project was developed and built by British firms only.[31] Fusing quasi-colonial, national, and corporate interests, Kuwait's modernization was unfolding as a transnational palimpsestic petroleumscape in which the plant (as built form and architectural design) emerged as a key representational structure.

In 1956, the *Architectural Review* dedicated an article to the power and desalination plant, illustrated with large black-and-white photographs of the three-part structure situated at the waterfront of Kuwait Bay (Figure 9.2).[32] The flat horizontal slab with the cantilevered window and French balcony, to the far left, contained the entrance hall on the ground floor and the control room above. This administration building adjoined "Power Station A," a large horizontal volume housing four turbines, pumps, switchgear, and additional offices, which opened onto a continuous covered balcony. Constructed from steel frames, reinforced concrete, and brick stone, the facades of these adjoined buildings displayed beautiful brick stones of different shades in geometric patterns. Noteworthy is the thin reinforced concrete roof with its overhangs that cast delicate zigzag shadows on the façade of the administration building. These were truly modern buildings in material, construction, and form, designed to showcase Kuwait's petro-fueled and water-dependent modernization.

The administration building and the power station faced an open courtyard with parking spaces that included a zigzag shaped pool with water fountains, filled with the precious material the plant was producing (Figure 9.3). The symbolic value of this dynamically shaped pool—generally considered the first lavish open-air water feature in arid Kuwait—cannot be overestimated. Its novel display of water in fossil-fueled movement recalls the Palais de l'Electricité at the 1900 Paris World Fair with its water cascades that translated the invisible

FIGURE 9.3 The water pool in the courtyard of the plant. *Source*: Unknown photographer, c. 1956, RIBA Collections, RIBA112509.

FIGURE 9.4 Oblique view of the water tower at Safat Square. *Source*: Unknown photographer, 1960. Image of the BP Archive, University of Warwick, ARC107058_16.

electric currents the building produced into an extravagant experience of waterworks in celebration of the electrical age.

Today, while the administration and the power plant are still in use, albeit for different purposes, the desalination plant, the third building, no longer exists.[33] The plant was a symmetrically stepped volume of reinforced concrete, clad with brick stones in geometric half-reliefs, especially on the main façade. Flat roofs made of extra-thin reinforced concrete slabs cantilevered the building for shading. The external chimneys connected to the water boilers, similar to those at the power plant, and framed the building. It housed offices, pumps, turbines, and switchgear and was connected to the evaporation units. Following the course of the freshly desalinated and remineralized water that was produced here leads us to the prominent water distribution scheme implemented together with the plant. It was not the administrative and industrial facility at Shuwaikh, but the four water towers located strategically across town that were the most visible manifestation of the city-state's new water infrastructure.

Putting the Waterscape on Display

"Water towers rise 80 to 100 feet above streets that never knew more water than could be carried in a goatskin," one British journalist reported in 1954.[34] Indeed, the water towers exceeded not only average building heights by far, but also common experiences of water availability and visibility in the city (Figure 9.4). The water towers needed to be in strategic positions to deliver water to the surrounding area efficiently and their height resulted from the need for a gravity feed for distribution. But their positions and town-wide visibility were in fact too preeminent to be merely practical. For example, one water tower was situated on a roundabout-turned-square and marked the entrance to Fahad Al Salem Street, the most modern and chic street at the time. Another water tower was erected on Safat Square, a big open space and the vibrant social and commercial heart of the city. Given the prominent locations and significant visibility, the towers were meant to be seen by all. Surprisingly, although the power and desalination plant was somewhat remote, it consisted of three beautifully designed buildings; in contrast, the water towers were extremely exposed and yet their architectural design seemed inadequate, almost unfinished.

The water towers' plain and utilitarian construction contradicted the prominent locations and the visibility the water towers were given around town. They consisted of two elevated water tanks, each supported by an open-braced steel structure and left unclad. The fine design of water works was considered an important task with a long tradition in Western architecture. The *Manual of British Water Supply*, which the British experts at work in Kuwait supposedly used, delineates: "As the greater part of the skillfully-designed works of a water undertaking is hidden underground it is not unreasonable to demand that those parts displayed to public view shall be of pleasing appearance."[35] In addition to being unadorned structurally, the coffered opaque white volumes of the tanks exposed its contents directly to the sun, as historical photographs show. Neither aesthetic nor purely functional, how can the architecture of the water towers be explained?

A former employee of John Taylor & Sons, who was involved in constructing the Kuwait water infrastructure, recalled that:

> each water tower [...] was to have been clad with architectural features, that in the city centre [Safat Square] being designed to be particularly lavish. The Ruler decreed,

however, that they should all be left unclad so that the population could see the benefits the oil had brought them![36]

By deliberately not cladding the elevated tanks, the construction put the liquid content on public display. The water towers provided visual proof of the new water source and showcased its immediate availability in each tank, thereby symbolically promising the urban ubiquity of these modern waters. Clearly, in the urban space, the water towers served as *pars pro toto* for the new water infrastructure of Kuwait, for which petroleum provided fuel, technological expertise, and financing. Water was publicly displayed as material evidence of technological success and a symbol of Kuwait's modernization. As Carola Hein has noted, "representation of technology is contingent to historical moments when technology is seen as a major societal achievement," which it was.[37] In fact, the government of Kuwait branded the new water infrastructure politically by integrating it into new national symbols. The first-ever series of definitive postage stamps (1959), issued by a Kuwaiti authority and with Kuwaiti motifs, included both a view of the power and desalination plant and of Safat Square with its unclad water tower.[38] In the following decades, a new generation of water towers continued to shape Kuwait's landscape in forms of even grander iconic architecture and with even bigger exposure.

A Monumental Sculpture to Water (and Petroleum): Desalination and Water Distribution in the 1970s

Petroleum continued to be the driving force behind Kuwait's water infrastructure, thereby perpetuating and consolidating the spatialization, but also celebration of the petroleum (scape) through the water(scape). In the 1970s, a new elaborate water scheme established a hitherto unprecedented sculptural iconicity of water infrastructure that paid tribute to petroleum as financial and technological motor. In 1965, the Ministry of Electricity and Water commissioned the Swedish construction and engineering company VBB (today SWECO) for a nationwide piped water distribution and storage scheme connected to the extended desalination facilities in Shuwaikh and Shuaiba. Commencing operation in 1976, the mushroom-shaped, light blue (sometimes gray) and white striped water towers became the spatial but also symbolic manifestation of the new large-scale water system (Figure 9.5). Looming above their flat surroundings, they left a lasting imprint on the landscape. Swedish architect Sune Lindström initially developed "The Mushroom" as a multifunctional water tower, including a transmitting station with antenna, a restaurant, and a viewing platform for the town of Örebro, Sweden.[39] For Kuwait, he adapted the design to function as a water tower only, exchanging the TV antenna with a flat roof and changing the color.[40] Placed in groups of varying numbers and heights at five strategic locations across Kuwait, the "mushroom" has become a landmark, an iconic element of Kuwait's urban and environmental landscape that today serves in color or shape as inspiration for souvenirs, stamps, and postcards.

The government of Kuwait continued to enhance the great symbolic value of water towers not only with the "mushrooms," but also with three water towers of even more elaborate design known as *abrāj al-Kuwayt* or the "Kuwait Towers" (Figure 9.6). This group of water towers was built as the final phase of the VBB's water scheme and is located at the eastern tip of Kuwait City's waterfront as part of an artificial leisure island. The Danish designer Malene Bjørn of Bjørn & Bjørn Design developed the unique architecture after Emir Jaber al-Ahmed

FIGURE 9.5 A group of mushroom-shaped Kuwait Water Towers built in the 1970s in the neighborhood of Abdullah Al-Salem, Kuwait City. *Source*: Photograph taken by author, February 2018.

al-Sabah (r. 1977–2016) demanded a different concept for this sixth group, which would be in sight of his residency—or so the story is usually told—and finally opted for Bjørn's design.[41]

The Kuwait Towers consist of three white shafts of different heights made of reinforced concrete and finished with silver spires of stainless steel that serve as lightning conductors. The highest tower (185 m) combines the shaft with two spheres covered with a multitude of enameled steel disks of blue and green shades that deflect solar radiation. The larger sphere contains a restaurant, a banquet hall, and an indoor garden in its upper half and a water tank in the lower, while featuring a horizontally extending shade much like a planetary ring. The upper, smaller sphere houses a rotating café and a viewing platform; the sphere's skin consists of dark blue triangular glass shapes to allow for a view. The second, smaller shaft (140 m) carries one sphere that serves as a water tank. Harnessed with lighting equipment, the third and smallest needle-like shaft (100 m), with no sphere, illuminates the other two and completes the ensemble harmoniously.[42]

Officially inaugurated in 1976 and 1977, respectively, the Water Towers (mushrooms) and the Kuwait Towers were awarded the first Aga Khan Award for Architecture 1980, and thus attracted much international attention to Kuwait. The award statement argued that the Kuwait Towers, an impressive technological achievement, references mosque architecture in its "minaret-like quality of the shafts" and green-to-blue steel disks, "recalling the tiled domes of historic mosques and shrines," and hence praised the architecture as a model for the "Islamic world of tomorrow," the designated scope of this award.[43] However, designer Malene Bjørn provided a different narrative of the design inspiration at the time:

> I asked myself: What's going on today, what's life, what's future? And so without really being conscious of it, there was "the globe" the earthy human warmth, and

FIGURE 9.6 Kuwait Towers at the waterfront of Kuwait City. *Source*: Photograph taken by author, February 2018.

> there were "the missiles" too, long sharp logic thoughts, fired against the heavens, and suddenly the globe was pierced and there were more globes, speared by the rockets. It was not my intention to make the towers look like minarets, but obviously they have something in common.[44]

The statement suggests that the design reflected the global political events of the 1960s and 1970s, like the Cold War, space exploration, and the first images of Earth from deep space. It contrasts sharply with the award committee's interpretation, which has remained influential. Even further, in light of Bjørn's concept and the ramifications of the "oil crisis" (1973) around the time of the towers' inauguration in the mid-1970s, the referentiality between water towers, oil rigs, and oil tanks is remarkable. Infrastructure of oil and water both stand out impressively from their surroundings; they mark violent processes of extraction, storage, and distribution of the Gulf countries' most precious goods and in the case of Kuwait, they come to "represent the rise of an economic power."[45] When Kuwait's new water infrastructure received an iconic architecture that achieved an unprecedented spatial presence and visual prominence, Kuwait's "fabulous development [...] from a small town to a modern capital" fueled by skyrocketing oil revenues had indeed received a monument to petroleum as well as water.[46]

Water Coolers: The Symbolic Power of Giving Water as Long as Oil Lasts

Petroleum has acquired the status of Kuwait's most precious property as a globally priced commodity at the heart of petroleumscape's value system. Meanwhile, the status of water, the once sought-after commodity, has also shifted throughout the twentieth century. Stabilizing its power as (oil) welfare state, the state deliberately "normalized" potable water as ubiquitous and free of charge. This has given way to new social practices around water. Today, in front of many luxurious villas (which constitute much of the urban fabric of Kuwaiti neighborhoods) and in front of some public and commercial buildings, a water cooler provides drinking water for free to anyone passing by (Figure 9.7). Connected to the building's water supply, some of these coolers have adopted various pop art-like sculptural designs, like plastic bottles or even the striped mushroom towers.

These water coolers, called *sabīl al-mā'*, are not just erected by the state, but also by Kuwaiti citizens, who consider their installation an act of charity in the memory of someone deceased. Literally, *sabīl* means "a road" or "a path," while its Quranic use "referring to a work done on behalf of God" probably explains its use for charitable fountains.[47] Water (*al-mā'*) in Islam is essential for ritual ablution and spiritual purity and framed as a life-bearing gift by God. The name might also derive from *'ābir sabīl*, meaning the one passing by.[48] Giving water to passing humans and animals, especially to pilgrims on their way to Mecca, is described as a Muslim charity in the Hadith.[49] Yet, in a country notorious for its high motorization, where even walking to the next kiosk can be socially questionable, usually only migrant laborers use the water dispensers. Also, many water coolers are neglected. Therefore, they serve a symbolic rather than a functional purpose, demonstrating the power to give water, but their complex, multivalent symbolism gives rise to many more questions: Does the social practice of the Kuwaiti *sabīl* echo the welfare state's function of providing its citizens with water? Are they miniature versions of the national water distribution system, yet also symbolic of religiously inspired historical practices? Do they serve as visible tokens of the enduring abundance of drinking water and therefore as symbols of petro-prosperity? Have they become nostalgic reminders of a once-powerful oil state and its reassuring water infrastructure, which has not received a contemporary update equally grand as the projects of the 1950s and 1970s?

In Modern Kuwait, Life without Oil Is Life without Water

A full-page advertisement published in 1988 in *Al-Kuwayti* (the KOC's Arabic language magazine established in 1961) depicts a drawing of a mushroom water tower in the desert framed by a desiccated tree, a historical oil tin water carrier, and the beaming sun (Figure 9.8). The water tower looms majestically above the landscape and—the way the viewer sees it from below—serves as a sculptural monument to technical, financial, and political power. The image brings to mind the modern history of potable water in Kuwait that has

FIGURE 9.7 Water cooler in Kuwait in the shape of a mineral water bottle located in Salmiyah, Kuwait City. *Source*: Author's photograph, taken January 2018.

manifested as the complex interweaving of the architectural, social, and symbolic layers of the petroleumscape and the waterscape—from early practices of distributing water in oil tins to the establishment of salt water desalination and the massive fossil-fueled plant built in 1953 and its unclad distribution towers, to the mushroom-shaped water towers and the sphere-shaped Kuwait Towers of the 1970s, and finally to today's outdoor water coolers. In the absence of petroleum as a fluid material, another commodified liquid, water, has been celebrated as the harbinger of modernity and prosperity. The water infrastructure laid the foundation for petroleum infrastructure and the commodification of petroleum as "precious property." The water architecture and the social and symbolic practices around it have spatialized oil, have fixed oil in Kuwait's urban space. The long-standing practice of physically erected structures, such as water infrastructure, has served to uphold the image of Kuwait's water and oil abundance, despite the finiteness of these precious properties. Still, given the water structures' spatial permanence and sculptural beauty, it is difficult to overcome the inherent oil celebration and therefore fossil fuel dependence.

FIGURE 9.8 Advertisement by the Ministry of Electricity and Water, Kuwait. *Source*: *Al-Kuwayti* magazine (July 1988), 2.

The 1988 advertisement also contained a caption. It reads, "Handle the water in accordance with its measure of scarcity because there is no life without it," and includes greetings from the Ministry of Electricity and Water. The government's advertisement called for moderate water use by imagining the water tower not in a booming modern urban environment, but instead in the desert, which still covers most Kuwait's territory today. This collage-like *Verfremdung* recalls the pre-oil, pre-desalination life world of Kuwait, asking the reader to be humble in view of Kuwait's history. But the image could also be read as a menacing future vision of Kuwait, which has one of the highest water consumption rates worldwide, without the fossil-fueled means of extracting and commodifying petroleum to maintain the current hydrological lifestyle inherent in the petroleumscape. A bitter taste of precisely such a future came with the first Gulf War, two years after the advertisement was published. The war turned Kuwait's oil fields into one big burning oil lake and brought the oil industry to a standstill. Several desalination plants were destroyed by Iraqi forces or severely damaged by crude spilling into the Gulf. It is said that there is life without oil, but not without water, but in modern Kuwait, it seems, there is no life without oil or without water; the two go hand in hand.

Notes

1 Political Agency, Kuwait, to Arabian Department, Foreign Office (FO), London, December 6, 1960, in Anita L. P. Burdett, *Water Resources: In the Arabian Peninsula 1921–1960*, two vols., Vol. 2: Kuwait (Slough: Archive Editions, 1998), 605.
2 British Embassy, Kuwait, to Foreign Office (FO), London, March 10, 1962, in Burdett, *Water Resources*, 611; see also P. Beaumont, "Water in Kuwait," *Geography* 62, no. 3 (July 1977): 190.
3 British Embassy, Kuwait, to Foreign Secretary, FO, London, September 26, 1962, in Burdett, *Water Resources*, 622.
4 Maria Kaika argues convincingly that modern city building always demands the transformation of nature, especially of water sources and waterways. Maria Kaika, *City of Flows: Modernity, Nature, and the City* (New York: Routledge, 2005).
5 Carola Hein, "Oil Spaces: The Global Petroleumscape in the Rotterdam/The Hague Area," *Journal of Urban History* 44, no. 5 (2018): 887–929, doi:10.1177/0096144217752460.
6 Ibid., 903. The academic debate on "oil urbanization" or "petro-urbanism," which started in the 1980s, usually situates the watershed moment between pre-oil and post-oil periods in the 1970s rather than in the 1950s. See, for example, Mohamed Riad, "Some Aspects of Petro-Urbanism in the Arab Gulf States," *Bulletin of the Faculty of Humanities and Social Sciences*, no. 4 (1981).
7 "Note on the Drinking Water Available in Kuwait and the Immediate Vicinity," Political Agency, Kuwait, December 1933, in Burdett, *Water Resources*, 4.
8 British Consulate-General, Basra, to British Embassy, Baghdad, November 17, 1950, in Burdett, *Water Resources*, 104. On APOC's development of Khorramshahr as petroleumscape, see Carola Hein and Mohamad Sedighi, "Iran's Global Petroleumscape: The Role of Oil in Shaping Khuzestan and Tehran," *Architectural Theory Review* 21, no. 3 (2017): 349–74, doi.org/10.1080/13264826.2018.1379110.
9 "Note on the Drinking Water," Political Agency, Kuwait, December 1933, in Burdett, *Water Resources*, 3.
10 Ibid.; Tom G. Temperley, "Kuwait's Water Supply," *Journal (American Water Works Association)* 57, no. 4 (April 1965): 419–20; Ministry of Energy, Electricity and Water, *Water and Electricity in the State of Kuwait* (Kuwait: Center for Research and Studies on Kuwait, 2005), 23–24.
11 "Note on the Drinking Water," Political Agency, Kuwait, December 1933, in Burdett, *Water Resources*, 4.
12 Temperley, "Kuwait's Water Supply," 419; Yūsuf 'Abd al-Muhsin al-Turkī, *Lamahāt Min Mādī Al-Kuwayt* ([Kuwait]: [s.n.], 1979), [Glances from Kuwait's Past], 54.
13 In contrast to Saudi Arabia, where water was of crucial agricultural importance, in Kuwait, potable water was an urban commodity in the first place. On Saudi Arabia, see Toby Craig Jones, *Desert Kingdom: How Oil and Water Forged Modern Saudi Arabia* (Cambridge: Harvard University Press, 2010).
14 Political Agency, Kuwait, to Political Residency, Bushire, November 13, 1912, in Richard Trench, ed. *Arab Gulf Cities: Kuwait City*, Vol. 4 (Cambridge: Archive Editions, 1994), 81. On the continuous use of kerosene tins, see Burdett, *Water Resources*, 4.
15 Sir Percy Cox, Political Residency, Bushire, to Secretary, Government of India, Foreign Department, Shimla, November 26, 1912, in Trench, *Arab Gulf Cities*, 79.
16 Trench, *Arab Gulf Cities*, 79. The "political resident" in the Persian Gulf was in charge of managing British India's relationship with the Gulf region.
17 Ibid.
18 See Laura Hindelang, "Photographing Crude in the Wild: Sight and Sense among Early Oilmen," in *The Life Worlds of Middle Eastern Oil*, eds. Nelida Fuccaro and Mandana E. Limbert (forthcoming).

19 Historical descriptions of discovering the first oil field in Kuwait and of geophysical prospecting of water indicate similar methods of surveying. Paul D. Foote, "The Discovery of an Oil Field," *Proceedings of the American Philosophical Society* 92, no. 1 (March 1948); Aubrey T. Hobbs, ed. *Manual of British Water Supply Practice* (Cambridge: W. Heffer & Sons, 1950), 123–26.

20 The concept of producing purified water by evaporating sea water can be traced back to Aristotle. The polymath Abu Mūsā Jābir ibn Hayyān (c. 721–815) wrote one of the first treaties on the distillation of liquids through evaporation. For an historical account of purification through distillation, see Moses Nelson Baker, *The Quest for Pure Water: The History of Water Purification from the Earliest Records to the Twentieth Century* (New York: American Water Works Association, 1948), 357–60.

21 Political Residency, Bushire, to Secretary, Government of India, Foreign Department, Shimla, November 26, 1912, in Trench, *Arab Gulf Cities*, 81.

22 Political Residency, Bahrain, to India Office, London, July 24, 1947, in Burdett, *Water Resources*, 72; Ministry of Energy, Electricity and Water, *Water and Electricity in the State of Kuwait*, 38.

23 Burdett, *Water Resources*, 72.

24 Mr. Kemp, London, to Middle East Secretariat, FO, London, May 2, 1950, in Burdett, *Water Resources*, 93.

25 Political Agency, Kuwait, to Eastern Department, FO, London, November 13, 1950, in Ibid., 105; FO, London, to Political Residency, Bahrain, December 6, 1950, in Ibid., 107; Gwilym Roberts and David Fowler, *Built by Oil* (Reading, MA: Ithaca Press, 1995), 103.

26 Mr. Kemp, London, to Middle East Secretariat, FO, London, May 2, 1950, in Burdett, *Water Resources*, 93; William Kitson, "Kuwait's Distillation Plant for Domestic Water," *The Times Review of Industry and Technology*, December 1951, 22.

27 Roberts and Fowler, *Built by Oil*, 104. Apparently, 1950 was the last year of Shatt al Arab water imports; see Fātima Husayn Yūsuf al-'Abd al-Razzāq, *Al-Miyāh Wa-L-Sukkān Fī Al-Kuwayt* ([Kuwait]: [s.n.], 1974), [The Waters and the Inhabitants in Kuwait], 153.

28 Political Agency, Kuwait, to Political Residency, Bahrain, July 5, 1947, in Burdett, *Water Resources*, 71.

29 English translation of Abdulla Mulla Saleh's speech, in Trench, *Arab Gulf Cities*, 96.

30 On the persuasiveness of development discourse since World War II, see Arturo Escobar, *Encountering Development: The Making and Unmaking of the Third World* (Princeton, NJ: Princeton University Press, 1995).

31 Most big development contracts of the time were given to British firms. This reflects the strong influence of the KOC and the British government, represented by the Political Agency in Kuwait, in the British quasi-protectorate Kuwait.

32 [s.n.], "Power Station at Kuwait, Architects Farmer and Dark," *Architectural Review* 120, no. 714 (July 1956).

33 I thank Bader Al-Anzi, Ahmad al-Azmi, Noura al-Deffeeri, Mr. Pallayat, Paul Frain and Fahad Alzuabi for organizing and facilitating my visit to the plants of Shuwaikh and Shammal Azzour in January and February 2018.

34 Elizabeth Monroe, "The Shaikhdom of Kuwait," *International Affairs* 30, no. 3 (July 1954): 278.

35 Hobbs, *Manual of British Water Supply Practice*, 420. Reference given by Gwilym Roberts, *Chelsea to Cairo, 'Taylor-Made' Water Through Eleven Reigns and in Six Continents: A History of John Taylor & Sons and Their Predecessors* (London: Thomas Telford, 2006), 363.

36 Roberts, *Chelsea to Cairo*, 364.

37 Hein, "Oil Spaces," 899.

38 See my work on Kuwaiti stamps: Laura Hindelang, *Iridescent Kuwait: Petro-Modernity and Urban Visual Culture in the Mid-Twentieth Century* (Berlin: De Gruyter, 2022), (forthcoming).

39 Developed by Lindström for VBB, the "Vattentornet Svampen" (water tower mushroom) opened in May 1958. It is vertically striped in two different shades of grey and its top is capped with a dome-shaped antenna.

40 Additional offspring of the mushroom series are the water towers in Riyadh and Jeddah, Saudi Arabia. VBB completed them in 1970 and 1977, respectively, for the Ministry of Agriculture and Water based on the Lindström design. Udo Kultermann, "Water for Arabia," *Domus*, no. 595 (June 1979).

41 Malene Bjørn, *The Light and the Airy: How It All Began in Sweden in 1945* (Växjö: Baltic Books, 2013), 24–35.

42 [s.n.], "Wassertürme in Arabien (VBB)," *Deutsche Bauzeitung*, no. 4 (1979); Udo Kultermann, *Contemporary Architecture in the Arab States: Renaissance of a Region* (New York: McGraw-Hill, 1999), 176–78.
43 Renata Holod and Darl Rastorfer, eds. *Architecture and Community: Building in the Islamic World Today* (Millerton: Aperture, for the Aga Khan Award for Architecture, 1983), 173–81 (Kuwait Towers), 180.
44 Ibid., 252.
45 Mokhless Al-Hariri, "Water Towers on-Site Review Report" (prepared for the Aga Khan Award for Architecture, 1980), https://archnet.org/publications/213. On oil and violence see Nelida Fuccaro, "Reading Oil as Urban Violence: Kirkuk and its Oil Conurbation, 1927–1958," in *Urban Violence in the Middle East: Changing Cityscapes in the Transition from Empire to Nation State*, eds. Ulrike Freitag, Nelida Fuccaro, Claudia Ghrawi, and Nora Lafi (New York: Berghahn, 2015).
46 Jan Östlin, "Architect's Record of Water Towers" (submitted to the Aga Khan Award for Architecture, 1980), https://archnet.org/system/publications/contents/214/original/FLS0218.pdf?1384747203.
47 Mostafa L. Saleh, "The Cairene Sabil: Form and Meaning," *Muqarnas* 6, no. 1 (1988): 34. Saleh establishes his argument based on Quranic textual sources and historical examples the *sabīl* as part of mosques.
48 I thank Deema Al-Ghunaim for this comment.
49 Saleh, "The Cairene Sabil," 34–35. See also Mandana Limbert's work on the moral, religious and social ideal of providing drinking water in Oman, sometimes in form of water coolers (as signs of development): Mandana E. Limbert, *In the Time of Oil: Piety, Memory, and Social Life in an Omani Town* (Stanford: Stanford University Press, 2010), 123–28.

10

DIMENSIONS OF THE PETROLEUMSCAPE IN THE PORT AND THE CITY OF HAMBURG

Christoph Strupp

At its annual press conference in 2017, the marketing company of the port of Hamburg presented slides with data on the development of the port during the previous year. Seventeen of the thirty-five slides focused on container handling. Bulk cargo, which includes petroleum, consisted of three slides and received the same limited space as cruise ship tourism.[1] Although Hamburg, as Germany's largest seaport, prides itself on its status as a "universal port" for all kinds of goods and services, the implicit message sent to customers and the public was that the port of Hamburg is primarily a container port. While containers indeed make up two-thirds of total turnover and generate the most value for the port, liquid cargo still accounts for 10 percent of total turnover. Of the 4,226 ha of land in the port area, about 12 percent is occupied by storage and production facilities pertaining to the oil industry, which remains a profitable pillar of the local economy. Headquarters of several oil companies are located in Hamburg, as are some oil lobby associations. Although local actors in the port economy and economic politics may have downplayed the significance of oil and oil handling in recent years, Germany's second-largest city remains an important node in the global petroleumscape.

The global petroleumscape, according to Carola Hein, consists of three interconnected layers (see also Hein's Introduction in this volume). The model emphasizes spatial dimensions of oil in the nineteenth and twentieth centuries and sheds light on the multitude of private and public actors involved in creating the petroleumscape.[2] This chapter explores the oil history in the port of Hamburg and focuses on the city's petroleumscape since the late nineteenth century, with an emphasis on the 1950s to 1980s. While Hamburg, like many other ports and port cities, has been serving as an important junction in the physical flow of oil around the globe,[3] and was publicly characterized as Germany's "oil city" in the interwar years and in the 1950s, the port and the city were not as intensively shaped by oil and the oil industry as, for example, Rotterdam. Therefore, the focus of "Hafen Hamburg Marketing e.V." on general cargo in its press conference in 2017 was no coincidence. Since emerging as one of Europe's leading overseas ports in the late nineteenth century, in Hamburg, port and trade have always been more closely linked than port and industry.[4] Emblematic of Hamburg as a city of trade was the "Speicherstadt" from the 1880s with its

DOI: 10.4324/9780367816049-13

brick warehouses for coffee, cacao, and general cargo, and the nearby "Kontorhausviertel," with its office buildings, from the 1920s to the 1930s.[5] This identity as a trading city was reinforced in the late 1960s, when the first container ships arrived in the port. Today, the Speicherstadt and the Kontorhausviertel are UNESCO World Heritage sites and tourist attractions, and three of Hamburg's four container terminals provide impressive views from the northern riverbank of the Elbe for locals and tourists alike. In contrast, most of the port's industrial sites are located out of sight in the southern part of the port (Figure 10.1).[6]

Expansion of the Industrial and Administrative Dimensions of the Petroleumscape from the Kaiserreich to the Third Reich

The spatial layer of the petroleumscape developed in Hamburg in the second half of the nineteenth century. The first barrels of oil arrived in the port from the US in 1861. In 1876, city officials' security concerns about the new flammable commodity resulted in the first expansion of the port—from the northern to the southern riverbank of the Norderelbe at Kleiner Grasbrook. Beginning in the late 1870s, a number of petroleum refineries were built in the port of Hamburg and the neighboring Prussian district of Harburg and Wilhelmsburg in the south. They included Ölwerke Ernst Schliemann (1879), Mineralölwerke Albrecht & Co. (1884), Ölwerke Stern-Sonneborn (1889), Johann Haltermann (1898), Ölwerke Julius Schindler (1908), and the Deutsche Vacuum Oil (1911), thus adding a new segment to the port-bound industry close to the waterfront. They benefited from the port's excellent hinterland connection to Berlin and the Saxon industrial area and—after 1888—from the port's newly granted free port status.[7]

In 1910, a new tank ship port was built at the western end of the port at Waltershof and, in the 1920s, additional storage and processing facilities for oil were installed at Waltershof, Kleiner Grasbrook, and in Harburg and Wilhelmsburg along the Süderelbe. These facilities consolidated Hamburg's position as oil center of the German Reich, even if the share of crude oil and oil products in total port turnover was limited (see Table 10.1). In the interwar years, up to 90 percent of German crude oil imports and most of the imported finished products such as gasoline passed through Hamburg. Contemporary aerial images show rows of storage tanks and production plants at several locations in the port area.

Hamburg was not only a central node in the industrial and infrastructure layers of the petroleumscape, but also an administrative hub. The company headquarters of ten of the largest German oil-processing companies—most of them in international hands—were located in the greater Hamburg area, accompanied by a large number of smaller companies based in various parts of the city. A commercial directory of Hamburg from 1922 lists almost 200 oil-related producers and distributors. Many of them had offices at prestigious addresses such as Rödingsmarkt, Hohe Bleichen, Mönckebergstraße, Glockengießerwall, and Alsterdamm.[8] Driven by the needs of local businesses, a separation of urban functions had begun in the nineteenth century and continued in the interwar years. Banks, trading houses, shipping companies, and industrial administrations clustered in the inner city.[9] In 1938, the Deutsch-Amerikanische Petroleum-Gesellschaft (DAPG) moved—in the words of the "Hamburger Nachrichten"—into a "new giant building at the Binnenalster"[10] that could easily match the other company headquarters there. The "Hamburger Fremdenblatt" emphasized that the new headquarters should "significantly influence the cityscape

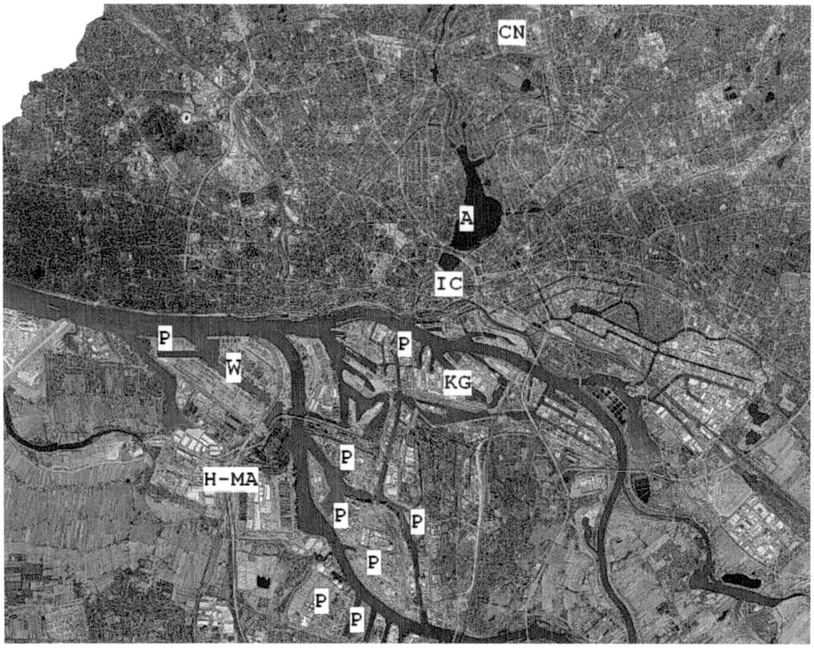

FIGURE 10.1 The city of Hamburg and the port area. P: Petroleum industrial sites; A: Alster; IC: Inner City; CN: City Nord; W: Waltershof; KG: Kleiner Grasbrook; H-MA: Hamburg-Moorburg and Altenwerder. *Source*: Creation by the author, based on an orthophoto of Stadt- und Landschaftsplanung Online (Planportal): https://www.hamburg.de/planportal/, Freie und Hansestadt Hamburg, Landesbetrieb Geoinformation und Vermessung, Data licence Germany – attribution – Version 2.0 (https://www.govdata.de/dl-de/by-2-0). Used by permission.

TABLE 10.1 Petroleum and Total Cargo Turnover in the Port of Hamburg, 1861–1938

Year	*Petroleum and Mineral Oil Import at the Port of Hamburg, in Tons*	*Total Cargo Turnover at the Port of Hamburg, in Tons*
1861	160	872,300
1871	40,000	2,663,000
1885	144,000	5,075,000
1900	375,000	14,432,000
1925	810,000	19,868,000
1938	3,700,000	25,760,000

Sources: Dieter Maass, *Der Ausbau des Hamburger Hafens 1840 bis 1910: Entscheidung und Verwirklichung* (Hamburg: Hansa, 1990), 78, 79; Andreas Kunz and Lars U. Scholl, eds. *Die deutsche Seeschifffahrt 1821–1989: Ein Datenhandbuch* (Bremen: H. M. Hauschild, 2011), 529–30; *Tabellarische Übersichten des hamburgischen Handels* (Hamburg: Herold, 1886, 1901); Klaus Weinhauer, "Handelskrise und Rüstungsboom: Die Wirtschaft," in *Hamburg im "Dritten Reich,"* ed. Forschungsstelle für Zeitgeschichte in Hamburg (Göttingen: Wallstein, 2005), 216.

according to modern architectural style."[11] Illustrated press reports such as these formed part of the representational layer of the petroleumscape, as did popular corporate media such as city maps with lists of sights to see and local gas stations.

In the Third Reich, in 1937, the Nazi regime merged the city of Hamburg with the neighboring Prussian cities of Altona, Harburg, Wilhelmsburg, and Wandsbek, which expanded Hamburg's territory from 415 to 745 km^2. The merger brought the port and industrial sites of Harburg and Wilhelmsburg, with its oil storage and refinery installations, under Hamburg's control and therefore further strengthened the industrial petroleumscape in the port (Figure 10.2). It also opened up new opportunities for a territorial adjustment of the port area, which would become relevant after 1945. While the official economic policy of the Nazi regime was guided by the ideal of energy autarky, motorization and rearmament led to a rise in demand for crude oil and petroleum-based products, which domestic sources alone could not meet. Oil handling in the port of Hamburg increased and the port-based local industry maintained and even expanded production levels in the late 1930s and early 1940s before targeted Allied air raids took place in June, August, and December of 1944, inflicting major damage to the port and adjacent industrial areas.[12]

FIGURE 10.2 Shell City Map of Hamburg with aerial photos of the Shell industrial facilities in Hamburg, Wilhelmsburg, and Harburg, mid-1930s. *Source*: Collection C. Strupp.

Industrial, Infrastructure, and Administrative Layers of the Petroleumscape in Hamburg in the Economic "Boom" of the 1950s

After World War II, the goal of making energy as cheap as possible soon dominated economic policies. The third quarter of the twentieth century saw a phenomenal rise in the use of oil, which replaced coal as the main energy carrier. In 1950, oil comprised 14 percent of energy demand in Europe; in 1970, its share was 60 percent.[13] Petroleum products were used in transportation, for heating and power generation, and more and more consumer products were based on oil derivatives. In the German Federal Republic, crude oil consumption rose from 4.5 million tons in 1955 to 83.6 million tons in 1965 and 110 million tons in 1973. The oil industry linked its product to economic progress and prosperity: "A revival and a gradual return to normal living conditions was unthinkable without a perfusion of the economy with oil," stated a 1951 brochure of Esso AG, the successor of the DAPG.[14]

In Hamburg, West Germany's hunger for oil—most of which had to be imported from oil-producing regions in the Middle East and North Africa—strengthened the port-based spatial layer of the petroleumscape, in particular industry, infrastructure, and administration. In the port, for two decades ever-larger tank ships and an expanding oil industry rather than the needs of general cargo would shape its spatial restructuring and expansion. In 1947, a new general development plan for the port included additional space for industrial plants by activating little used areas of port land along the Süderelbe.[15] However, in the immediate postwar years, priority was placed on the rebuilding of the destroyed facilities in the port and the reestablishment of the existing storage and refinery capacities at Waltershof, in Harburg, and Wilhelmsburg. In 1947, the Harburg asphalt plant of the DAPG/Esso AG restarted production. The Oelwerke Julius Schindler rebuilt their original refinery for oil products, and in 1949 expanded their facilities with a pipe distillation plant to process crude oil. Additional facilities for lube oil and sodium were added in the 1950s. In 1951, the Oelwerke became a subsidiary of the Anglo-Iranian Oil Company, the later BP. In November 1949, the rebuilt Harburg plant of Deutsche Shell AG was inaugurated with a festive event.[16] In 1959, Shell opened a refinery with 520 storage tanks at Hohe Schaar in Wilhelmsburg. All major oil corporations were represented in Hamburg and all expanded their plants in the 1950s, although primarily on land that had already been part of the port—the area of Neuhof, Kattwyk, and Hohe Schaar in the west of Wilhelmsburg and the Harburg port area further to the south. Ten companies maintained storage tanks in the port, mostly at the petroleum port at Waltershof.

For ports all over the world, the rising influx of crude oil initially meant a significant increase in total turnover. In Hamburg, the total turnover increased from 1.8 million tons in 1945 to 10.9 million tons in 1950, 24.0 million tons in 1955, and 30.8 million tons in 1960. Crude oil turnover rose from 2.1 million tons in 1950 to 6.3 million tons in 1955 and 10.1 million tons in 1960, thus now accounting for about one-third of total turnover (see Table 10.2). Another statistic underlines the importance of oil for the port: in the late 1950s, the quota of general cargo turnover had reached only 89 percent of the 1936 level, while bulk goods stood at 158 percent and oil at 225 percent of what it had been in 1936.[17]

The industrial and infrastructure layers of the petroleumscape expanded. Oil increased local traffic at the port and strengthened the industrial sector in Hamburg; it provided jobs in the shipbuilding industry and stimulated economic recovery in general. This economic importance was mirrored in the representational petroleumscape: in the 1950s, Hamburg was called the German "oil city" and the "oil center of the Federal Republic."[18] Local media reported regularly on the presence of the first mayor or the minister for economic affairs at the opening ceremonies for rebuilt or newly constructed production plants.

In 1954, at the fourth Hamburg "Overseas Day"—a lobbying event for business and politics, which takes place yearly around the "harbor birthday" of May 7—on the grounds of the Esso refinery in Hamburg-Harburg, the guest of honor, German Chancellor Konrad Adenauer, gave a friendly welcome speech (Figure 10.3). Later in the year, German President Theodor Heuss, during his state visit to Hamburg, also visited Esso. A press report characterized the pipes of the Esso refinery as "towers of the twentieth century," made of steel and aluminum instead of stone.[19]

In the mid-1950s, the industry employed approximately 5,500 employees each in the industrial plants and in administrative functions.[20] While the storage tanks and refineries were located in the port area, the administrative layer of the petroleumscape was concentrated in the inner city. As in the interwar years, the Hamburg-based company headquarters of Deutsche Shell AG, Deutsche BP AG, Esso AG and others were housed at prominent addresses such as Neuer Jungfernstieg, Alsterufer, Steinstraße, Hohe Bleichen, or Glockengießerwall. Trade associations such as the Mineralölwirtschaftsverband e.V. founded in Hamburg in September 1946 or the Uniti-Vereinigung deutscher Kraftstoffgroßhändler e.V. founded in July 1947 and specialist publishers were also located in Hamburg.[21]

TABLE 10.2 Crude Oil, Mineral Oil Products and Total Cargo Turnover in the Port of Hamburg, 1945–2005

Year	*Crude Oil Turnover, in Millions of Tons (Import Only)*	*Mineral Oil Products Turnover, in Millions of Tons (Import and Export Over Sea)*	*Total Cargo Turnover, in Millions of Tons*
1945	0.9 [1946]		1.8
1950	2.1		11.0
1955	6.3		24.0
1960	10.2		30.8
1965	13.1		35.1
1970	12.12	7.6	47.1
1975	10.85	7.05	48.6
1980	9.85	8.19	63.1
1985	3.68	9.42	59.52
1990	4.16	9.6	61.36
1995	4.82	7.27	72.12
2000	4.28	5.96	85.09
2005	4.42	6.75	125.74

Sources: Behörde für Wirtschaft und Verkehr, *Hafen Hamburg 1945–1965: Zwanzig Jahre Aufbau und Entwicklung* (Hamburg: Behörde für Wirtschaft und Verkehr, 1965), 43–46; Klaus-Jürgen Juhnke, "Die Bedeutung der Mineralölindustrie des Hamburger Hafens für Kosten und Erträge des Hamburger Staates" (Thesis, University of Hamburg, 1962), 56; Werner Klugmann, *Tatsachen, Daten und Zahlen über den Hamburger Hafen* (Hamburg: Okis, 1966), 131; Statistikamt Nord, *Schiffahrt und Außenhandel Hamburgs 1970 bis 2009* (Hamburg: Statistisches Amt für Hamburg und Schleswig-Holstein, 2010), 6.

FIGURE 10.3 German Chancellor Konrad Adenauer visits the Esso Refinery in Hamburg-Harburg, May 1954. *Source*: State Archive Hamburg. Used by permission.

The import of crude oil replaced some of the cargo volume lost in the East because of the division of Europe, and on-site processing of oil brought added value for the city. From a spatial perspective, however, it provided Hamburg—a city-state with fixed political boundaries and an upstream port about 120 km away from the Elbe estuary—with two major challenges: providing land for the industry and keeping up with the rapidly growing size of the ships. For both problems, port planners developed solutions that tried to meet the spatial needs and the interests of the oil industry without neglecting the interests of other established players in the port, particularly general cargo handling.

Since the early 1950s, the Ministry for Economic Affairs was in talks with representatives of the oil companies to accommodate their expansion plans. The city purchased agricultural land in Moorburg in the southwest, adjacent to the existing industrial sites, ever-rising prices because the farmers were aware of the value of their property, and then leased it to the oil industry for new production facilities.[22] Michael Miller has described how multinational companies Shell, Esso, and the others exercised "relentless pressure and blackmail" to dictate their conditions to port cities all over Europe.[23] In Hamburg, too, hardly any of the correspondence of the German CEOs with the ministry or the first mayor concerning the size of the locations, division of costs, or timetables lacked a reference to the allegedly favorable settlement conditions in the Rhineland.[24] The city was confronted with effects of the global character of the petroleumscape: local authorities were dealing with the German branches of multinational companies with production facilities in various locations for whom it did not necessarily matter to keep administration and production units together. The global petroleumscape threatened to overwrite local ties and Hamburg's politicians could not count on sentimental feelings toward their "oil city."

The director of Hamburger Hafen- und Lagerhaus AG (HHLA), a publicly held company in charge of much of the port operations, and then-Minister for Port Affairs Ernst Plate, a key player in Hamburg's port activities in the 1950s and 1960s, pointed out in parliament in June 1956:

> During the past weeks and months you have all read the frequent good press reports about the new anchoring of the mineral oil industry in the West German region, in Cologne. This development, I want to say openly as a Hamburg citizen, has not been very pleasing for us [...]. In this situation, I believe I can take the view that we are all the more interested in making available to the local oil industry all the facilities it needs to further expand its capacity.[25]

For this position, Plate could count on the support of all political parties and the local media.

Since the available land for industrial development in the port was gradually depleted at the end of the 1950s, the Port Development Act of 1961 designated 2,500 ha of land in Hamburg-Moorburg and neighboring Altenwerder, west of the Süderelbe, as future port area and subjected the inhabitants to expropriation and resettlement.[26]

The growing size of ships, which owed much to oil, posed another challenge for many ports after World War II.[27] In Hamburg in 1950, only two out of ninety-two tankers had a draft of more than 9.8 m, while in 1959, 108 of the 346 arriving tank ships needed this water depth.[28] In 1953, at the highly publicized Hamburg launch of the tanker Tina Onassis, the press mused whether an economic limit had been reached with its 35,000 tons.[29] In May 1962, the Howaldtswerke in Hamburg celebrated with prominent guests and thousands of spectators the launch of the 88,000-ton tanker Esso Spain, the largest cargo ship built to date in Europe. Within months, the Esso Deutschland and the Esso Bayern, both capable of carrying 91,000 tons of oil, broke this record (Figure 10.4).[30]

In the second half of the 1960s, ship sizes had increased to 276,000 tons, and in 1972 the first tanker with a capacity of 447,000 tons was under construction. In light of Hamburg's upstream location, the scenario of large tankers being unable to reach the port fed the fear that the port would not be able to keep up with competitors Bremen, Rotterdam, and Antwerp. This fear had been a driving force behind Hamburg's economic policy decisions since 1945.[31]

After the initial focus on rebuilding and expanding existing structures, the spatial demands of oil led to broader redesign of parts of the port by local authorities. In August 1956, the city's Ministry for Economic Affairs issued a memorandum, "On the Question of Hamburg's Tank Ship Ports," which predicted a further increase in oil turnover, a significant increase in ship arrivals, and a need for new processing plants.[32] The port had to be adapted accordingly. The city government and the parliament adopted a comprehensive expansion program, which involved modernizing the petroleum port at Waltershof, constructing several new tank ship ports (Waltershof, Neuhof, Kattwyk, Schluisgrove), and deepening the Elbe to 11 m. This would create 10 berths for large tankers, 17 berths for medium-sized tankers, and additional berths for smaller inland tankers and cleaning places. The cost of DEM 65.8 million would be offset by using the dredging material for landfilling in other areas of the port and amortize itself within a few years with additional revenue from the oil industry. This expansion program was completed in 1963.[33]

The dredging of access routes and the construction of larger berths were conventional measures to adapt the port's spatial layout to a changed economic environment. In response to European-wide debates about entirely new ways of importing oil—such as discharging

FIGURE 10.4 Beer mat of a local brewery celebrating the 775th port anniversary in 1964 with a drawing of the Esso Bayern. *Source*: Collection C. Strupp.

tankers at sea at floating port facilities or specialized oil ports at geologically suitable sites—in November 1960, the city government revealed plans to build an offshore port for large bulk carriers at the island of Neuwerk, a Hamburg exclave at the Elbe estuary; but such an outward growth of the petroleumscape in the end never happened. Due to economic and technological changes in the 1960s and 1970s, and after years of fruitless planning and discussions, the Neuwerk plan was abandoned in 1979.[34] In Hamburg, the industrial footprint within the spatial layer of the petroleumscape did not grow beyond the established port area and the port did not expand beyond city limits.

Clouds over Hamburg: Spatial and Representational Dimensions of the Petroleumscape in the 1960s

The Neuwerk plan and the restructuring of parts of the port to accommodate larger tank ships came at a time when the general conditions of the oil supply of the Federal Republic and the status of Hamburg as an "oil city" had already started to change. A few months before the memorandum of 1956 predicting the need to increase Hamburg's oil-handling capacity, the West German Federal Government decided to fund a specialized oil port in Wilhelmshaven on the coast of Lower Saxony. A pipeline to refineries near Cologne in the Rhineland went into operation at the end of 1958; as early as 1960, Wilhelmshaven replaced Hamburg as the leading oil port.[35] In addition, in July 1960, a pipeline from Rotterdam to the Rhine-Ruhr area became operational, and later pipelines from Mediterranean ports started to serve Southern Germany. Before the construction of the pipelines, in 1957, 100 percent of West German crude oil imports had been discharged directly in the ports. This rate fell to 22 percent within a decade.[36] In terms of production facilities, the landscape diversified. New refineries were built in western and southern Germany, close to main areas of consumption. They received their crude oil at competitive prices from Rotterdam via the pipelines and from German extraction sites. Even in 1957, the capacity of the eight refineries at the Rhine and Ruhr outnumbered that of the Hamburg plants. In 1960, the share of Northern German refineries in Hamburg (BP, Shell, Esso, Oelwerke Schindler), Bremen (Mobil Oil), and Heide (Texaco) amounted to 27.1 percent of the total capacity of the Federal Republic, but dropped to 14.9 percent within a decade.[37]

Hamburg's politicians argued that the pipelines were no competition since they transported oil to regions outside of Hamburg's trading area in Northern Germany.[38] However, this trading area had shrunk and the increase in oil turnover at the port was flattening. Experts of the Hamburg Ministry for Economic Affairs admitted this as early as 1960.[39] It was also reflected in a debate on the status of the tank ship expansion program in parliament in 1960, but members reaffirmed that "in the context of the European Economic Community Hamburg has to make sure not to be pushed aside by the so-called Europoort position which certain ports claim for themselves."[40] This was aimed at Rotterdam, Europe's leading oil port, which expanded its port toward the North Sea and added almost 6,000 ha of space between 1952 and 1973.[41]

In the second half of the 1960s, Hamburg had more than thirty berths for tank ships in three of its seven port groups (Waltershof, Kattwyk-Hohe Schaar, and Harburg), the largest of which were able to handle ships with a capacity of 60,000–70,000 tons. However, the Howaldtswerke in Kiel had already filled orders for ships of up to 171,000 tons and in Japan ships with more than 200,000 tons were under construction. In the Federal Republic of

Germany, only Wilhelmshaven could keep pace with this trend; that port gradually expanded to accommodate ships with up to 250,000 tons by 1975.

In terms of sales volume, in the early 1960s the mineral oil industry in Hamburg remained at the forefront of the port industries. However, while the shipyards employed some 33,000 people, the oil industry provided only around 8,600 jobs (out of a total of 235,000 industrial workers in Hamburg).[42] While the time of major enlargements of storage and processing capacities in the port in Hamburg was over—the newly designated port areas of the Port Development Act of 1961 were never used for the oil industry—and Esso, for example, now celebrated the opening of new refineries in Karlsruhe (1963) and Ingolstadt (1964), the oil cluster in the port continued to benefit from the unabated demand for oil and oil products. In 1968, Esso announced an investment of DEM 150 million to expand the capacity of its Harburg refinery from 3.6 to 5.5 million tons of crude oil annually.[43] The small Oelwerke Julius Schindler built additional processing units for lube oil products at its refinery in Neuhof (1963–1967).[44] A signal of confidence in the long-term future of oil in Hamburg followed in 1969, when Esso and Deutsche BP announced the founding of a company to study the construction of a pipeline from Wilhelmshaven to Hamburg. The pipeline, which in the end was not built until 1981, would have strengthened the position of the local refineries but would have meant for the port a yearly loss of several million tons of cargo. Ministry for Economic Affairs consoled itself with the added value the expansion of the capacities of the refineries would bring.[45] In the early 1970s, crude oil imports in Hamburg had become local traffic that, in contrast to other goods, had no competition with other ports.

In the second half of the 1960s, the needs of the oil industry no longer dominated port planning in Hamburg, but planners focused on a broader range of services and goods.[46] As in many other ports, the introduction of intermodal containers revolutionized general cargo handling. This fundamental change in maritime shipping posed new challenges to port authorities, port cities, and their strained budgets, but offered also new business opportunities. In terms of added value for the ports, oil compared unfavorably to other bulk and general cargo. The occupation of large areas of port property by the oil industry also came under critical review. From an economic point of view, it seemed to be more appropriate to concentrate on general cargo and to promote "modern" industries. Hamburg shifted its economic strategy in this direction with a comprehensive plan for promoting the "industrialization of the lower Elbe region." Profitable aluminum, steel, and chemical plants were to be attracted to Hamburg and the Elbe region and supplied with cheap energy via nuclear power stations.[47]

In the late 1960s, the industrial dimension of the spatial layer of the petroleumscape in the port of Hamburg stagnated but persisted—not the least because of the strong infrastructural path dependencies created by oil storage and production facilities. However, the administrative petroleumscape evolved and expanded into an entirely new area of the city, further away from the port and the prestigious inner city. The oil industry reaffirmed its commitment to Hamburg when four major companies relocated their headquarters to spacious modern office buildings in the "City Nord." Werner Hebebrand, Hamburg's director of urban development, had planned this cluster of large functional office buildings north of the city center on former garden land in the late 1950s, when congestion in the inner city increased and several companies in Hamburg demanded additional space for their administrations. Renowned contemporary architects designed the decidedly modern buildings. Esso moved into its new building in 1968, Deutsche BP in 1971, Deutsche Shell AG in 1974, and finally Texaco in 1977 (Figure 10.5).[48] Groundbreaking ceremonies and the openings of the buildings always took place in the presence of high-ranking members of the local government. Private institutions such as the prestigious Overseas Club, in which almost all the CEOs and other senior staff were members, also provided opportunities for political networking.[49]

The spatial petroleumscape in Germany was adjusted in the 1960s with the rise of Wilhelmshaven and the Rhine region, the refineries in Bavaria and Baden-Württemberg, and the introduction of the pipelines as an alternative to shipping. Hamburg remained a prominent knot in this web because of the combination of storage and production sites in the port and administration headquarters in the city.

Coping with Crises: Adjustments of the Petroleumscape in Hamburg Since the 1970s

In the 1970s, the question of the energy supply of Western Europe increasingly changed from a technical to a policy issue and moved from national to supranational levels.[50] At the beginning of the decade, Rotterdam had developed into the central hub of the petroleum sector in Europe. Wilhelmshaven, Hamburg, and Antwerp together accounted for only one-third of Rotterdam's oil turnover.[51] Regardless of the flattening growth rates of crude oil in Hamburg in the 1960s, optimism still prevailed at the Elbe. In September 1971, the Department for Economic Affairs signed a contract with Dutch Paktank N.V. on the construction and operation of a large storage tank on the Hohe Schaar, which was supposed to bring an additional 2–3 million tons of crude oil to the port annually. In October 1972, Esso reopened its expanded facility in Harburg and other refineries expanded their capacities as well.[52] Hamburg's refineries had a combined capacity of more than 13 million tons, but Deutsche BP, Esso, and Deutsche Shell all operated larger facilities at other locations.[53]

The first "oil crisis" of 1973 was, in political and psychological terms, a shock to contemporaries. It is regarded as a fundamental divide in the history of the second half of the twentieth century.[54] In the port of Hamburg and the local oil industry, however, the crisis

FIGURE 10.5 Office building of Deutsche Shell AG in the "City Nord," designed by the architect Meinhard von Gerkan; Ornament at the entrance of the building. *Source*: Wikimedia Commons, https://commons.wikimedia.org/wiki/File:Ex_shell_überseering_35_06.jpg/https://commons.wikimedia.org/wiki/File:Ex_shell_überseering_35_02.jpg, Author: Dirtsc, Date: July 18, 2015, CC BY-SA 4.0 (https://creativecommons.org/licenses/by-sa/4.0).

did not leave deep traces. The total turnover in the port recovered within a year (see Table 10.2). The import of crude oil dropped slowly from 12.12 million tons in 1970 to 9.85 million tons in 1980. In addition, finished oil products of about 4–5 million tons annually were handled in the port in the 1970s. Nevertheless, in 1976 oil companies stopped all future expansion plans for their coastal sites.[55]

The second "oil crisis" of 1979–1980 with its massive price increases and a 25 percent reduction in consumption within four years drove the port of Hamburg and the local oil industry into a fundamental crisis.[56] From 1982 to 1983, the port experienced a record drop in total turnover from 61.9 to 50.8 million tons. In the German oil industry, production capacities were cut in half within a few years, after capacity utilization had dropped to 57 percent in the early 1980s (see Tables 10.2 and 10.3). The crisis affected northern Germany in particular because the market there could be supplied cost-effectively with finished oil products from abroad because of its access to transport routes. In the port of Hamburg, the import of finished products grew from 5.3 million tons in 1980 to 7.7 million tons in 1990, but could not compensate for the loss of crude oil.[57]

While the first "oil crisis" had hardly affected the spatial layer of the petroleumscape in northern Germany, the second "oil crisis" changed it significantly. In Wilhelmshaven, Mobil Oil closed its refinery in March 1985.[58] In Hamburg, Deutsche BP reduced the capacity of its own refinery and of the Oelwerke Julius Schindler by a total of 2.1 million tons in 1979 and 1981. In 1983, crude oil processing was completely abandoned and in 1985, the refinery closed. Esso first reduced its capacities in 1985 and completely stopped production on January 12, 1987. The refinery was sold to Holborn Europa Raffinerie GmbH, which reopened the plant in 1988 on a smaller scale and obtained crude oil via the new pipeline from Wilhelmshaven. Only Deutsche Shell AG benefited from the closings of its competitors and expanded its Hamburg location because a more favorable product structure—less

TABLE 10.3 Mineral Oil Consumption in the Federal Republic of Germany and Crude Oil Distillation Capacity in Northern Germany, 1950–2005

Year	*Mineral Oil Consumption in the Federal Republic of Germany, in Millions of Tons*	*Crude Oil Distillation Capacity in Northern Germany (Schleswig-Holstein, Bremen, Hamburg), in 1,000 Tons*	*Crude Oil Distillation Capacity in Northern Germany, Percentage of West German/ German Total Capacity*
1950	4.052	2.635	49.2
1955	9.746	6.570	44.3
1960	28.730	10.970	26.4
1965	74.278	16.470	18.9
1970	124.447	17.910	13.6
1975	129.599	21.350	12.4
1980	138.948	20.630	12.1
1985	121.595	13.250	12.1
1990	122.811	11.930	11.8
1995	130.204	12.700	12.1
2000	125.032	14.250	12.6
2005	117.502	14.250	12.3

Source: Mineralölwirtschaftsverband e.V., *Jahresbericht Mineralöl-Zahlen 2005* (Hamburg: Mineralölwirtschaftsverband e.V., 2006), 27, 51.

gasoline and heating oil, more petrochemical basic materials—made it possible to adapt to the changed market conditions.[59]

Port and port city officials had little influence on the location decisions of international corporations. Two of the three potential strategies to overcome the crisis—modernization, partial withdrawal, or closure—would inevitably lead to a loss of turnover and possibly to abandoned land in the ports, which could not easily be reused. This was the case in Hamburg, for example, with the BP site at Finkenwerder.[60] However, the industrial layer of the petroleumscape in the port of Hamburg remained strong and the oil industry is to this day a factor in the local economy. With the exception of the old petroleum port at Waltershof, almost all the oil sites—from the small area next to Kleiner Grasbrook at the Norderelbe, today occupied by Sasol Wax and Deutsche Shell, to the Harburg-Wilhelmsburg cluster, today occupied by H&R Oelwerke Schindler, Vopak Dupeg storage tanks, Haltermann, Nynas specialty oils, Holborn, and others—are still in use. This confirms Carola Hein's observation of the "staying power" of refineries.[61]

In the 1990s and early 2000s, with some delay, the administrative layer of the petroleumscape also changed again. The major oil companies maintained their presence in Hamburg, but abandoned the spacious office buildings in the City Nord in favor of smaller headquarters throughout the city. The Mineralölwirtschaftsverband e.V. moved to Berlin in December 2008, as did the Uniti in 2009, but other lobby groups such as the Verband Schmierstoffindustrie e.V. stayed in Hamburg. In public life, oil remained visible in various ways: cultural aspects of the representational layer of the petroleumscape were manifest, for example, in the sponsorship of the major league soccer team Hamburger Sport-Verein by Deutsche BP (1979–1987) and the Hamburg-Marathon by Deutsche Shell (1991–1999) (Figure 10.6).

Conclusion

This chapter has provided an overview of dimensions of the petroleumscape in Germany's biggest seaport city Hamburg since the early 1860s. In the port, petroleum shaped much of the expansion process of the port area in the Kaiserreich and the interwar years. In the 1950s and 1960s, further port growth was limited by the political borders of the city-state and its upstream location. Port planners accommodated the spatial demands of the oil industry through the reallocation of port land and some adjacent areas. The introduction of pipelines and the oil crises of the 1970s ended new projects in Hamburg, but the oil industry maintained most of the established storage and production sites, underlining the path dependencies that large-scale industrial investments create.

The administrative dimension of the petroleumscape in Hamburg with its company headquarters and additional oil-related institutions and associations had originally formed a tight cluster in the inner city, not far from the port and close to the city hall and the headquarters of shipping companies, trading houses, and banks. In the late 1960s, this cluster disintegrated when more than two dozen companies, including the four major oil companies, moved their headquarters to the newly built City Nord. Since the 1990s, the administrative layer of the petroleumscape has changed again and further dispersed throughout the city.

For the port of Hamburg, the economic importance of oil grew in the interwar period and reached its peak in the late 1950s and early 1960s. Crude oil strengthened local traffic

FIGURE 10.6 Autographed card of a Hamburg soccer player with the BP logo on his sports shirt and on the stadium wall, mid-1980s. *Source*: Collection C. Strupp.

and partially made up for the loss of cargo in Eastern Europe after the war. For the city, the oil industry has been a profitable branch of the industrial portfolio throughout the twentieth century. Representations of Hamburg as Germany's "oil city," however, had to compete with established perceptions of Hamburg as a city of trade whose image was shaped by colonial goods and later on by containers.

Notes

1 "Hafen Hamburg Jahrespressekonferenz 2017," https://www.hafen-hamburg.de/downloads/media/dokumente/hhm-jpk-2017-16zu9.pdf.
2 Carola Hein, "Oil Spaces: The Global Petroleumscape in the Rotterdam/The Hague Area," *Journal of Urban History* 44, no. 5 (2018): 887–929.
3 Carola Hein, "Between Oil and Water: The Logistical Petroleumscape," in *The Petropolis of Tomorrow*, eds. Neeraj Bhatia and Mary Casper (New York: Actar, 2013), 436–47.
4 Arnold Kludas, Dieter Maass, and Susanne Sabisch, *Hafen Hamburg: Die Geschichte des Hamburger Freihafens von den Anfängen bis zur Gegenwart* (Hamburg: Kabel, 1988); Wolfgang Michalski, *Hamburg: Erfolge und Erfahrungen in der globalisierten Welt* (Hamburg: Murmann, 2010).
5 Ralf Lange, *Hamburgs Welterbe: Speicherstadt und Kontorhausviertel* (Hamburg: Junius, 2016).
6 Oliver Lieber, *Hafen versus Stadt: Konfliktanalyse der Flächenkonkurrenz zwischen Hafenwirtschaft und Stadtentwicklung in Hamburg* (Wiesbaden: Springer Fachmedien, 2018), 126.
7 Ernst Hieke, *Wilhelm Anton Riedemann: Anfang und Aufstieg des deutschen Petroleumhandels in Geestemünde und Hamburg, 1860–1894* (Hamburg: Verlag Hanseatischer Merkur, 1963); Dieter Maass, *Der Ausbau des Hamburger Hafens 1840 bis 1910* (Hamburg: Hansa, 1990), 78–81, 125–30, 192–93; Matthias Rademacher, *Die Geschichte des Hafen- und Schiffahrtsrechts in Hamburg, vol. 4:*

Die Entwicklung des Hamburger Hafens und Hafenrechts von der Jahrhundertwende bis zur Gegenwart (Hamburg: Rademacher, 1999), 120–126; Rainer Karlsch and Raymond G. Stokes, *"Faktor Öl": Die Mineralölwirtschaft in Deutschland, 1859–1974* (Munich: C. H. Beck, 2003), 30–36, 56–62, 84–90.

8 *Handels–Adreßbuch Hamburg: Ausgabe 1922* (Hamburg: Rößler & Delbeck, 1922). See also Wilhelm Ehricht, "Hamburgs Mineraloel-Handel und –industrie: Seine Bedeutung für Deutschland" (PhD diss., University of Hamburg, 1925).

9 Carola Hein, "Hamburg's Port Cityscape: Large-scale Urban Transformation and the Exchange of Planning Ideas," in *Port Cities: Dynamic Landscapes and Global Networks*, ed. Carola Hein (London: Routledge, 2011), 184–91.

10 "Neuer Riesenbau an der Binnenalster," *Hamburger Nachrichten*, March 2, 1936.

11 "Der Neubau des DAP-Verwaltungsgebäudes," *Hamburger Fremdenblatt*, October 8, 1936.

12 Karlsch and Stokes, *"Faktor Öl,"* 156–202, Klaus Weinhauer, "Handelskrise und Rüstungsboom: Die Wirtschaft," in *Hamburg im "Dritten Reich,"* ed. Forschungsstelle für Zeitgeschichte in Hamburg (Göttingen: Wallstein, 2005), 202, 215–18.

13 Robert Millward, *Private and Public Enterprise in Europe: Energy, Telecommunications and Transport 1830–1990* (Cambridge: Cambridge University Press, 2005), 212.

14 *Fünf Jahre Wiederaufbau: Tätigkeitsbericht der Esso AG 1945–1951*, ed. Esso AG (Hamburg: Hanseatische Druckanstalt, 1951), 25. All translations from German source quotes by the author.

15 Arved Bolle, "Der Generalplan für den Ausbau des Hamburger Hafens von 1947," in *Die Neugestaltung des Hamburger Hafens*, ed. Verwaltung für Wirtschaft und Verkehr (Hamburg: Verwaltung für Wirtschaft und Verkehr, 1947), 20–27.

16 "215 Tanks in einer Mondlandschaft," *Hamburger Abendblatt*, November 5, 1949.

17 Behörde für Wirtschaft und Verkehr, Amt Strom- und Hafenbau, *Grundlagen für den künftigen Ausbau des Hafens Hamburg* (Hamburg: Behörde für Wirtschaft und Verkehr, 1960), 7a, 8.

18 "Ölstadt Hamburg," *Die Zeit*, July 21, 1949; Sven von Müller, "Ölstadt Hamburg. Drei Säulen: Selbstvertrauen, Wille und Können," in *Neues Hamburg, vol. X: Die Wiederaufrichtung Hamburgs 1945–1955*, ed. Erich Lüth (Hamburg: Hammerich und Lesser, 1955), 87–88; "Hamburg, das große Ölzentrum der Bundesrepublik," *Hamburger Abendblatt*, May 7, 1954, addendum.

19 "Türme des 20. Jahrhunderts," *Hamburger Abendblatt*, April 29, 1954.

20 Müller, "Ölstadt," 88; Klaus-Jürgen Juhnke, "Die Bedeutung der Mineralölindustrie des Hamburger Hafens für Kosten und Erträge des Hamburger Staates" (Thesis, University of Hamburg, 1962), 41–42.

21 Hansgeorg Günter, *Mineralöl-Handbuch* (Hamburg: Christen, 1950); Curt Weber, "Mineralölindustrie: Rohöle aus Venezuela und dem Mittleren Osten," in *Hamburg als Industrieplatz*, eds. Handelskammer Hamburg and Behörde für Wirtschaft und Verkehr (Hamburg: Hammerich und Lesser, 1952), 68–69; "Hamburg, das große Ölzentrum der Bundesrepublik," *Hamburger Abendblatt*, May 7, 1954, addendum; "Hamburg: Deutschlands Ölhafen Numer eins," *Hamburger Abendblatt*, May 7, 1957, addendum.

22 Notes and Correspondence of the Ministry for Financial Affairs, State Archive Hamburg 311-3 I: 196: Erweiterung der Ölindustrie im Hamburger Hafen, 1954–1955; 311-3 I, Abl. 1981: 410–18/69: Erweiterung der Ölindustrie im Hamburger Hafen, Vol. 1–2, 1955–1960 (1962, 1964).

23 Michael B. Miller, *Europe and the Maritime World: A Twentieth-Century History* (Cambridge: Cambridge University Press, 2012), 311.

24 Notes and Correspondence of the Ministry for Economic Affairs, State Archive Hamburg 371-16 II: 4001: Firmen der Mineralölwirtschaft (Beispielakte), Deutsche Shell AG, 1951–1961.

25 *Stenographische Berichte über die Sitzungen der Bürgerschaft zu Hamburg im Jahre 1956* (Hamburg: Auerdruck, 1956), 760.

26 Rademacher, *Geschichte*, 418–19.

27 Miller, *Europe*, 307–308.

28 Behörde für Wirtschaft und Verkehr, Amt Strom- und Hafenbau, *Grundlagen für den künftigen Ausbau des Hafens Hamburg* (Hamburg: Behörde für Wirtschaft und Verkehr, 1960), 7–8; Behörde für Wirtschaft und Verkehr, *Hafen Hamburg 1945–1965: Zwanzig Jahre Aufbau und Entwicklung* (Hamburg: Behörde für Wirtschaft und Verkehr, 1965), 13–14.

29 E. Förster, "Warum so große Tanker?," *Hamburger Abendblatt*, December 9, 1953.

30 "Ein Festtag für den Hamburger Hafen," *Hamburger Abendblatt*, February 20, 1963.

31 Hans-Henning Kroll, "Sie sind zu groß für Europas Küsten: Tanker-Giganten finden keine Häfen mehr," *Hamburger Abendblatt*, September 27, 1966; Lu Seegers and Christoph Strupp, "Hafen- und Handelsstadt oder Stadt der Industrie? Wirtschaftspolitik und Deutung des Strukturwandels in Hamburg," in *Wirtschaftsräume und Lebenschancen: Wahrnehmung und Steuerung von sozialökonomischem Wandel in Deutschland 1945–2000*, eds. Stefan Grüner and Sabine Mecking (Berlin: De Gruyter Oldenbourg, 2017), 205–22.

32 Behörde für Wirtschaft und Verkehr, *Zur Frage der Hamburger Tankschiffhäfen* (Hamburg: Behörde für Wirtschaft und Verkehr, 1956); Hans Laucht, "Planung und Bau neuer Tankschiffhäfen in Hamburg," *Handbuch für Hafenbau* 3 (1957): 233–39.

33 Juhnke, "Bedeutung," 81–99, 124–52.

34 Rademacher, *Geschichte*, 413.

35 Jörn Eiben, *Industriestädte und ihre Krisen: Wilhelmshaven und Wolfsburg in den 1970er und 1980er Jahren* (Göttingen: Wallstein Verlag, 2019), 50–51; Marten Boon, Multinational Business and Transnational Regions: A Transnational Business History of Energy Transition in the Rhine Region, 1945–1973 (New York: Routledge, 2018) Boon, *Multinational Business*, 108–15; Juhnke, "Bedeutung," 154–56.

36 Boon, Multinational Business, part II; Michael Poppe, *Integration von Infrastrukturen im historischen Vergleich, vol. 5: Öl- und Treibstoffpipelines* (Baden-Baden: Nomos, 2015); Miriam Bader-Gassner, *"Pipelineboom:" Internationale Ölkonzerne im westdeutschen Wirtschaftswunder* (Baden-Baden: Nomos, 2014).

37 Boon, *Multinational Business*, 64, 85–92; Wolfgang Viernow, *Mineralölpipelines: Ihre Funktion und ihre Stellung im Verkehrssystem der Bundesrepublik* (Cologne: DVWG, 1971), addendum 3.

38 Hubert van Drimmelen, "Mineralölzentrum Hamburg wird sich behaupten," *Hamburger Journal* 8, no. 5 (1960): 43–44; Klaus Kramer, "Für Hamburg keine Konkurrenz," *Hamburger Abendblatt*, August 17, 1963.

39 Uwe Christiansen, "Hamburg im Seehafen-Wettbewerb," *Wirtschafts-Correspondent* 14, no. 9 (1960): 14.

40 *Stenographische Berichte über die Sitzungen der Bürgerschaft zu Hamburg im Jahre 1960* (Hamburg: Auerdruck, 1960), 481.

41 Hein, "Oil Spaces," 908, 912, 917–20; Boon, *Multinational Business*, 154–58; Ferry de Goey, *Ruimte voor industrie: Rotterdam en de vestiging van industrie in de haven, 1945–1975* (Delft: Eburon, 1990).

42 Juhnke, "Bedeutung," 37–42.

43 *100 Jahre Dapolin: Dapolin-Benzin, Standard, Esso 1890–1990*, ed. Esso AG (Hamburg: Esso AG, 2009), entry for 1968.

44 Theo Müller and Annette Schlapkohl, *100 Jahre Schindler: Chronik einer Hamburger Firma* (Husum: Husum Druck– und Verlagsgesellschaft, 2008), 46–48.

45 Hans Spilker, "Keine Prestige-Investitionen! Zum Bericht der Tiefwasserhäfen-Kommission," *Wirtschaftsdienst* 52 (1972): 204; Wilhelm Voss, "Hamburg als Massenguthafen," *Hansa* 110 (1973): 832–833.

46 M. S. Husain, "Influences on Development Policy in the Port of Hamburg," in *Cityport Industrialization and Regional Development: Spatial Analysis and Planning Strategies*, eds. Brian S. Hoyle and David A. Pinder (Oxford: Pergamon Press, 1981), 232, 239.

47 Husain, "Influences," 232; Helmuth Kern, *Ein Modell für die wirtschaftliche Entwicklung der Region Unterelbe* (Hamburg: Behörde für Wirtschaft und Verkehr, 1970).

48 Sylvia Soggia, *City Nord: Europas Modellstadt der Moderne* (Munich: Dölling und Galitz, 2009); Carola Hein, "City Nord – eine Geschäftsstadt im Grünen," in *Das ungebaute Hamburg: Visionen einer anderen Stadt in architektonischen Entwürfen der letzten hundertfünfzig Jahre*, ed. Ulrich Höhns (Hamburg: Junius, 1991), 200–9.

49 Übersee-Club e.V., Mitgliederverzeichnis, Mai 1963 (Hamburg: Übersee-Club e.V., 1963).

50 Martin Czakainski, "Energiepolitik in der Bundesrepublik Deutschland 1960 bis 1980 im Kontext der außenwirtschaftlichen und außenpolitischen Verflechtungen," in *Energie – Politik – Geschichte: Nationale und internationale Energiepolitik seit 1945*, eds. Jens Hohensee and Michael Salewski (Stuttgart: F. Steiner, 1993), 17–34; Rüdiger Graf, *Öl und Souveränität: Petroknowledge und Energiepolitik in den USA und Westeuropa in den 1970er Jahren* (Berlin: De Gruyter Oldenbourg, 2014).

51 Johannes Marx, "Der Hamburger Hafen als Energieumschlagplatz," *Hamburg in Zahlen*, no. 4 (1974): 98–100.

52 "85 Millionen für den Hafen," *Hamburger Abendblatt*, September 3, 1971; Klaus Repenning, "Die Zukunft der Petrochemie," in *Hamburg auf dem Weg in das Jahr 2000: Prognosen*, ed. Ernst Schmacke (Düsseldorf: Droste, 1970), 224–35.
53 Erich Schieweck, "Wirtschaftliche und politische Aspekte des Erdöls in den siebziger Jahren," *Glückauf* 106 (1970), 807.
54 Jens Hohensee, *Der erste Ölpreisschock 1973/74: Die politischen und gesellschaftlichen Auswirkungen der arabischen Erdölpolitik auf die Bundesrepublik Deutschland und Westeuropa* (Stuttgart: F. Steiner, 1996); Stefan Göbel, *Die Ölpreiskrisen der 1970er Jahre: Auswirkungen auf die Wirtschaft von Industriestaaten am Beispiel der Bundesrepublik Deutschland, der Vereinigten Staaten, Japans, Großbritanniens und Frankreichs* (Berlin: Logos, 2013).
55 K.-D. Fischer, "Stagnation der westeuropäischen Raffineriekapazität," *Annuaire européen du pétrole* (1976): 44–50.
56 Hans Michaelis, "Die Energiewirtschaft der Bundesrepublik Deutschland 1970 bis 1990," in *Energie – Politik – Geschichte: Nationale und internationale Energiepolitik seit 1945*, eds. Jens Hohensee and Michael Salewski (Stuttgart: F. Steiner, 1993), 52–53; Hohensee, *Ölpreisschock*, 78.
57 Bernhard Hnat, *Strukturwandel in der mineralölverarbeitenden Industrie der Bundesrepublik Deutschland: Untersuchung über das Anpassungsverhältnis der Mineralölgesellschaften als Folge der Ölkrisen* (Göttingen: Vandenhoeck & Ruprecht, 1992), 61; Susanna Degrassi, "The Seaport Network Hamburg" (Dissertation, University of Hamburg, 2001), 281–82.
58 Eiben, *Industriestädte*, 225–28.
59 Rademacher, *Geschichte*, 470–72; Hnat, *Strukturwandel*, 109, 115.
60 Carola Hein, "'Old Refineries Rarely Die': Port City Refineries as Key Nodes in the Global Petroleumscape," *Canadian Journal of History* 53 (2018): 476–79; David A. Pinder and Sohail Husain, "Deindustrialisation and Forgotten Fallow: Lessons from Western European Oil Refining," in *Revitalising the Waterfront: International Dimensions of Dockland Redevelopment*, eds. Brian S. Hoyle, David A. Pinder, and M. S. Husain (London: Belhaven, 1988), 232–46.
61 Hein, "'Old Refineries'," 450.

11

"PRODUCTION FIRST, LIVELIHOOD SECOND"

The Life and Death of Worker-Peasant Model Villages in a Chinese Oil Field

Hou Li

European and American researchers often associate oil with consumerism and middle-class suburban lives. However, China's entry into the hydrocarbon age was marked by revolutionary, anti-imperialist, and nationalist struggles. The Chinese official oil narrative depicts an emerging Chinese working class and a young socialist state that prioritized production over consumption—as in the slogan "production first, livelihood second" deployed by the leaders of Daqing Oil Field, the largest oil field ever discovered in the People's Republic of China (PRC). In 1963, three years after the field's discovery in northeastern China, the country gained oil independence. At the time, the People's Republic had barely recovered from the nationwide Great Famine (1958–1962) and was threatened with war by both world superpowers—the US and USSR. The distinctive feature of oil exploration and production in China is that it was state-sponsored and centrally planned.

China's hydrocarbon modernity was inevitably tied to a nationalist narrative and national power struggles. The process of oil discovery, which is technologically and financially demanding, requires the ability to mobilize resources and navigate political institutions (in this book, see also Peyerl).[1] To guarantee the timely completion of state industrialization goals, the central authorities channeled national resources in a way that concentrated on key projects and strategic locations. Through regional planning, the socialist regime sought to integrate projects of oil exploration and industrial construction. This enabled the fast expansion of an industrial landscape in China's oil-rich basins in the northeast in the 1960s, then to the hinterland in the 1970s. Furthermore, during these two decades, the great success of the petroleum industry empowered oil cadres to reform the formerly Soviet-modeled state planning machine.

In this sense, the oil industry not only shaped the "petroleumscape"[2] of oil exploration, extraction, transportation, and refining, but also reshaped national standards of ancillary "nonproductive" construction—such as public housing, schools, and hospitals—that profoundly impacted people's daily lives as well as the national landscape.

DOI: 10.4324/9780367816049-14

Wars, Campaigns, and the Discovery of Oil

Transnational corporations became important actors in the history of oil before the establishment of China's communist regime and long before oil was discovered in mainland China in the twentieth century. The use of kerosene for illumination had been introduced to China in the mid-nineteenth century, after the first Opium War, when China was forced to open its domestic market to foreign trade. In 1870, the Standard Oil Company extended its business into China not long after it was founded by John D. Rockefeller in Cleveland, Ohio. It quickly stepped up sales to become the largest American company operating in China. A few decades later, Texaco Petroleum Company—another American company—and Asiatic Petroleum Company, the Far Eastern branch of the Royal Dutch-Shell Company, became the two other most competitive companies to enter the Chinese market.[3] By the end of the nineteenth century, imported kerosene had become part of Chinese everyday life in both the cities and the countryside.

As the Chinese state and society realized the value of crude oil, they began the search for domestic oil fields. However, the spatial distribution of oil deposits in China is very uneven. Early discoveries were all in the Northwest, thousands of miles from China's industrial centers, which posed problems regarding transportation and communication in a country lacking a modern infrastructure. The Japanese invasion of Manchuria in 1931 marked a turning point with respect to China's oil consumption and exploration. By the 1930s, although private ownership of cars was rare, gasoline had replaced kerosene as China's most important petroleum product.[4] Oil had become not just a part of people's everyday lives, but a strategic national commodity. At the time, China's consumption of oil was still highly dependent on the three global companies.

Imported oil was first stored in the coastal treaty ports, including Shanghai, Tianjin, Ningbo, and Guangzhou, and then was delivered to the hinterlands mostly by ship. The Yangtze River was the main transportation corridor for the oil trade. After the Japanese army seized control of key cities along the major communication links, including the Yangtze River, China's foreign oil supply was almost completely cut off. The National Defense Planning Commission, later renamed the National Resources Commission, was established to make national defense preparations that involved rigorous efforts to obtain a domestic supply of oil.[5] In the late 1930s, the commission succeeded in discovering oil deposits sufficient for industrial production, first in Yanchang and then Yumen (Figure 11.1).

These early discoveries helped China survive its wartime energy crisis, but could hardly meet the country's needs for a massive societal transformation. At the start of the communist regime, the state's plans for industrialization mandated heavy use of energy, and a tense international situation required self-sufficiency in oil production. In the 1950s, the Ministry of Petroleum Industry turned oil prospecting into labor-intensive campaigns and imported modern geological surveying technologies and equipment. The 1960 discovery of Daqing's rich oil deposit right at the center of China's most industrialized region resulted from years of effort and careful strategy, but it was also a matter of luck. In the late 1960s and the early 1970s, the ministry discovered other major oil fields—Shengli, Dagang, Zhongyuan, and Huabei in North China, Jianghan in the Southwest, Liaohe and Jilin in the Northeast—one after another (Figure 11.2). From 1968 to 1978, China's crude oil production maintained an astonishing annual increase of 20 percent. Within a decade, China was transformed from an oil importer to an oil exporter.

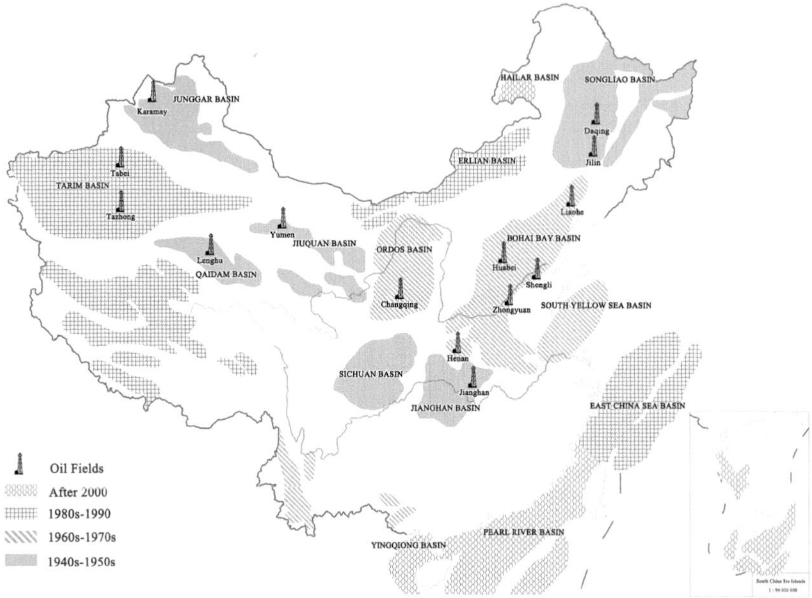

FIGURE 11.1 The spatial distribution of oil fields in China. Drawn by the author.

FIGURE 11.2 The Yumen Oil Refinery, located in the mountains, with a dormitory at the middle right, year unknown. Photo courtesy of China Petroleum Corp., Limited (Republic of China).

Planning the Daqing Petroleumscape: The Pursuit of Egalitarian Industrialism

By the end of the first decade of the People's Republic, the Chinese economy faced a series of setbacks and complications. The difficult task of building a balanced relationship between industry and agriculture, production and livelihood, had seemingly come to a dead end. Shortages of both capital and labor inputs limited the growth of agricultural production, which in turn curtailed industrial construction. At the same time, the implementation of the state's industrialization plans triggered extensive urban growth. Massive movement took place back and forth between the cities and the countryside. The deficiency of the newly established planning economy eventually led to extreme scarcity of construction materials and also, in the end, to fatal food shortages.

In this situation, the ancillary facilities, which were considered a constituent part of "nonproductive construction," such as housing and urban amenities, were strictly controlled and often were the first to be cut. The state spent most of its revenue reinvesting in "productive" units. Capital accumulation, referring to the revenue returned to production instead of using for urban construction, was maintained at a high rate and urgent social needs were ignored. Chinese urban planners faced waves of criticism for their "excessive spending" of state investments in urban construction, which left little for the construction of urban housing, transportation, education, or other aspects of public welfare.[6] The state promoted frugality, austerity, and self-sufficiency and people were asked to curb consumption and leisure in order to achieve higher rates of economic growth.

In 1960, as a pragmatic solution, the Ministry of Petroleum Industry built mud huts for oil workers and their families. On the Daqing prairie, winter comes as early as October and the ministry could not fulfill its mission if the labor force had to retreat to nearby cities for the long winters. The ministry was hoping to put the oil field into production within a year. In 1960, the shortage of oil had slowed industrial production and threatened national security. China was importing more than half of its crude oil, mostly from the Soviet Union. With increasing tension between the PRC and USSR, further imports from the Soviets were not guaranteed, and the state faced significant financial difficulties due to its huge foreign currency payments. For fuel, the military relied almost entirely on imports, so large-scale operations that required substantial supplies of fuel became increasingly difficult to maintain. The serious shortage of fuel even influenced pilot training and the daily operation of the air force.[7]

Daqing Oil Field was surrounded by an unusually fertile area that was perfect for grain production. In the famine years, many oil field workers had to share their limited grain with wives and children who had followed them to the field. Housewives who came from peasant backgrounds picked up their hoes and began reclaiming the nearby land in order to ease the burden on their families and to provide food for their hungry children. More families began taking part and their actions were welcomed by the Daqing leaders. The Oil Field Headquarters organized the unemployed women in collective farms, and before long most mud hut villages were surrounded by farmland. In Daqing's exploratory years, the petroleum workers tended to be quite young, either fresh graduates or the veterans back from the Korean War. The harsh conditions encouraged the young men and women to live communally; they not only worked and lived together, but they also built their houses and worked together to obtain food. The physical and human constraints and scarcities contributed to the community's solidarity. Many first-generation Daqing people remembered this period as a utopian moment, representing a primitive form of communist life.[8] Every individual was mobilized and integrated as part of the group, becoming both objects and agents of revolutionary change. Although the nationwide famine cast a huge shadow on the young People's Republic, the struggling central leadership perceived the Daqing campaign as a source of light and Daqing's unique landscape an ideal representation of the resilient Chinese socialism.

At the regional planning level, Vice Minister of Petroleum Sun Jingwen, formerly the head of the state's Urban Construction Bureau, decided against building an oil city in Daqing. Drawing on lessons learned in the 1950s, Sun insisted on a dispersed, production-oriented development model. He further summarized the pattern as one that represented the ideal of an "integration of industry and agriculture, integration of workers and peasants, integration of town and countryside."[9] Such a philosophy of construction was backed by the other ministers of oil, Yu Qiuli and Kang Shi'en, and then was highly praised by the central leaders, first Prime Minister Zhou Enlai, later President Liu Shaoqi, and then Chairman Mao. They saw the approach as having helped achieve a level of industrial productivity needed to conquer material scarcity. Furthermore, it could embody the ideal communist society described by Karl Marx because of its efforts to eliminate the three great gaps: between town and countryside, between workers and peasants, and between mental and manual labor.

The Daqing settlement construction plans were both pragmatic and ideologically appropriate at a time when petroleum leaders' influence was rising. Consequently, Daqing's petroleumscape became the hallmark of the new socialist country's industrial spirit. Architects in Daqing were required to develop a "scientific gandalei" model for a rammed earth house that combined local building techniques and modern innovations (Figure 11.3). In addition to sun-dried pounded blocks of mud and straw, the Daqing gandalei design team used other locally available construction materials. Small pieces of wood provided support for a curved roof. Bitumen (also known as asphalt), a by-product at the oil refinery, was mixed with the mud and hay for insulation and to provide a water-resistant layer for the roof. One of the difficulties was to design a foundation for a house that could be built quickly and function well in the northern climate. The region's traditional houses frequently encountered structural failures as a result of humidity or thawing in the spring. After several tests, the foundation was finally set at 30 cm deep, covered by a 5 cm layer of bitumen and mud mixture at a ratio of 1:5, along with a building skirt around the outer wall. Ventilation and heating systems were carefully designed to avoid fire hazards and to introduce adequate fresh air. Compared to the typical local mud houses, the walls of the "scientific gandalei" houses were thinner and the roofs were elevated so as to increase the internal space.[10]

In the spring of 1964, large-scale construction began on decentralized settlements in the vicinity of the Daqing oil field (Figure 11.4). Architects and workers labored together to build the first worker-peasant village in Saertu. The headquarters hosted an on-site conference on gandalei building. All the work units sent representatives to study the new model and to begin their own construction projects. Communal building housed the majority of the oil labor force, more than 200,000 people, under this single design. The locations of the settlements followed the expansion of oil exploration and production. Master plans of the oil settlements called for barrack-like structures with unadorned empty ground, ideally surrounded by farmland. The entire Daqing landscape became a flowing assembly line for

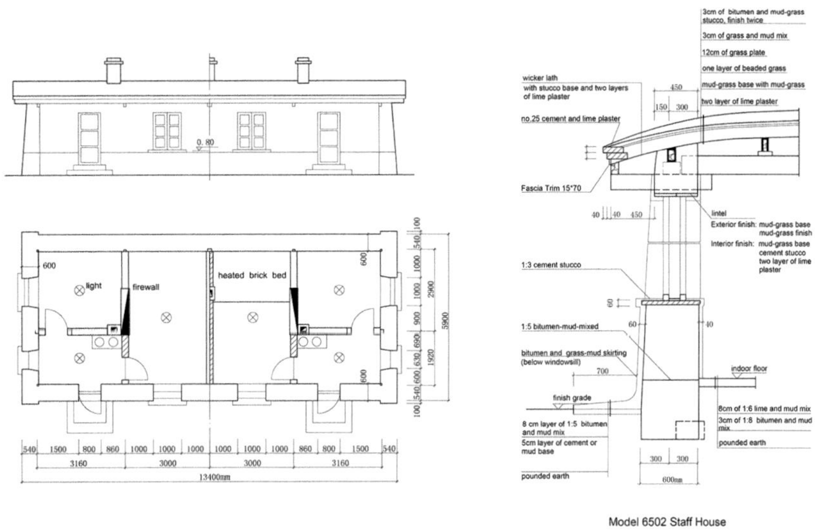

FIGURE 11.3 Scientific gandalei house: floor plan and height, Daqing Design Institute, 1964. Original drawing by the Daqing Design Institute, redrawn by the author.

FIGURE 11.4 The first scientific *gandalei* village, Hongweixing (Red Satellite) Village, 1964. Tongji Planning Department Archives.

industrial and agricultural production. Ornamentation and decoration were considered a waste of state resources. The petroleum leaders saw urban consumption and demands for a higher quality of life as an undesirable by-product of industrial growth that would compromise state development goals.

In the early years, the dispersed and highly varying appearance of Daqing's physical landscape was not entirely a planned product. The results of improvisation eventually became a policy endorsed by top leadership. Mobilization for oil production in a national crisis and the self-sacrifice of the model oil workers became enshrined in the humble mud-straw houses as well as the agricultural-industrial integrated lifestyle. The decentralized villages with their identical houses were also seen as indicative of an equal society, but the rules were imposed from above. With the rise of Daqing as a national industrial model, in December 1964, Minister of Petroleum Yu Qiuli, with Mao's backing, replaced State Planning Commissioner Li Fuchun and began drafting state planning strategies. During the last several days of 1964, oil cadres in Daqing received Yu's instructions that all headquarters—meaning all administrative offices—were to remain in their humble gandalei houses, and the brick buildings were only to be used for public services. Unlike in most countries, the administrative centers in Daqing Oil Field were devoid of monumentality and did not have higher standards than the workers' living compound. Now, no more multistory buildings were to be constructed. Planning for the Daqing headquarters in Ranghulu came to an abrupt halt. The economic logic of Daqing's petroleumscape had become politicized. The struggle between different development strategies was represented as mud versus brick or single-story buildings versus multistory buildings.

With the introduction of the Daqing model nationwide, the late 1960s and 1970s saw continuous diffusion of the urbanization that had begun in the 1950s. Industries along the coast were moved inland. Investment that had been concentrated in the north was diverted to the south. Population growth in the large cities was strictly controlled, even leading at times to negative growth. The 1960s and 1970s witnessed the emergence of numerous mining districts and factory complexes that statistically were not large enough to be counted as cities and whose populations were engaged in more than just mining and manufacturing. This dispersed pattern of development marked the physical landscape throughout the country, accompanied by a quickly increasing national population—it grew 300 million in fifteen years—and the extension of inhabited areas. The decentralized model was also favored because the country faced the threat of war: under this model, it would be more difficult for the enemy in the air to distinguish urban from rural areas, industry from agriculture, centers from peripheries.[11]

The Settlements

The Oil Headquarters divided Daqing's belt-shaped oil field, stretching 140 km in a north-south direction, into several oil extraction fields, following the sequence of their development and a division of labors. No. 1 Oil Extraction Factory was the area centered around the Saertu Railway Station. No. 2 Oil Extraction Factory was located several miles to the south of Saertu, and No. 3 Oil Extraction Factory was to the north. No. 4 and No. 5 Extraction Factories were farther to the south, near Datong, where the region's oil was first discovered. No. 6 Factory was located in the far northwest of Saertu, at the Lamadian Oil Field. Other ancillary departments were located relatively close to Saertu, but still at some distance. For example, the construction headquarters was in Hongweixing (literally, Red Satellite) Village, northwest of Saertu, and the transportation headquarters was in Chengfeng (literally, Riding on the Wind) Village, 20–30 km to the south of Ranghulu.

The Master Planning Division at the headquarters and the production units picked the settlement sites together. Then, the workers and their families built the houses with some assistance from professional building teams. There were three types of locational choices. First, for those work units that had fixed but dispersed working sites, such as an oil extraction factory, scattered "settlement points" were selected close to their production unit.

Second, for work units whose locations were fixed and concentrated, such as oil or water pump stations or research institutes, larger-scale "central villages" were built together with the work unit. Remote settlements in the peripheral areas might be built to meet the farming needs of the dependents in these work units. Third, for highly mobile construction and drilling teams, a living base was built near the oil exploration frontier. There was no viable map for locating these settlements until 1966, so planners determined locations by investigating on-site. They usually chose higher places (if available) near the work units, as every spring the plains encountered serious flooding problems.

Under the industrial master plan, these villages and settlement points became linked to roads and electric grids as oil field construction continued on the plains. Every settlement had its own water station in the middle, providing piped and hot water, and a public earth toilet on the periphery. The central village might have a few more daily service facilities, such as a grocery shop, a grain station, a barbershop, a bathhouse, or a clinic. Daqing planners added elementary and middle schools in the villages in the late 1960s and early 1970s, as the first baby boomers in the oil field reached school age. As rammed earth houses could not meet the requirements of large public spaces for school buildings, professional teams made schools out of brick. The brick school buildings became conspicuous landmarks, surrounded as they were by clusters of low mud houses. The so-called "scientific gandalei" house was the standard housing type in Daqing for all residents regardless of occupation, age, or family size. Young couples and singles usually shared two bedroom units, whereas families with children might occupy an entire unit. With the adoption of this standard form, architects could save their time and energy for designing the oil field's industrial structures.

By the late 1960s, the petroleumscape of the Daqing Oil Field consisted of (a) three small towns—Saertu, Ranghulu, and Longfeng—in the north along the Binzhou Railway, all within 5–10 km of each other; (b) an intermingling of oil wells, dwellings, and farmland in the belt-like oil field stretching from the railway to the south; (c) dispersed industrial facilities associated with oil extraction and processing; and (d) a grid pattern of roads linking the gandalei settlements and the production points. According to the "three-integration" planning principle, the industrial, administrative, ancillary, and infrastructural components were closely integrated. This landscape presented a decentralized cellular pattern, in which each cell was self-sustained and multifunctional. For more than a decade, Daqing People's Hospital and Daqing Design Institute were the only multistory buildings, serving as landmarks for new arrivals—and perhaps convincing them that they were no longer in the middle of the countryside.

The First Model Village: Red Satellite

Red Satellite Village (Figure 11.5) was Daqing's first model village. It was the site of the Daqing Oil Field Construction Headquarters—in charge of all of the oil field's building and construction—as well as a base for workers and their families. The site comprised one central village and four settlement points (Nos. 2, 3, 4, and 8). The central village had about 600 households, and each settlement point varied from one hundred to 300 households. The settlement points centered around the village at almost equal distances; Nos. 2, 3, and 8 were all within 1 km, whereas No. 4 was about 2 km from the village. The central village was developed at the intersection of two main roads, with branches of the road linking the settlement points, while the settlement points also had their own links to the roads. The

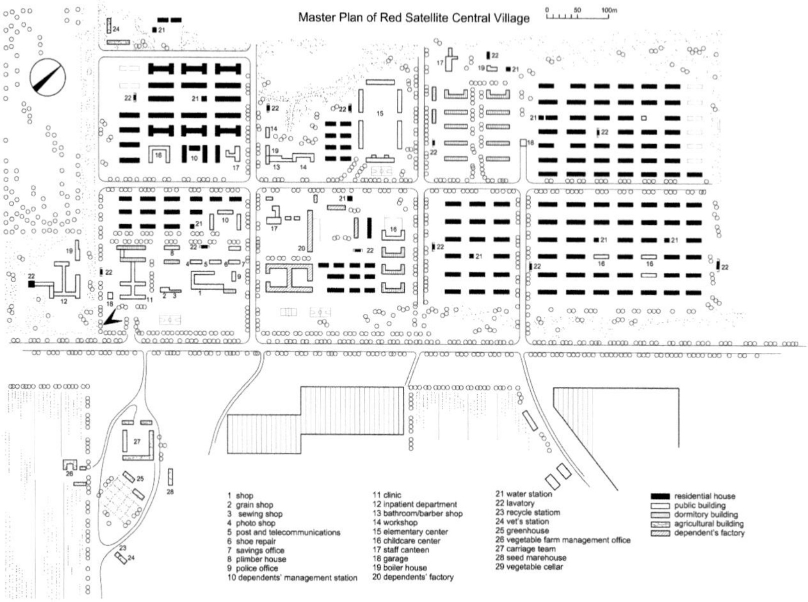

FIGURE 11.5 Master plan for Red Satellite Central Village. Daqing Design Institute, 1964.

offices of the three Basic Construction Teams belonging to the Oil Field Construction Headquarters were located either at the periphery of the village or facing the village across the road. Public services could be found at the center of the village, and the schools occupied a relatively quiet corner, away from traffic. Structures serving agricultural production were located midway between the village and the farmland.

Settlement layouts consisted of a simple and universally applicable matrix. Identical mud houses were lined up at equal distances. Black curved oil-painted roofs, brown mud walls, blue wood eaves, doors, window frames, and green, naturally grown poplars surrounding the village constituted the uniform landscape for all the settlements. Inside the settlements, there were no paved roads. Instead, roads consisted of pounded earth. A grid of electricity poles, cables, and gas pipelines provided every household with electricity and natural gas. A public water station provided cold and boiled water to the residents. This was also where the residents did their laundry and engaged in conversation. Sewage was collected through an open trench that led to local rivers and ponds, similar to the situation in other rural areas. The planned villages were similar to traditional rural ones in the Northeast, except they featured an orderly design and better infrastructure. Despite its designated status of being a "worker-peasant village," in 1978, Red Satellite provided accommodations for 7,628 residents, a scale equivalent to that of a small town at the time.[12]

The "Dependents": Pioneer Village

In Chinese, "dependents" refers to family members, including housewives, without permanent jobs and those households classified as agricultural. The oil extraction industry was dominated by young men. In 1964, the female staff constituted only 10 percent of the total employees at Daqing Oil Field, representing a reality in many heavy industries.[13] In Daqing, farming not only solved the problem of the food crisis, but it also helped to organize rural women into agricultural collectives, transforming them from "consumers" at home to "producers" in the field. During the later years of the Daqing Exploration Campaign, an increasing number of dependents were persuaded to join the collective agricultural teams. Among them, the wives of oil drilling workers from Pioneer Village often were portrayed as role models for dependents coming from rural backgrounds.

Pioneer Village, a model worker-peasant village built mainly by family dependents, was an early settlement built in Daqing. It was named after its residents, who were the pioneers of collective farming during the explorative years. The site was 35 km from the Saertu headquarters, and was built as a base for drilling workers and their dependents. In many ways, this was a female-dominated community, because 80 percent of the husbands in these families worked on the oil-drilling frontier and would only occasionally return to the village for a short break. While the men were away, their wives took over building and running the homes. Most of the gandalei buildings in Daqing were built with the dependents providing homemade mud bricks and then professional male teams helping to erect the houses. Some women in the village were reported to have become skillful carpenters, tilers, welders, or glaziers—jobs traditionally performed by men.[14] The women also received military, political, and technical training. With the rapidly growing need for laborers, many of the younger dependents were recruited as temporary workers, building infrastructure and houses, serving in the canteens, and working in the shops. Some daring young women who were willing to challenge the limits of human endurance and to compete with the men worked on the oil front as "iron girls," mostly engaging in oil extraction. And the elderly women set up cooperative nurseries to offer basic childcare for the working mothers (Figure 11.6).

Pioneer Villages, like Red Satellite, consisted of a cluster of settlements, with one central village and four smaller settlements: Pioneer Village No. 1 to No. 4. The average distance

FIGURE 11.6 Dependents walking on the unpaved main road in Pioneer Village, with houses on either side, 1977. Daqing Archives.

between the central village and the other settlements was about 1.5 km, with farmland in between. The central village was located at the geometric center and linked to the four settlements by paved radius roads. Until 1978, the Pioneer Villages as a whole had 2,255 inhabitants in 667 households, about one-fourth the population of Red Satellite near the central headquarters.[15] Pioneer Village No. 1 and No. 2 each had about 100–200 households, whereas the other two settlements were slightly smaller. The central village was a miniature version of Saertu: the courtyard of the headquarters was located to the north of the main east-west road, facing a grid of mud houses to the south. The open space bordering the public buildings was reserved for large public meetings and for use as an open-air theater.

Far away from the base town, the Pioneer Village complex attempted to achieve a high level of self-sufficiency. It provided cradle-to-grave services to its inhabitants so as to limit the need to commute. Every settlement had its own childcare facilities, a barefoot medical station, boiling water stations, and a public toilet. The central village hosted most of the public services, including an administrative office, a health clinic, elementary, middle, and high schools, a general store, a grain shop, a financial services team (which also served as a post office, a place to make deposits, and a reading facility), a general services team (barber, bathroom, tailor, and repair shop), and a food-processing workshop (producing baked goods, soy sauce, vinegar, and wine). In terms of agricultural production, each village was a full-scale cooperative commune. It provided grain to all dependents, and livestock, fishery ponds, orchards, and vegetable fields offered extra nutrition for both the families and the drilling workers on the front. The village also offered agricultural production services, such as stations for agricultural research, collective agricultural machines, a weather observation station, and veterinary services.

The Daqing dependents performed their productive lives in the new socialist culture, both in reality and on stage. "The Rising Sun" is one of the most well-publicized dramas performed by the Daqing dependents; it was directed by Sun Weishi, chief director of the Central Experimental Theater, to audiences of tens of thousands in Beijing.[16] In the local gazette *Daqing Women*, most of the dependents who lived in No. 2 Pioneer Village claimed they had not left the village for years, asserting they had been able to meet all their needs locally.[17] Pioneer Village and Red Satellite Village were the most visited and propagandized gandalei villages in Daqing. Visitors became part of the villagers' daily lives in the 1960s and 1970s. Images of these mud houses arranged in such an orderly fashion in the midst of oil wells and farming land had become the symbol of an exceptional socialist way of life after China deviated from the Soviet model of socialist construction.

A Basic Production Unit: Strengthening Village

Strengthening Village is the living quarters for the No. 2 Oil Extraction Factory. It is typical of the settlements built parallel to the oil production sites. The No. 2 Oil Extraction Factory was established in November 1964, one of the earliest. The numbering of the factory represented a development sequence; this was the second production site developed in Daqing. The oil extraction terrain was a bit south of the Saertu railway station. As soon as the location of the factory headquarters was finalized in April 1964, about 150 workers were drawn from the construction team of the No. 1 Factory to build one hundred units of gandalei houses, totaling 7,965 m^2, within a period of six months. By November, the cadres and workers had begun to move in. There were attempts to build multistory offices and headquarters. For example, during the winter of 1964, the factory hired the Heilongjiang No. 3 Construction Company to build offices, dormitories, a canteen, and a bathhouse made of brick and wood in the middle of Strengthening Village, estimating a total of 4,966 m^2. In the middle of the construction, Daqing was promoted as a national model. Then the Minister of Petroleum Industry ordered the factory to stop constructing brick office buildings and to cover the completed brick buildings with a mud facade.[18] All ancillary buildings under construction were provided with curved mud roofs, conforming to the existing gandalei houses in the village. The professional construction team was replaced by local peasants. The administrative office (or "the headquarters," as Daqing widely used the military term) remained a humble courtyard house.

The No. 2 Factory was further divided into several mining units. The size of mining units varied from place to place, depending on the labor force required for oil production. The central villages contained three hundred to five hundred resident households. New Village, built in 1965 where the oil storage tank was located, consisted of twenty-two gandalei buildings. The seats of the No. 1 and 9 Mining Districts consisted of ten mud houses. The most remote settlements might consist of only a single family, which would watch over the so-called "husband-and-wife wells." In the mining field, housing and utilities were free, and public services, such as childcare and the canteens, charged minimum fees. The gandalei houses required yearly maintenance and the residents assumed full responsibility for such maintenance—they could apply for a one-week leave per year for housing maintenance. The demands and features of oil production defined the spatial pattern of the village and the daily life of the villagers, and would continue to do so in the following years.

Redefining the Purpose of Socialist Production

The oil field saw its first great baby boom in the early 1970s and with it an increased demand for housing. Large-scale housing construction was prohibited. Housing increased with self-built additions and piecemeal expansions of existing settlements. With the improved economic situation and fewer political mandates, the villages grew larger, with new houses primarily made of brick instead of mud and straw. Many of the remote single-household settlements were abandoned when shuttles became available. Schools were filled with children, and hospitals, shops, and buses were ever more crowded.[19]

Riding the wave of the dramatic rise in world petroleum prices, oil became China's most important source of foreign revenue. From 1975 to 1978, 10 million tons of crude oil were shipped abroad, which accounted for about one-tenth of national extraction.[20] Nevertheless, despite the profit that Daqing produced for the country, in Daqing, the ban on building multistory housing remained. It had become the symbol of the Daqing spirit. The self-reliant, decentralized, "new socialist industrial-mining districts," which integrated industry with agriculture, became a model for all types of Chinese work units. Even in Beijing, one could find gandalei houses and urban agriculture in the very center of the city. Large-scale planned state housing construction was replaced by cheap self-built housing with the assistance of the work units.

In early 1978, the Office of Political Research, led by economist Yu Guangyuan, conducted a survey of living conditions among Chinese workers.[21] The survey report, which argued that poor living conditions had seriously dampened labor productivity, was circulated to the central leaders. According to data from the Ministry of Building Construction, the average living space per person (of state employees) had dropped from 4.5 m^2 in 1952 to 3.6 m^2 in 1978. In 182 Chinese cities, 35.8 percent of urban households were in urgent need of housing.[22] After twenty years of the "production first" strategy, cities were suffering greatly from a lack of investment. In the field of philosophy, rising humanist concerns were voiced, criticizing the practice of treating people as a means to achieve state goals rather than as an end in themselves.[23] *People's Daily* published articles by Yu Guangyuan and many others advocating a more balanced relationship between production and livelihood. Three major oil papers—*Daqing Daily*, *North China Oil Paper*, and the *Petroleum Industry Newsletter*—reacted strongly to such criticisms. Two articles were published on November 17 and November 19, 1979, respectively, written by the Investigation and Research Office of the Ministry of the Petroleum Industry: "Less Empty Talk, More Solid Work" and "Improve the People's Livelihood on the Basis of Developing Production."[24]

By the end of 1979, despite the great pressure for continuous growth, crude oil production in China had stagnated. No major new discovery of oil fields occurred after that point. Total output dropped for the next two years. Contracts for crude oil exports could not be met. On August 26, 1980, Minister of the Petroleum Industry Song Zhenming wrote a self-criticism that appeared in all the major newspapers taking responsibility for covering up the sinking of an oil rig, Bohai No. 2, during a storm in November 1979, and he was relieved of his position. Kang Shi'en, the former commander-in-chief of the Daqing Oil Campaign, was also relieved of his position as vice premier and was reprimanded. Vice Premier Yu Qiuli was shuffled out of the State Planning Commission. This marked a halt in the rising power of the "Petroleum Group" in the Chinese central government.

The Disappearance of the Daqing-Modeled Petroleumscape?

In the early 1980s, large-scale housing construction resumed throughout Chinese cities. Mid-rise planned residential districts with brick structures have gradually replaced former villages; urban farming and chicken coops in the cities have gradually disappeared, along with the end of Daqing-modeled construction throughout China. After four decades of extensive urban construction, the older petroleumscape represented by gandalei housing is scarcely visible today. Many of the dispersed settlement points have been abandoned. Daqing is seldom mentioned. The Northeast, the land with rich natural resources, the most important heavy industry base for Japan's colonial empire and then the young socialist republic, has now become China's rust belt.

The historic metamorphosis of Daqing's petroleumscape marked an unusual episode for China and epitomized the major threats and changes experienced by the Chinese state and society since the 1950s. The political aspirations of the socialist regime have shaped the meanings of petroleum production and spaces in many ways as well as many aspects of the everyday lives of ordinary Chinese. The discovery of Daqing Oil Field contributed to creating a domestic oil supply in China for the first time in the modern era. However, with increasing globalization and the marketization of oil industries in China, China has become the No. 1 crude oil importer of the world. But still, the spatial pattern of Chinese oil industries is heavily influenced by the regional planning that took place in the Daqing era when oil infrastructure was laid out: pipes, ports, refineries, oil institutions, and skilled laborers. It is especially evident in North China. The Daqing-modeled petroleumscape has not disappeared.

Notes

1 Li Hou, *Building for Oil: Daqing and the Formation of the Chinese Socialist State* (Cambridge, MA: Harvard Asia Center, 2018).
2 Carola Hein, "Oil Spaces: The Global Petroleumscape in the Rotterdam/The Hague Area," *Journal of Urban History* 44, no. 5 (2018): 887–929.
3 Sherman Cochran, *Encountering Chinese Networks: western, Japanese, and Chinese Corporations in China, 1880–1937* (Berkeley: University of California Press, 2000), 12–43.
4 William C. Kirby, "Engineers and the State in Modern China," in *Prospects for the Professions in China*, eds. William E. Alford, Kenneth Winston, and William C. Kirby (New York: Routledge, 2011), 283–313.
5 William C. Kirby, *Germany and Republican China* (Stanford: Stanford University Press, 1984), 78.
6 Cao Hongtao 曹洪涛 and Chu Chuanheng 储传亨, eds. *Dangdai Zhongguo de chengshi jianshe* 当代中国的城市建设 (*Contemporary China's Urban Construction*) (Beijing: Zhongguo shehui kexue chubanshe, 1990).
7 Wen Houwen 温厚文 et al., eds. *Kang Shi'en zhuan* 康世恩传 (*Kang Shi'en: A Biography*) (Beijing: Dangdai Zhongguo chubanshe, 1998).
8 From interviews conducted by the author in 2007 and 2008 in Daqing.
9 Xu Zhen and Li Wiechen, "Sun Jingwen," in *Zhongong Dang Shi Ren Wu Zhuan*中共党史人物传 (*Biographies of CCP Leaders*), eds. Qi Wang and Zhiling Chen, Vol. 75 (Beijing: Zhongyang wenxian chubanshe, 2000).
10 Daqing Oil Field Design Institute, "Gandalei fangwu de sheji yu shigong" 干打垒房屋的设计与施工 (The Design and Building of *Gandalei* Houses), *Jianzhu xuebao* 建筑学报 (*Architectural Journal*), nos. 4–5 (1966): 30–32.
11 *Xia Fei* 霞飞, *"Mao Zedong zai shangge shiji liushi niandai de yi ge zhongda zhanlue juece"* 毛泽东在上个世纪六十年代的一个重大战略决策 (Mao Zedong's Important Strategic Decisions in the 1960s). *Dang shi zong heng* (*Party History*), no. 1 (2008): 42–45.
12 *Chengzhen juzhuqu guihua shili* 城镇居住区规划实例 (*Models of Urban Residential District Planning*) (Beijing: Zhonggong jianzh gongye chubanshe, 1979), 2.
13 "Daqing funü zhi" bianzuan weiyuanhui, ed. *Daqing funü zhi* (大庆妇女志) (*Gazetteer of Daqing Women*) (Beijing: Fangzhi chubanshe, 1995).
14 *Zhanbao* (战报) (*Battlefield Newsletter*), December 10, 1964, 2.
15 *Dui Daqing jingyan de zhengzhi jingjixue kaocha* 对大庆经验的政治经济学考察 (*A Political Economy Examination of the Experience of the Daqing Model*) (Beijing: Renmin chubanshe, 1979), 322.
16 Sun Weishi (孙维世), "Daqing tongxin" (大庆通信) (Correspondence from Daqing), *Hongqi* (红旗) (*Red Flag*), no. 12 (1965): 41–49.
17 Daqing funü zhi" bianzuan weiyuanhui 大庆妇女志编纂委员会, ed. *Daqing funü zhi* 大庆妇女志 (*Gazetteer of Daqing Women*) (Beijing: Fangzhi chubanshe, 1995).
18 This occurred at the same time that Ranghulu's plan for a three-story headquarters was abandoned. Based on interviews conducted by the author in 2007 in Daqing.
19 Qin Zhijie 秦志杰. "Cong fensan dao jizhong" 从分散到集中 (From Dispersion to Aggregation), in *Wushinian huimou: Xin Zhongguo de chengshi guihua* 五十年回眸: 新中国的城市规划 (*Fifty Years Review: City Planning in New China*), ed. China Academy of Urban Planning (Beijing: Shangwu yinshuguan, 1999).
20 Data estimated from several sources. Li Xin 李昕, "Import and Export of China Oil Since 1949 (1949年以来中国石油进出口地位演变)," *Journal of Southwest Petroleum University* 16, no. 1 (2014): 1–6.
21 Yu Guangyuan于光远, Wo yi Deng Xiaopeng 我忆回忆录 (*My Memories of Deng Xiaoping*) (Hong Kong: Shidai guoji chuban you xian gong si, 2005).
22 Cao Hongtao (曹洪涛) and Chu Chuanheng (储传亨), eds. *Dangdai Zhongguo de chengshi jianshe* (当代中国的城市建设) (*Contemporary China's Urban Construction*) (Beijing: Dangdai Zhongguo chubanshe, 2009).
23 Wang Ruoshui (王若水), *Hu Yaobang xia tai de beijing: Rendao zhuyi zai Zhongguo de mingyun* (胡耀邦下台的背后: 人道主義在中國的命運) (*The Background to Hu Yaobang's Stepping Down: The Fate of Humanism in China*) (Hong Kong: Mingjing chubanshe, 1997); Li Honglin (李洪林), *Si zhong zhuyi zai Zhongguo* (四种主义在中国) (*The Four Isms in China*) (Beijing: Shenghuo, Dushu, Xinzhi Sanlian shudian, 1988).

24 See various articles in *Daqing Zhan Bao*, November and December, 1978, such as "Shaoshuo konghua, duogan shishi" (Less Empty Talk, More Solid Work), November 17, 1978; "Zhi neng zai fazhan shengchan de jichu shang ganshan shenghuo" (We Can Only Improve Our Livelihood by Developing Production), November 19, 1978; "Gao sihua bixue xian shengchan hou shenghuo" (Production First and Livelihood Second is Necessary for the Four Modernizations), November 21, 1978; and Zheng Xuansong, "Xian gai lou zai na you, haishi xian na you zai gai lou?" (Multistory Housing First or Oil Production First?), December 6, 1978.

PART III

Oil Ecologies and Imaginaries

12

ENERGY HUMANITIES AND THE PETROLEUMSCAPE

Imre Szeman and Caleb Wellum

Recent debates around fossil fuels in Canada have brought home the importance of energy resources to the life of the country. They have also shown the importance of space to disputes over how, why, and whether fossil fuels should be used. The contentious function of oil in linking space and governance has been made clear in the extended struggle over the construction of pipelines from northeastern Alberta (close to the extraction site of the Athabasca tar sands) to the coasts, especially to the West Coast. For the Alberta provincial government and the Canadian federal government, pipelines have the potential to expand markets for Canadian oil, most of which goes to the US[1]; for the provincial government of British Columbia, through whose territory they must pass, and for many First Nations in the region, the structures pose more threat than opportunity. Some of the disputes over energy in Canada are based on environmental principles and the argument that oil must be left in the ground. In the case of the pipelines and in the case of recent challenges by conservative governments in Saskatchewan and Ontario to the federal carbon tax, the struggle is over politics rather than principles, and it often plays out in the construction and representation of space. Conservative governments want the business of fossil fuels to continue unabated, even as British Columbia's current left-of-center government seems intent on mitigating the outcomes of fossil fuel use, including the fires that have plagued its forests over the past several years.

Despite the fact that Canada's history and the space of its political sovereignty have been shaped in relation to resource extraction, it is safe to say that such practices have only recently animated vigorous public debate about their short- and long-term viability as well as their environmental implications. Yet these debates have only scratched the surface of the social and cultural commitments Canadians have made to their resource culture. This isn't to say that there aren't politically committed environmentalists in Canada, such as author Naomi Klein, whose contributions include her involvement in the 2015 Leap Manifesto, a document that demands restructuring Canada's economy to end fossil fuel use. Instead, it is to say, even as environmentalists push Canada to change its sources of energy, there has been less of a demand for an accompanying social transition. All too often, ending the use of

DOI: 10.4324/9780367816049-16

fossil fuels is imagined as leaving almost everything the same, *except* for the source of energy that powers the country. It is assumed that current modes of work, economic organization, and daily life would continue unchanged. This limit of imagination doesn't plague Canada alone. We have all barely begun to discuss the profound changes to the existing petroleumscape that must accompany any transformation in global fossil fuel use.

For the interdisciplinary field of study known as the "energy humanities," there is a deep link between the energy on which society depends and the character of its cultural, social, political, and spatial forms. To put it directly, far from being just a neutral input into our society, energy plays an *essential* role in shaping who we are and how and where we live. The history and shape of modernity is deeply connected to the character of the energy that has been dominant over the past 150 years—fossil fuels—and the petroleumscape it has created. Since the mid-nineteenth century, changes in the forms of energy that we employ have altered what we are capable of, how we imagine ourselves, and the spaces we inhabit. The structure of transportation systems, the overall shape and design of cities, and the expansion of global trade are linked to the ease of access and affordability of fossil fuels, which has created durable structures such as refineries and highways that will have to be repurposed in a post-oil future. Much of the psychological, affective, and political characteristics of modernity—individuality, autonomy, self-development, and freedom, for example—are also connected to the ever-expanding use of fossil fuel energy and the spaces it has produced (Figure 12.1). This is especially the case since World War II, a period that has been described as the "Great Acceleration" by some scholars and as the "Great Derangement" by others.[2]

We are not claiming that there is a one-way cause-and-effect relationship between the expanded per capita energy use enabled by fossil fuels and the socio-spatial developments that characterize modernity. The relationship is dialectical: the advent of cheap and accessible

FIGURE 12.1 "Oil Slick Surrounds the Statue of Liberty in New York Harbor." *Source*: EPA-Documerica—Chester Higgins. Record Group 412, National Archives II, College Park, MD.

sources of energy helped to generate new forms of technology, infrastructure, and social behavior, which in turn lead to an ever-greater use of fossil fuels in a process Jeff Diamanti and Mark Simpson have termed "energy deepening."[3] As Jean-Claude Debeir, Jean-Paul Deléage, and Daniel Hémery argue in their iconic history of energy:

> while there is no energy determinism there is a powerful energy determination at work in all societies ... the energy determination is itself determined: it is the result of the interplay of economic, demographic, psychological, intellectual, social and political parameters operating in the various human societies.[4]

The process of a century and a half of energy deepening with fossil fuels, a continuous doubling down of social bets on this substance, makes it very difficult to imagine a quick, easy turning-off of the taps. If it were somehow possible to switch immediately from fossil fuels to (say) solar energy, much of global society would nevertheless remain a fossil fuel society, thanks in large part to the inertia of the global petroleumscape. Car-dependent suburbs cannot be quickly retrofitted for walkable density, nor can oil refineries easily be converted to other uses. Debeir, Deléage, and Hémery remind us that a true transition would require "a radical change in the key economic choices which shape civilization over long periods. What is required is a decisive broadening of political and social democracy, a profound change in individual behavior and educational systems."[5] In other words, energy transition requires something akin to a political revolution, a new articulation of the commons to undergird and support the introduction of solar panels, electric transport, and sustainable configurations of space.

The intervention of the energy humanities is (i) analytic and conceptual, and (ii) speculative and political. With respect to the first, the fundamental goal of the energy humanities is to argue for the need to include energy in accounts that the human sciences offer of *any* subject whatsoever, with an emphasis on the history of modernity—the period during which rapid increases in energy use transformed human life experience. Accounts that foreground energy have already begun to re-narrate our understanding of key moments in the development of contemporary culture, society, and politics, as well as the spaces that shape them. The work of political theorist Timothy Mitchell stands out as one of the most powerful examples of this approach. In *Carbon Democracy: Political Power in the Age of Oil*, Mitchell connects (among other things) the shift from coal to oil as having a major impact on unions. As forms of energy moved by pipeline became dominant over coal, the capacity of unions to shut down energy flows from coal mines declined. The bigger takeaway from Mitchell's influential book is the link between the growing reliance on oil for energy and the emergence of what might be described as "democratic governmentality," that is, a form of governance that divides the commons to limit "claims for greater equality and justice."[6] This governmentality consigns particular concerns to distinct spheres, such as to nature or the private sphere, that are subject to specific rules that limit certain kinds of political claims. One of the most significant of these concerns is the topic of this volume: the claim over and right to construct the petroleumscape, and to do so largely in relation to the imperatives of government and business. Understanding energy in relation to the construction of space is essential to grappling with its political and socioeconomic force. While space is not always foregrounded in the wide range of inquiries collected under the umbrella term of "energy humanities," our description here of key moments in the constitution of this field should be seen as in the dialog with the spaces of petroculture.

Like Mitchell, the work of Swedish historian Andreas Malm provides a reimagining of another key moment in the history of the present. The Industrial Revolution is often seen as a period in which liberal political ideology and technological advances improved the lot of humanity, a development signaled in the switch from water mills to more efficient and powerful factories and engines run on coal to ones fueled by petroleum, a foundational development in the history of the global petroleumscape. In *Fossil Capital: The Rise of Steam Power and the Roots of Global Warming*, Malm argues that the transition from water to steam was actually a form of class warfare.[7] Coal was *not* cheaper than water, but it was more mobile, allowing factories to concentrate in cities in order to exploit unspecialized workers for the benefit of the factory owners. Other analyses have produced similar insights into the character of the present, including Amitav Ghosh's account of the inability of the modern novel to address climate change and Dipesh Chakrabarty's reframing of the terms on which historical research is carried out in light of global warming.[8]

The speculative and political dimensions of the energy humanities are an equally important element of the work it carries out. The absence of energy from our accounts of modernity has impoverished our understanding of it and limited our sense of the forces and imaginaries shaping the present, including the role of the petroleumscape in shaping perception. It has also made it difficult to respond effectively to one of the consequences of intense fossil fuel use: the production of ever-increasing levels of carbon dioxide that have induced global warming. The analytic work of the energy humanities produces an expanded sense of the social and infrastructural commitments that have been made to fossil fuels—commitments that constitute the "energy deepening" or "energy determination," so aptly captured in the petroleumscape concept. But when it comes to energy use, it is not enough simply to generate knowledge or insight into processes or protocols. It is vital to develop political responses at the right scale and at optimal sites, in order to begin to unravel the tightly coiled temporality of energy deepening.

Building on existing discussions in the energy humanities,[9] we want to point to some ways in which including energy in our investigations generates insights into where we have been and where we might be going. Just how does the approach of the energy humanities generate new insights into the *historical* practices, processes, and spaces of the modern? And how, too, does it give us tools for revamping the work of what has come to be known as *theory*?

Energy, History, Modernity

The history of energy is often practiced as a form of economic or environmental history or as a historical subfield. Energy historians tend to examine the exploitation of different energy forms and technologies, highlighting their role in production and consumption more broadly as well as their social and environmental consequences. J. R. McNeill's environmental history of the twentieth century, *Something New Under the Sun*, Vaclav Smil's substantial scholarly corpus on the history of energy technologies and use, and E. A. Wrigley's work on the energy dynamics of the Industrial Revolution are key works in this vein.[10] Each generated debate about the economic, political, and environmental consequences inherent in the history of energy and demonstrated that energy is an important factor in historical change. The historical claim of the energy humanities, however, is deeper and more radical: energy is a fundamental category of historical analysis to be added to the triumvirate of class, race, and gender. Like those categories, paying closer attention to energy in our histories will transform how we understand our societies and the spaces they inhabit. If the claim that there is a deep link between energy and the social and political is to stand and to be analytically useful, then the work of historicizing cultural imaginaries, social practices, and political formations in relation to energy must be at the core of historical praxis.

This approach is particularly apt for the history of modernity, which hinges on radically altered human capacities and imaginaries. Its purview includes revolutionary scientific and technological shifts (mobility, health, and communication); the emergence of nation-states and new political ideologies (democracy, communism, fascism); unprecedented population growth; mass production and consumption; mass migration; secularization; and the spread of new norms and values. The fact that these changes are unevenly distributed in space and time does little to detract from their historical and environmental importance. It only complicates potential responses to the challenge of climate change and imparts greater urgency to the creation of new narratives of modernity. We must now ask: To what extent is the history and status of modernity tied up with fossil fuels as well as the tendencies and imperatives that their use entails? How might new histories deepen our understanding of the commitments to fossil fuels that we have made? We will focus on the era of fossil fuels from the late eighteenth century to the present to suggest some ways that an analytical concern with energy opens up new understandings of the contingency and materiality of modernity and some of its fundamental categories, concepts, and spaces, that is, the complex web of distinct elements that constitute the petroleumscape.

Attention to energy, in the first place, can open up the histories of historical consciousness and crisis. This point begins with the basic insight that different forms of energy generate unique time and space relations. Chopping and burning wood for heat tends toward different temporalities than flicking on a natural gas furnace or stove. Drawing on bodily energies to walk and burning gasoline to drive similarly entail different experiences and expectations of time and space. Such distinctions may seem banal, but they suggest energy's role in deeper shifts in social and cultural awareness. The emergence of modernity revolved in large part around such reconfigured human relations to time and space *and* overlapped significantly with the eighteenth- and nineteenth-century energy transition from flows of solar energy (primarily expended via bodies) to the stock energies that powered machines and fueled—and continue to fuel—aspirations for a workless utopia.

According to the German historiographer Reinhart Koselleck, by the end of the eighteenth century modernity contained a "peculiar form of acceleration" generative of expanded historical consciousness—a sense that the future (the "horizon of expectations") could *and should* differ from the past (the "space of experience").[11] In this accelerating divergence of future from past lay the origin of "crisis," a transcendental placeholder that imparted meaning to history and demanded a new future.[12] By the twentieth century, the ideology of "progress" was firmly entrenched, while thinkers like Walter Benjamin critiqued the culture of modernity in terms of speed and shock.[13] Crisis (in this volume, see also Introduction, Strupp and Fuccaro)—a concept at the heart of the experience of modernity and of modernist critique—proliferated alongside and within the pervasive sense of acceleration most often symbolized by industrial machinery, trains, and automobiles that were, in turn, a condition of possibility to the world wars that scored the first half of the century. These fossil-fueled technologies then continued to generate further ruptures in the experience of time, remaking space along the way to accommodate their expansion.[14]

The energy humanities claims that this modernity of acceleration, abstraction, and crisis must be understood as a "petromodernity"—that is, as fueled by and coextensive with the mass consumption of fossil fuels that the global petroleumscape both embodies and abets. Its roots are therefore significantly material and historically contingent. This claim opens up all kinds of questions about the relationship between energy, experience, and capitalism. It encourages us to reexamine alternative orientations to time and struggles over time, such as those studied by E. P. Thompson and Anson Rabinbach in the context of the labor history of Western Europe in the seventeenth to the nineteenth centuries, to understand how the exploitation of fossil energies has remade time and space.[15] It also demands that we historicize more recent structures of time and space. For instance, Malm has argued that the prerogatives of steam power within capitalism contributed mightily to the abstract reconfiguration of time and space to maximize productivity, separating human labor, bodies, and minds from natural spaces and temporalities.[16] Cultural theorists Jonathan Crary and Dominic Pettman suggest that this process of abstraction and fragmentation has only intensified as capital turns from mining the resources of the earth to human attention and creativity as new sources of energy and profit.[17] All these developments require enormous inputs of fossil fuel energy and carry social, economic, and political consequences that will reverberate well into the future. The key task—and analytical opportunity—of foregrounding energy in our accounts of modernity is to see how time and space are shaped by it, and to trace the multiple and intersecting consequences of that shaping.

Attention to energy also has the potential to illuminate the primary political forms of modernity: nationalism and the nation and their spatial dimensions. The history of nationalism parallels the mass availability of energy, flourishing in the nineteenth and early twentieth centuries alongside the fossil-fueled Industrial Revolution. Although its histories are complex and multi-causal, nationalism presupposes awareness beyond the local and a sense of a shared, national public sphere. Such cultural and psychological shifts, in turn, require mass mobility and communication technology to sustain the sense of a shared space of nationhood. Nationalism then presupposes large energy inputs to power communication and mobility, which in turn rely on infrastructures that knit spaces together and shrink the expanse of time, helping disparate communities and regions understand themselves as parts of larger wholes. Indeed, energy infrastructures, from hydroelectric dams and oil refineries to railways and highways, are often loaded with nationalist discourses of modernity, progress, and common purpose (in this volume, see Ramos, Uyttenhove, Peyerl, Couling, Scotto, Hindelang, and Hou). Fossil-fueled transportation technologies and manufacturing techniques also enabled mass mobilization at the scale of the nation to engage in nationalist conflicts, as both of the twentieth century's world wars attest.[18] These ideas are speculative, but point to the possibility of formulating a theory of the nation that would reframe its history in terms of access to energy.

It is also useful to frame post-national one-worldist and "think global" politics, prevalent in globalization, multicultural, and climate change discourse, as deeply embedded in the energy abundance of the twenty-first century. These ideologies, which seek to supersede the nation in the interests of equality, peace, and a sustainable global future, emerged in response to the global conflagrations of the twentieth century, but they have also been enabled by energy-intensive infrastructures of global trade, mobility, and communications. What will be their future in a post-oil world? How might seeing the capacities and politics of energy resources at the heart of both nationalist and post-national imaginaries change our understanding of them?

The historicizing claim of the energy humanities extends from modernity's configurations of time and space to concepts often taken to be fixed and scientific. Of these, perhaps the most important and under-analyzed is the concept of energy, which is typically defined as power or the ability to do work. The concept of energy is of course much more historically situated and complex than that, imbued with different social and political meanings embedded in time and place. Nearly thirty years ago, Rabinbach linked nineteenth-century notions of energy within physics and thermodynamics to the management and politics of labor through his analysis of the human motor metaphor.[19] Since that time, little work has been done to historicize energy in the twentieth century, despite the rapid acceleration of energy production, consumption, and crisis that marked the era of mass automobility, nuclear power, and digital technologies. Since these changes have pervasively intensified our control over and reliance on non-bodily energies, we must reflect on the ways in which they may have reshaped the notions of energy that we presently rely on to understand the past and to imagine a world after oil.

One place to start is with the science of ecology, which developed its core concepts of the ecosystem and energy flow within the systems science research of the mid-twentieth-century nuclear-industrial complex in the US as well as in response to the environmental catastrophes visited upon the world by automobiles and nuclear testing. The work of US ecologists Eugene and Howard T. Odum and their colleagues is particularly important for understanding how the nineteenth-century conception of energy as a universal force flowing through the laboring body was abstracted to fit the era of systems science and control.[20] Working on sites of nuclear testing and power generation with support from the Atomic Energy Commission, systems ecologists developed a cybernetic idea of energy as *flow through systems*. This new notion of energy was as effectively mobilized in efforts to colonize outer space and to develop vast digital networks of consumption and surveillance as it was mobilized by the environmental movement. Ecology's focus on energy as an animating feature of systems paralleled the increasing influence of systems thinking in understanding the ecology of the earth and in the development of digital technologies. It has helped us to make climate change legible, but has also underwritten the acceptance of illiberal concepts in popular environmentalism, such as "spaceship earth."[21] It is worth thinking more deeply about these historical links to better understand the questions and opportunities that our current notions of energy open up and, perhaps more importantly, render untenable. Is our notion of energy separable from twentieth-century abundance or is it hindering how we think about pathways beyond our present impasse?

Although systems ecologists tended to favor holism and equilibrium over growth, the hegemonic political ideologies of the twentieth century pursued economic growth as the holy grail of peace and progress after a half century of global conflict. The analytical emphases of the energy humanities can help us to re-narrate this commitment too, opening up difficult questions about the viability of dominant approaches to economics, such as Keynesianism and neoliberalism. Since the late 1930s, the center-left has favored a Keynesian approach to economic policy. Its dominant narrative is that Keynesian stimulus spending paired with a welfare state generated the postwar "Golden Age of Capitalism" characterized by steady economic growth and relative social equality. This narrative concludes that capitalism can be regulated to reign in its worst excesses and generate sustained growth to benefit a significant majority, *as only capitalism can*. As Mitchell argues, however, the postwar economic boom depended on precipitous increases in oil consumption in the US and the industrialized West. One of Mitchell's signal contributions is to show that "the economy" as a distinct sphere of discourse and intervention is of relatively recent vintage, having emerged out of the interwar crisis to underwrite the continued viability of nation-states and the international system that they comprised. Crucially, the emergence and growth of "the economy" also depended on the increasing availability of oil.[22] Flows of cheap oil allowed economies to grow (volume of exchange) without physical expansion (of land) and made the Bretton Woods system possible. This dependence eventually came into view in the 1970s, when the specter of energy scarcity wreaked economic havoc on now heavily oil-dependent nations.

Mitchell's insight about the relation between oil, the economy, and the Keynesian paradigm casts a shadow over the economic achievements of Keynesianism and of the twentieth century more broadly. If Keynesianism owed much of its success to the energy regime in which it was formulated, where does the center-left turn after oil? Would there have been a Golden Age of Capitalism without oil? Liberal capitalism's late-century triumph suddenly seems fragile. Moreover, given the historic reliance of communist states on industrial production and fossil-fueled imaginaries of abundance—histories that are still largely unwritten—does the Left have any clear sense of how to achieve just living standards without oil?

The Marxist literary theorist Raymond Williams, writing in 1958, predicted that the central challenge of the twentieth century's second half would be the equitable distribution of wealth and resources to create a "good common culture." "The means to a good, abundant economy," he said, "we already understand."[23] Williams was writing in the early days of the Great Acceleration, when growing energy use was largely taken for granted. He recognized the role of "steam power, the petrol engine, [and] electricity" in relieving the backbreaking labor and poverty of rural life, and embraced them for that reason.[24] When we bring energy into our analyses, however, we must confront whether or not we really do understand the "means to an abundant economy" that is fair *and* sustainable. This kind

FIGURE 12.2 The environmental costs of the good life in the 1970s US. The original caption reads "Clark Avenue and Clark Avenue Bridge, looking east from West 13th Street, obscured by industrial smoke, in Cleveland, Ohio in July 1973." *Source*: EPA-Documerica—Frank Aleksandrowicz. Record Group 412, National Archives II, College Park, MD.

of analysis and the insights it generates also pose tough questions, as Ghosh and others have pointed out, about other twentieth-century concepts, such as the standard of living and individual freedom, as we understand and live them today (Figure 12.2).[25]

Although Keynesianism remains influential and is periodically revived in contemporary political discourse, neoliberalism has surpassed it as the dominant socioeconomic paradigm. Neoliberalism's history, suppositions, and future transmogrify in light of energy. For instance, the ascendency of neoliberalism, which is typically dated by the elections of Margaret Thatcher and Ronald Reagan in 1979–1980, owes much to the 1970s energy crisis. Free market intellectuals and economists used that decade's energy woes to argue that overregulation, rather than excessive consumption, caused the problems of energy scarcity and expense, and that deregulating energy production and pricing would drive innovation to generate future energy abundance. Almost paradoxically, while neoliberals decried the "scarcity paradigm" of environmentalists who interpreted the energy crisis as a manifestation of the "limits to growth," the neoliberal economy of the 1980s sought economic growth and value generation in financialized abstraction, hence the proliferation of financial instruments and the growth of service industries. All of this relied, of course, on the offshoring of energy-intensive manufacturing, where the costs of labor could be reduced, and on an extensive network of container ships, transport trucks, and air freight to cover the world with endless flows of cheap commodities.

Attending to energy helps us to see neoliberalism as a response to the limits to growth and a denial of limits that deepened reliance on abstract energies. Its most recent iteration—the "New Economy" of "innovation" and digital technology—has only intensified these tendencies. It relies on techno-utopian visions of energy-efficient economies of creativity and information that rely not only on global networks of manufacturing, but also on the energy-intensive infrastructures that undergird the internet, "the cloud," and the batteries needed to run it all. Like Keynesianism, neoliberalism continues to prioritize—even to obsess over—economic growth.[26] It also denies scarcity and assumes that the free market will drive the technological and economic innovations that we need to address our socio-technical problems, from climate change to housing and beyond. Neoliberalism presupposes energy abundance even as it fetishizes efficiency. Consequently, its values of radical individualism, choice, and growth must be understood in light of the energy conditions that surrounded their birth.

As fossil fuels tightened their grip on the world in the twentieth century, the cultural importance of their capacities and crises deepened. What might it mean, finally, to reimagine postmodernity—as a cultural and intellectual movement—in terms of energy, as a post-energy crisis phenomenon? The 1970s energy crisis threatened the material foundations of mid-twentieth-century life and caused many to question the viability of their petroculture and its spaces. When oil prices receded in the 1980s and the US reestablished geopolitical primacy in the Middle East, however, energy transition retreated to the margins. In its place roared an avalanche of discourse about the elimination of time and history and the primacy of space, and fascination with 24/7 televisual and digital connection, while globalization dominated the pages of the *New York Times* and the *New Left Review*. Is it possible to establish parallels between post-1973 financialization and postmodernity as denials of the limits to growth and escape into abstraction? How should a contemporary critical theory that faces the threat of climate change understand the intellectual legacy of the 1980s and 1990s? To what extent is the endless refashioning of the subject through digital interfaces an effect of fossil-fueled neoliberalism and its ideologies of entrepreneurial flexibility?

If the energy humanities is to help us to "return to the flow" or to realize some other future, it must also expand its purview to other forms of energy. This is very difficult in light of our contention that energy shapes us. How can we, as analysts peek above the fog? The critique of fossil fuels and of nuclear power will remain central, particularly since we are nowhere near transitioning away from them. But the generation of other possibilities and understanding of other limits will mean tracing the pasts of many forms of energy, including natural renewables (food and muscle), renewable technologies (solar, wind, geothermal), and others. We need to understand how these different forms have shaped and might still shape nature, social relations, and imaginaries. Present energy transition debates prioritize the maintenance of current forms of modernity, but without fossil fuels. The challenge that we face is to reimagine a modernity without nearly our present levels of fossil fuel use. Doing so will mean dismantling and repurposing the global petroleumscape that to date has preserved oil's power, as Dominic Boyer suggests.[27] Perhaps the past can help in that respect as well.

Theory After Energy

The claim that forms of energy shape culture, society, and politics also has implications for (critical) theory. It demands the reinterpretation of the history of theory, particularly in the fossil-fueled twentieth century, and the reimagining of theory in light of this new understanding. Malm has recently made a similar point about climate change, namely that it is now a "shared litmus test" for contemporary theory, which must be able to make sense of the fossil economy that drives global warming.[28] The energy humanities pose a similar but broader challenge to theory. Whereas Malm focuses on the immediacy of grasping climate change and the fossil economy, the energy humanities extends the challenge to all theory formulated over the past century and to all forms of energy. Although it shares Malm's sense of urgency about climate and fossil fuels, the energy humanities proposes that the material realities of all forms of energy carry social and political implications, as do the imaginaries that are often linked to those forms of energy.

In this section, we want to talk briefly about some of the implications of energy for how we understand theory, focusing on two broad areas: (1) politics and society, and (2) subjectivity. For the most part, the issues that energy raises for theory in these areas have yet to be worked out in detail; there are only a few works that link energy to theory explicitly.[29] The role of energy in theory goes beyond politics and society and theories of subjectivity; there is an obvious need, for example, to develop the points of connection and disjuncture of the energy humanities from post-humanisms and new materialisms, including object-oriented ontologies. What follows is an initial map of the issues that might shape some aspects of theory after the introduction of energy into accounts of the cultural, social, and political—zones of experience marked deeply by energy, even if to date they have rarely been explored as such.

Politics and Society

Dipesh Chakrabarty has provocatively suggested that "The mansion of modern freedoms stands on an ever-expanding base of fossil fuel use. Most of our freedoms so far have been energy-intensive."[30] The development and extension of liberal freedoms (such as they are, and where they exist at all) since the Enlightenment is rarely narrated in relation to the expansion of the use of energy. Reading energy not only into the apparatus of freedom—in the liberal tradition, the operation of law to secure freedoms for and against—but also into our conceptual understanding of it would generate new insights into the character and materiality of political freedom. Re-narrating post-Enlightenment practices and verities in relation to energy would draw attention to the environmental implications of the political as such, and do so in a manner that pushes past the meek gestures that have constituted the majority of responses to global warming to date (e.g., carbon taxes—a continuation of the use of tax as a mechanism of liberty that dates back to Adam Smith). It might also tell us something about the threats to freedoms that lie just over the horizon, as access to abundant and cheap fossil fuels begins to wane, either as a result of the depletion of supplies or due to the environmental necessity of moving away from their use.

In addition to the lack of focus on the role of energy in liberal understandings of freedom, there has been virtually no attention to the role of energy in the operations of contemporary power, politics, and the spaces of governance, and whether these operations are figured in the language of standard political science or understood through the lens of the work of thinkers such as Michel Foucault. The careful elaboration of the rationale and mechanics of governmentality that Foucault outlines in his later lectures *never* describes the import of energy to the management of modern populations. It is not just that the operations of governmentality require energy and that access to more energy in turn animates new modes of governmentality, but that population is itself a phenomenon linked to energy. The expansion of modern populations that accompanied the increasing scale and intensity of energy use is deeply conjoined with rising levels of energy use; the capacity of states to intervene at the scale of the population to exercise governmentality also depends on access to ever-greater levels of energy. In his later lectures, Foucault also examined the development of neoliberalism as an aspect of modern governmentality.[31] As we described earlier, one dimension of neoliberalism that is commonly stressed is the downloading of state responsibilities to the market. Here, too, there are practices in the flow of modern energy that can be seen as complicating this accepted narrative of neoliberalism. The externalization of the costs of ever-expanding car culture and suburban infrastructure to governments (a now global phenomenon) have helped markets stung by stalled or slow growth post-1973 to continue to support a belief in the growth of capitalist markets and of indefinitely increasing levels of energy.

Despite the obvious import of fossil fuel energy to the shape of contemporary politics and society, very few theorists have attempted to make certain that energy is part of the larger story of power. This is as true of Marxist theory as it is of other modes of political theory (the work of Bellamy and Diamanti is a notable exception). Marx's "Fragment on Machines" from the *Grundrisse*, which has played an important role in the reimagining of contemporary political theory on the Left, describes a world in which advanced technology frees labor from production. Even were technology able to set individuals free from work so that they might engage in self-development, these machines would still require energy to function. It is necessary to understand the energetic and environmental implications of the manner in which we frame political transformation. It is not only our understanding of capitalism that is impeded when we fail to factor energy into social theory, but also our imaginings of the character of social and political emancipation.

Some theorists have begun to refigure social and political theory with energy in mind via the use of new categories such as energy commons, energy ethics, or solar communism.[32] A great deal of work remains to be done, insofar as these interventions continue to be mostly viewed as asides to the main narrative through which we theorize the politics and society of modernity.

Subjectivity

If one had to summarize the past century and a half of theories of subjectivity—that is, when such theories overlap historically with the era of fossil fuels—what emerges as central is an insistence that subjects are not self-identical or self-present. In other words, the claim advanced in work from Marx to Freud and beyond is that there is no fixed, identifiable subject per se. To be a subject is to be something other than oneself, whether as a result of the dynamics of ideology and power or the operations of language, the psyche, and gender. While the phenomenological experience of being a subject is to understand oneself as oneself, the reality of being a subject is that one can never fully map the social, linguistic, and psychological forces that produce one's subjectivity—including the very sense that one is a subject qua subject.

What would it mean for our understanding of the subject if we were to recognize energy as a key element of subjectivity—as powerful a force of decentering the subject as language or ideology? As subjects have expanded their energy use over the course of modernity, they have extended their capacities (from slowly walking to flying over oceans in a matter of hours), created new spatial relations, and have molded their sense of selfhood in relation to the abilities that energy affords. Energy might thus be viewed as something like the discourse networks explored in the work of Friedrich Kittler—akin to the mechanisms of communication (pens, typewriters, film, etc.) that are key to the production of subjectivity.[33] Yet, fossil fuel energy goes beyond communication systems in its import to subjectivity. In "The Resources of Fiction," literature scholar Graeme Macdonald writes: "Oil's emancipatory role in habitual experience is repeatedly vaunted in this incorporating system of petro-acculturation: how *could* we live without it?"[34] The deep, unspoken, and largely untheorized connection between the subject and energy—one of the core relations in the construction of the petroleumscape—can be witnessed in the difficulty of imagining how one could be a subject *at all* in the absence of energy. Allan Stoekl notes that "In speaking of the finitude of energy supplies, we are only speaking of the limits to the human, the fundamentally limited availability of ordered energy capable of doing 'work' for Man. We are speaking, in other words, of death, of the incommensurability of intimate Nature."[35] Subjectivity has become fossil fueled; this is likely one of the key reasons why it is so difficult to imagine (much less actually undertake) a transition away from oil and the spaces that oil has birthed around the world.

Theorists interested in adding energy to subjectivity have frequently gestured to Patricia Yaeger's identification of an "energy unconscious." Her identification of this unconscious is provocative, even if in her work she is speaking not of a mode of subjectivity but of a method of reading for energy (an extension of Fredric Jameson's *The Political Unconscious* [1981]).[36] Nevertheless, the intent of this notion of an "energy unconscious" is to position energy at the core of subjectivity in a way that must be accounted for, both with respect to the subject of oil *and* to whatever subject might be brought into being via energy transition. Herbert Marcuse captures the notion of energy unconscious in his description of the subjective depths required for genuine social change:

> Our world emerges not only in the pure form of time and space, but also, and *simultaneously*, as a totality of sensuous qualities—object not only of the eye (synopsis) but of *all* human senses (hearing, smelling, touching, tasting). It is this qualitative, elementary, unconscious, or rather preconscious, constitution of the world of experience, it is this primary experience itself which must change radically if social change is to be radical, qualitative change.[37]

FIGURE 12.3 "Oil waste on a barren hillside." *Source*: EPA-Documerica—Gene Daniels. Record Group 412, National Archives II, College Park, MD.

Jordan Kinder's remapping of Louis Althusser's notion of ideology and ideological state apparatuses by adding energy performs a similar function: to insist that the modern subject is an energy subject and the operations of subjectivity cannot be understood without considering energy (Figure 12.3).[38]

As with our discussion of political and social theory above, it is important to insist here too that the aim of theorizing subjectivity in relation to energy is not to create a subfield of academic analysis of petro-subjectivity. It is, rather, to insist that *all* theories of the subject have to be alert to the impact and import of energy. The Lacanian subject, for instance, should always already be imagined as a Lacanian petro-subject; this addition of energy to our understanding of the subject might in turn cause us to reimagine (in this instance) the operations of the psyche.

Conclusion: The Politics of Energy in the Twenty-First Century

The political struggles and debates over the petroleumscape in Canada are taking place in many parts of the world. Do we continue to build more pipelines and other fossil fuel infrastructures and the spaces associated with them to support a growing planetary population, or is this the moment to put on the brakes? How do we minimize fossil fuel use even while extending to much of the planet's population access to more energy (and with it, some of the undeniable benefits of petromodernity)?

Answering these questions requires that we begin to ask—and answer—the questions that the energy humanities prompt about the complex ways in which we have shaped ourselves into creatures of petromodernity inhabiting a global petroleumscape. Anything approximating energy transition requires that we commit ourselves to refiguring the codes, practices, and infrastructures of oil modernity—a challenge that goes far beyond any insistence that, from now on, we just leave the stuff in the ground. The struggle to bring about a fossil fuel transition demands that we more fully understand the shape, form, character, and history of the petroleumscape we inhabit today—a planetary space of oil that we continue to mistake for the norms and patterns of everyday life. As the various contributions to this book have shown, we can no longer afford to make this mistake.

Notes

1 See Government of Canada, "Crude Oil Facts," https://www.nrcan.gc.ca/energy/facts/crude-oil/20064, accessed January 7, 2021.

2 J. R. McNeill and Peter Engelke, *The Great Acceleration: An Environmental History of the Anthropocene Since 1945* (Cambridge, MA: Harvard University Press, 2016); Amitav Ghosh, *The Great Derangement: Climate Change and the Unthinkable* (Chicago, IL: University of Chicago Press, 2016).

3 Jeff Diamanti and Mark Simpson, "Five Theses on Sabotage in the Shadow of Fossil Capital," *Radical Philosophy* 2, no. 2 (2018): 3–13, 3.

4 Jean-Claude Debeir, Jean-Paul Deléage, and Daniel Hémery, *In the Servitude of Power: Energy and Civilization through the Ages*, trans. John Barzman (New York: Zed Books, 1991), 13.

5 Ibid, 237.

6 Timothy Mitchell, *Carbon Democracy: Political Power in the Age of Oil* (New York: Verso, 2011), 9.

7 Andreas Malm, *Fossil Capital: The Rise of Steam Power and the Roots of Global Warming* (New York: Verso, 2016).

8 Ghosh, *The Great Derangement*; Dipesh Chakrabarty, "The Climate of History: Four Theses," *Critical Inquiry* 35, no. 2 (2009): 197–222.

9 See, e.g., Imre Szeman and Dominic Boyer, "Introduction: On the Energy Humanities," in *Energy Humanities: An Anthology*, eds. Imre Szeman and Dominic Boyer (Baltimore, MD: Johns Hopkins University Press, 2017), 1–13; Jennifer Wenzel, "Introduction," in *Fueling Culture: 101 Words on Energy and Environment*, eds. Imre Szeman, Jennifer Wenzel, and Patricia Yaeger (New York: Fordham University Press, 2017), 1–16.

10 J. R. McNeill, *Something New Under the Sun: An Environmental History of the Twentieth-Century World* (New York: W.W. Norton & Company, 2000); Vaclav Smil, *Energy and Civilization: A History* (Cambridge, MA: MIT Press, 2017); E. A. Wrigley, *Energy and the English Industrial Revolution* (Cambridge: Cambridge University Press, 2010).

11 Reinhart Koselleck, *Futures Past: On the Semantics of Historical Time*, trans. Keith Tribe (New York: Columbia University Press, 2004), 11.

12 See Janet Roitman, *Anti-Crisis* (Durham, NC: Duke University Press, 2013), 9ff.

13 Walter Benjamin, "On Some Motifs in Baudelaire," in *Illuminations*, ed. Hannah Arendt, trans. Harry Zohn (London: Jonathan Cape, 1970), 157–202.

14 See Christopher Wells, *Car Country: An Environmental History* (Seattle: University of Washington Press, 2012); Carola Hein, "Oil Spaces: The Global Petroleumscape in the Rotterdam/The Hague Area," *Journal of Urban History* 44, no. 5 (September 2018): 887–929.

15 E. P. Thompson, *The Making of the English Working Class* (New York: Vintage, 1966); Anson Rabinbach, *The Human Motor: Energy, Fatigue, and the Origins of Modernity* (Berkeley: University of California Press, 1992).

16 Malm, *Fossil Capital*.

17 Jonathan Crary, *24/7: Late Capitalism and the Ends of Sleep* (New York: Verso, 2014); Dominic Pettman, *Infinite Distraction* (Cambridge: Polity, 2016).

18 See Stephen Kern, *The Culture of Time and Space, 1880–1918*, 2nd ed. (Cambridge, MA: Harvard University Press, 2003).

19 Rabinbach, *Human Motor*.

20 Joel B. Hagen, *An Entangled Bank: The Origins of Ecosystem Ecology* (New Brunswick, NJ: Rutgers University Press, 1992).

21 Peder Anker, *From Bauhaus to Ecohouse: A History of Ecological Design* (Baton Rouge: Louisiana State University Press, 2011).

22 Timothy Mitchell, "Carbon Democracy," *Economy and Society* 38, no. 3 (August 2009): 399–432 and "Fixing the Economy," *Cultural Studies* 12, no. 1 (1998): 82–101.

23 Raymond Williams, "Culture is Ordinary," in *Cultural Theory: An Anthology*, eds. Imre Szeman and Timothy Kaposy (Oxford: Wiley-Blackwell, 2010), 56–57.

24 Ibid.

25 Bob Johnson, *Carbon Nation: Fossil Fuels in the Making of American Culture* (Lawrence: University Press of Kansas, 2014).
26 Philip Mirowski, *Never Let a Serious Crisis Go to Waste: How Neoliberalism Survived the Financial Meltdown* (London: Verso, 2013).
27 Dominic Boyer, "Revolutionary Infrastructure," in *Infrastructures and Social Complexity: A Companion*, eds. Penelope Harvey, Casper Jensen, and Atsuro Morita (New York: Routledge, 2017): 174–86.
28 Malm, *Fossil Capital*, 16.
29 Brent Bellamy and Jeff Diamanti. *Materialism and the Critique of Energy* (Toronto: MCM Prime, 2018).
30 Chakrabarty, "Climate of History," 208.
31 Michel Foucault, *The Birth of Biopolitics: Lectures at the College de France, 1978–79*, ed. Michel Senellart, trans. Graham Burchell (New York: Palgrave Macmillan, 2008).
32 George Caffentzis, "The Petroleum Commons," *Counterpunch* (December 15, 2004), www.counterpunch.org/2004/12/15/the-petroleum-commons/, accessed September 12, 2018; Giovanni Frigo, "Energy Ethics, Homogenization, and Hegemony: A Reflection on the Traditional Energy Paradigm," *Energy Research and Social Science* 30 (2017): 7–17; Patricia S. Mann, "On the Precipice with Naomi Klein, Karl Marx and the Pope: Towards a Postcapitalist Energy Commons and Beyond," *Radical Philosophy Review* (August 2016), https://www.pdcnet.org/radphilrev/content/radphilrev_2016_0999_8_8_64; David Schwartzman, "Solar Communism," *Science & Society* 60, no. 3 (1996): 307–31; Jessica Smith and Mette High, "Exploring the Anthropology of Energy: Ethnography, Energy and Ethics," *Energy Research and Social Science* 30 (2017): 1–6.
33 Friedrich Kittler, *Gramophone, Film, Typewriter*, trans. Geoffrey Winthrop-Young and Michael Wutz (Stanford, CA: Stanford University Press, 1999).
34 Graeme Macdonald, "The Resources of Fiction," *Reviews in Cultural Theory* 4, no. 2 (2013): 11. Emphasis in original.
35 Allan Stoekl, *Bataille's Peak: Energy, Religion and Postsustainability* (Minneapolis: University of Minnesota Press, 2007), 200.
36 Patricia Yaeger, "Editor's Column: Literature in the Ages of Wood, Tallow, Coal, Whale, Oil, Gasoline, Atomic Power, and Other Energy Sources," *PMLA* 126, no. 2 (2011): 305–26, 3; Fredric Jameson's *The Political Unconscious* (1981).
37 Herbert Marcuse, "Nature and Revolution," in *Counter-Revolution and Revolt* (Boston, MA: Beacon Press, 1972), 63. Emphasis in original.
38 Jordan Kinder, "The Coming Transition: Fossil Capital and Our Energy Future," *Socialism and Democracy* 30, no. 2 (2016): 8–27.

13

ANTWERP'S PETROLEUMSCAPE

Imagining the Carbon Age

Pieter Uyttenhove

Since the nineteenth century, oil has been a major source of economic growth in the port city of Antwerp. It has been stored, transported, refined, and used to produce petrochemicals. Most of these activities once took place in close proximity to the city center, including along the quays of the River Scheldt. Today, immense surfaces are devoted to activities related to oil, but in new port zones far outside the city's residential areas. In fact, the sum of the surfaces of petroleum-related sites is bigger than the size of the city of Antwerp inside the Ring Road. Although you can still sense its presence at night when the horizon lights up with flares from the petrochemical port, the oil industry is much less visible than it was a century ago.

The port has dominated the urban space of Antwerp for many centuries, just as it has dominated the way citizens imagine the city. The port has long been considered a more important part of the city than the parts of the city where people live. This is partly because of the power of the oil industry served by the port. The relationship between the city and the physical presence and expansion of the oil industry is clearly a matter of the imaginary. Inhabitants have been ready to offer parts of their urban comfort and tranquility and accept being shoved and hustled by the ruthless machinery of a port's expansion and transformation in exchange for their port city gaining in aura and influence. This is true in general in the era of the oil industry's first activities inside the city as well as in the city's immediate surroundings, when it gave form to the harbor city of Antwerp. The construction of quays and canals pushed out neighborhoods, docks replaced fortifications. This is also true, on a regional scale, when in the twentieth century, the port was extended to the Dutch border, sweeping away villages and agricultural fields. Oil was not a consequence of the development of modern Antwerp, but the driving force of it: it shaped the city's deepest structures.

Throughout this period, oil has also shown its dark sides in Antwerp, as a driving force for wars, as a cause of explosions and fire, and as a source of environmental pollution. Indeed, oil seems to have made a strong impression on inhabitants when it comes to its tremendous powers of destruction at different time scales, whether in the short term by fires and explosion or in the long term by the exhaustion of natural resources, the pollution of

DOI: 10.4324/9780367816049-17

the human biotope, or the breakdown of Earth's renewability. The impact of petroleum's dark sides on various levels of a city is, in a port like Antwerp, clearly observable. Since the first years of its industrial presence in Antwerp, petroleum has been the cause of accidents, terrifying the city's inhabitants and decision makers. One result of these accidents was that industrial activities linked to oil were progressively moved away from the city. Worldwide, oil became a geopolitical weapon and this made a petroleum port like Antwerp particularly vulnerable. Environmental pollution due to industrial oil activities has only recently began to be considered unacceptable, at the same time as the petroleum industry's architectural and urban heritage is being recognized as authentic and worthy of preservation. As Antwerp takes stock of and reimagines its relationship with oil, I propose that oil be regarded as one essential part of what has been an age dominated by carbon.

Wealth, Growth, and Development

In the development of Antwerp, the size of the port and the city's wealth both grew as a result of oil. Historically, Antwerp developed first on the right bank of the river, with canals entering deep inside the city, and later as a large port for cargo with docks, cranes, warehouses, and industrial plants. Since the Austrian and Napoleonic harbor extensions of the eighteenth and early nineteenth centuries—and actually since the first urban extensions by Gilbert Van Schoonbeke in the sixteenth century—the Antwerp port has grown mainly to the north. Although oil arrived relatively late in the city's history, it had a radical impact. In the second half of the nineteenth century, long-haul trade to Africa and Asia was starting, and trade through Antwerp with the German hinterland—including the trade in oil—was booming.[1] During a long period of public works between 1875 and 1911, when Antwerp was emerging as a European hub and bursting at the seams as a trading city, the city government took on the task of straightening the quays of the Scheldt, at the city's waterfront. The old canals, or fleets (*vlieten*), running perpendicular to the river into the city were filled in. New docks were built in the northern part of the city to satisfy the needs of the growing trade.

From the last quarter of the nineteenth century on, newspapers, magazines, and books advertised how fast the production of oil was continuously increasing and how oil was one of the main factors behind the booming of Antwerp's port activities. In his 1920 book *Petroleum*, A. J. Hendrix, for instance, pointed out that during the year before World War I, the world produced more than 700 million hL of petroleum.[2] In 1910, more than 2 million hL were imported to Antwerp, with more than half originating from the US and the rest coming from countries such as Romania, Russia, and Galicia. Of this imported petroleum, more than three quarters was consumed in Belgium, one quarter was again exported, mainly to the Netherlands and Germany, with only a small part going to other countries. Throughout the first half of the twentieth century, the progression continued. In 1928, the big Kruisschans Lock, better known as the Van Cauwelaert Lock, was inaugurated. It was 270 m long and 35 m wide. The lock was situated closer to the North Sea than the already existing locks and allowed huge vessels and oil tankers to enter much earlier into the docks and make an economically important gain in time, although the sluicing itself was very time consuming.

After World War II, Antwerp became Europe's biggest chemical cluster, thanks to the Marshall Plan, the American reconstruction plan for Europe. Antwerp's harbor had suffered less destruction than other large Western European harbors during the war, despite important damage. Notably, Petroleum Zuid, the petroleum port in the south of Antwerp, was almost completely destroyed by numerous bombings. Nevertheless, Antwerp was one of the only ports that was able to resume normal operations very soon after the war. In 1949, the government authorized the construction of a large Total refinery on the right bank of the Scheldt. It was located south of the Kruisschans Lock, near the Marshall Dock built for inland shipping, and west of the new Petroleum Dock for the international maritime transport of oil. The site opened in 1951 housed a large refinery, built in association with BP. In the mid-1950s, the first chemical installations, specializing in the manufacture of plastics, were erected near the refinery, shortening the transportation and production chain, and increasing Antwerp's industrial diversity.[3] Initially, the site of the refinery was a marshy polder, but using 5 million m^3 of hydraulic backfill, it was raised by 5 m to put it above the high tide sea level of the Scheldt.[4] A port city like Antwerp has to protect extensive international investments. Pushed by global industrial profit games, port cities function on power mechanisms that go beyond national political players and policies. Survival logics of big ports overrule national powers and force companies and port authorities to look for solutions unimaginable for local national actors. For instance, in order to continue to exist as a global player, Antwerp was forced to look for technical solutions solving the problem of its geographical confinement.

Despite the huge dock expansion projects it inaugurated in the 1950s, the port of Antwerp continued to suffer from a lack of accessibility. This became more and more problematic due to the increasing draft and tonnage of ships, especially petroleum tankers. Although a maritime port, Antwerp, located near the northern border of Belgium, is situated on the Scheldt, which, from the Belgian border to the North Sea, flows over Dutch territory for about 80 km. For decades, the responsibility for the maintenance and the dredging of the Scheldt has been a point of discussion between both countries. Tankers up to 80,000 tons can reach the Antwerp refineries. When the Antwerp oil installations risked becoming inaccessible, both the Dutch and the Belgian governments authorized the construction of a 102-km-long pipeline transporting crude oil from Rotterdam to Antwerp. The pipeline was built in the second half of the 1960s and began functioning in 1971.

To keep up with the continuous increase in shipping capacities, ever-bigger docks, quays, and surfaces were built for transshipping, storing, and distributing goods. Additions included the Delwaide Dock, opened in 1979, the Europa Terminal in 1990, and later the North Sea Terminal in 1997. In the 1970s, the port of Antwerp was progressively expanding to the north, until eventually it would reach the Dutch border. Consequently, the decision was made to cross the Scheldt and start developing port areas in the polders of the left bank. In 1989, the Berendrecht Lock was opened to allow giant ships to access the docks efficiently and safely.

Mental Landscapes of Numbers and Symbols

Showing the port's effective growth implicitly prefigures its future expansion as it affirms and prepares the mindset for it. Since the last quarter of the nineteenth century, Antwerp's mental landscape included images undergirding the paradigm of unstoppable modernization. In this, oil plays a central role. This mental landscape—in a way, the representational petroleumscape—is in constant dialog with the physical reality of buildings, urban forms, and infrastructures, with symbolic gestures, allegorical sculptures, and paintings, and with names of places and institutions.

But, besides ornament and architecture, oil's evocation of magnitude, wealth, and power is also expressed through data and statistics. In an international context, harbors are often compared and ranked according to size, growth, and quantity—and oil plays an important role in those rankings. Administrators, industrialists, economists, and politicians like to promote the international importance of big ports like Antwerp by using numbers in a manner apt to overwhelm a lay audience. Numbers are used to emphasize the dynamic aspects of flows and stocks visually evoked by the presence of reservoirs, pipelines, and other technical infrastructure. The use of grandiloquent discourse and quantitative data is part of the port's communication routine. It concerns not only economic and technological aspects, but also figures concerning the size, surface area, and scale of docksides, shipping, and port infrastructure.

In 1952, for instance, the Société Industrielle Belge de Pétroles published a brochure on the construction and functioning of its brand-new Kruisschans Refinery, where "5,000,000 man-hours have been devoted to the erection of this industrial complex with an annual processing capacity of approximately 1,800,000 tons."[5] The refinery's impressive size and productivity were communicated using imageries of quantification:

> In the first eight months of the refinery's operation, about 80 ocean-going vessels unloaded more than 1,200,000 tons of crude oil, and 27 ocean-going tankers loaded more than 400,000 tonnes of product. More than 600,000 tons have already been shipped by around 1,300 barges and the same number of tank cars have taken over 35,000 tons of finished products.[6]

The text was accompanied with wide aerial photos of the installations, details of complex construction works, maps of various scales, and day and night views of the splendid high-tech installations after completion.

The Berendrecht Lock was also praised for its gigantic size. Newspapers and promotional publications repeatedly stated that the lock was the biggest in the world because it could handle more than 100 million tons of goods in a year. Its size of 500 m in length was illustrated in architectural terms by explaining that it could contain "four times the length of Antwerp's Onze-Lieve-Vrouw cathedral."[7] Citing its place in the world's port rankings and confirming that the port is home to the biggest petrochemical cluster in Europe and to the second-largest complex in the world, after only Houston, Texas,[8] have been frequent rhetorical devices intended to evoke pride in the city.

The representational petroleumscape also takes shape in other ways, including spectacular architecture. In 2007, the decision of the Antwerp Port Authority to build an enormous new administrative building in the middle of the port was not just a strange whim. It was meant to serve as a city landmark that could be used for marketing the city and improving its visibility. This branding policy for city and port was a sign of its time more than a need, and part of a contemporary trend of "star architecture" serving economic and commercial goals. The new project for the Port Authority administration was to be added to the existing fire station building, a big building from the early twentieth century that was designed to resemble the seventeenth-century Antwerp City Hall. The location of the so-called Port House was chosen because it was deep in the harbor and close to an elegantly designed viaduct, planned to be part of Antwerp's northern ring road. At that spot, it would have been visible from passing cars and trucks. Unfortunately, the plans for the viaduct were canceled.

FIGURE 13.1 The Havenhuis, housing the Port Authority Administration at the Kattendijk Dock in Antwerp in 2016. The extension designed by Zaha Hadid Architects refers both to the hull of a boat and to a cut diamond reflecting sunlight. Photograph by Torsade de Pointes.

Zaha Hadid's winning design for the building (Figure 13.1) illustrates, probably in the clearest way possible, the city government's desire for imagery reflecting the past, present, and future of the city's grandeur. Hadid's design sits on top of the fire station. Like the city hall itself a long time ago, the fire station once burned down and was then restored to host the Port Authority administration. It is located near the Hout Dock in the northern part of the old harbor of Antwerp, dating from the early days of the Industrial Revolution, where formerly, until the end of the nineteenth century, the oil ships, barrels, and tanks were concentrated. Hadid's spectacular building was built between 2009 and 2016, the year of her death. It resembles a huge vessel navigating the clouds, while its glass and metal skin is faceted like a cut diamond, a symbol of Antwerp's diamond industry, which like the port, is ranked among the world's leaders.

Oil, a Geopolitical Weapon

Besides the fact that it has exerted an important material and spatial impact on Antwerp's port and city development and has generated a mental petroleumscape of data and symbols referring to wealth, modernity, and grandeur, oil is also a means of power. Its global economic dynamics and its infiltration deep into our society's daily life and technology have generated an efficient geopolitical power in the hands of governments and industrial groups. Petroleum's power is based on its essential presence, its wide diffusion, and its diversification at all levels and in all activities of society.

This power was explained by the Belgian Louis Gérard Nauwelaerts, author of two books on oil: his 1936 reference work *Petroleum*, written in Dutch and translated into German, Hungarian, and Polish, and his 1939 *Het groene goud ("petroleum")* ("The green gold petroleum," with "green" referring to the natural color of extracted oil), a collection of four of his lectures on oil.[9] For Nauwelaerts, oil has replaced gold as a global power value, the main reason being that the concentrated ownership of gold has lost its symbolic power.

In recent times, real power is based on the use and ownership of the means of production.[10] In that context, petroleum is probably the purest form of power. Nauwelaerts declares that oil was only at the beginning of its global ascent.[11] He explains that petroleum, as a result of its various forms and levels of integration in society, is resistant to crises because its economic demand is generated by basic human needs like mobility, transportation, and heating. The improvement of social welfare in various nations, he argues, relies on the stability and continuous development of the international petroleum industry—a compelling illustration of the feedback loop as described by Carola Hein.[12] According to Nauwelaerts, oil is the "biggest conquest of the 20th century" and "one of the most important milestones of History," and has only begun to improve modern life.

From the start, petroleum had two primary qualities: it was technologically a multifaceted substance and it was immensely attractive in economic terms. Based on these qualities, it is not difficult to understand how it has been used, in many ways, as a geopolitical weapon and why it has played a primary role in international political strategies. Antwerp had a long tradition of using its geographical and political situation to its advantage.[13] Under various occupiers—Spanish, French, Dutch—the citadel of Antwerp served as a place of protection and defense, but even more, it was a place built for controlling the citizens of Antwerp. On a crossroad between the Netherlands, France, and Germany, it was a highly strategic stronghold and an important port, although completely dependent on free passage over the Scheldt to the North Sea. Napoleon's decision to build new docks and shipyards for the military development of Antwerp was inspired by his vision, expressed through his well-known sally, that the harbor on the Scheldt was "a pistol pointed at the heart of England."

Petroleum's importance in geopolitical affairs can be observed before World War II when its power was made effective by means of embargoes. At the end of the 1930s, when the threat of war was high, nations like the US, Japan, and Germany were turning their normal commercial and industrial activities into war economies. When Mussolini invaded Ethiopia in 1935 for his own domestic political reasons, oil proved a valuable international weapon: the Society of Nations decided to take sanctions against Italy by closing down the supply of oil products as was suggested by many media. The world understood this would have an immediate effect on Italy: "L'embargo sur le Pétrole, c'est la paix!"[14]

An example of oil's geopolitical power in connection with the port of Antwerp occurred when the Arab Petroleum Exporting Countries proclaimed an oil embargo in 1973, in reaction to Israel's Yom Kippur War. It set off a raw materials boom which would have long-term policy consequences.[15] The big oil-producing Soviet Union formed a storage terminal company called Nafta, together with some European investors. Nafta built a large oil storage terminal in Antwerp, with a capacity of about 1 million tons, and there was even talk, for a while, of building a refinery. From the Antwerp terminal, petroleum was shipped to buyers all over Western Europe and the US.[16] Nafta set up its own network of wholly owned and financed filling stations. The US hoped that this and other developments would keep the Soviet Union from reverting to autarchy, but the consequences were minimal.

Fears of Devastation

Alongside the imaginary of the urban wealth and magnitude of the Antwerp port city development in relation to oil, a darker side of oil also existed. From a historical perspective,

Antwerp's urban planning and physical organization was partly shaped in response to fires and explosions related to oil's risky presence close to the city.

Historically, even before petroleum, Antwerp had already suffered quite a lot of urban violence by fires and explosions. The fire consuming in flames the rich and beautiful City Hall, mentioned earlier, was like a harbinger of the city's fate. Finished only ten years before, City Hall burned down in 1576 during the Spanish Fury, by a fire ignited by revolting Spanish soldiers (Figure 13.2). It was also by means of fire that the Dutch army at the end of the sixteenth century tried to attack the Spanish stronghold of Antwerp, more precisely the pontoon bridge blocking off the Scheldt from all harbor traffic. It steered ships with explosives, so-called "infernal machines" or "hellburners," onto the bridge, which caused heavy loss of life after exploding. These events symbolize the violent context of the port city, its huge commercial and military challenges as well as the fragile reality of wealth and political power. A complete and terrible destruction by bombardment and fire was also the fate of the Antwerp citadel when the huge, originally Spanish, fortification was occupied by the Dutch in the early nineteenth century and attacked by the French, who did not leave two stones on top of one another.

A more recent example of Antwerp's difficult relationship with fire can be found in the opening event of Antwerp 93: Cultural Capital of Europe, which featured artistic pyrotechnics by Pierre-Alain Hubert.[17] The French fire artist was asked to open the European cultural year with fireworks on the Scheldt. His show was conceived as an ode to

FIGURE 13.2 In 1585, during the Siege of Antwerp carried out by the Dutch and the English, Giambelli conceived the "hellburners," burning ships loaded with explosives that were used to destroy the pontoon bridge. The Spanish Alessandro Farnese, the Duke of Parma, had the pontoon built to block off the Scheldt. Frans Hogenberg, eighteenth century. Rijksmuseum.

Antwerp's connection with the Scheldt, to its harbor and industry, taking inspiration from the dramatic glow of bright flares of the port's petrochemical industry in the background. In addition to burning oil drums, small torches, and flaming fagots along the Scheldt, the choreography consisted of firing projectiles horizontally over the water, instead of vertically in the air. Except for nice reflections on the water, many found the idea of the event distasteful. People used to impressively crackling, crepitating, and thunderous skyrockets were unable to watch the spectacle.

The story of petroleum and how it is imagined in Antwerp brings to the fore the reality of accidental fire, devastation, disaster, and the risks of handling and keeping oil in the vicinity of the city. As is the case for many other port cities, Antwerp's harbor history in the nineteenth and twentieth centuries is marked by destructive fires and explosions, to the point that one could state that its urban layout was partially ruled by security questions. Also, as a result of an often-tragic learning process, dramatically destructive events led to radical decisions affecting the relationship between the city and the port. In the mind of civil society, oil and the petrochemical industry constituted undoubtedly the biggest area of concern regarding dangers of fire, explosion, and pollution.

The import of the first oil barrels from the US in 1861 was the start of Antwerp becoming the leading European harbor for the discharge of American oil. It served as a distribution hub for Europe and remained the main European petroleum harbor for decades until World War I. The earliest major disaster in Antwerp caused by oil occurred on August 10, 1866, in a building for petroleum storage on Zak Street, situated in the historic heart of Antwerp.[18] Several houses went up in flames. On August 28, 1882, a fire ignited in the storehouses in the Ferdinandus polder—north of the actual Houtdok—where a number of warehouses had recently been built containing wood, a sawmill, and guano, but also oil barrels. The fire started in a bar and a strong wind blew it over to the storehouses of wood, menacing the masonry warehouse holding 6,500 barrels of oil. Thanks to the instantaneous demolition of parts of the buildings, the fire was contained.[19]

Until the end of the nineteenth century, oil was generally kept in warehouses in different parts of the city, but oil-related activities were soon regulated and gathered outside the city limits, around the Amerika Dock. Inaugurated in 1887, the dock was officially named Petroleum Dock and it housed Antwerp's oil activities until 1903. On September 6, 1889, an incredible explosion, followed by a huge column of white smoke and another of black smoke, marked the beginning of a fire covering 2 ha of the Amerika Dock[20] (Figure 13.3). A workplace dedicated to the disassembly of cartridges exploded and set fire to the neighboring petroleum store of thousands of oil barrels belonging to a private company and to the petroleum reservoirs of the city of Antwerp. The blast heavily damaged several ships in the harbor. Due to the shock of the explosion, the hydraulic power station producing power for all the engines in the harbor collapsed, ruining all its machines (Figure 13.4). The air was filled with the detonations of barrels exploding, one after the other, while the glow of the fire could be seen as far away as Malines, halfway to Brussels, and up to 50 km north of Antwerp. The burning oil flew into the Scheldt, setting fire to a number of ships and their freight. The first explosion was so powerful that its destructive effects reached the city center. Fragments of cartridges were found in a radius of several hundred meters. In the area around the Amerika Dock, not one of the fifty houses was left intact. At the site of the cartridge workplace, a funnel of several meters deep with a perimeter of 500–600 m was what remained.[21] Three hundred people died, many of them in atrocious circumstances.

FIGURE 13.3 Informative illustration showing an overview of the site of the disaster that occurred on September 29, 1889. A legend explains the different locations of the tragedy. From Le Petit Moniteur illustré, September 29, 1889.

FIGURE 13.4 The site after the explosion of a workplace for the disassembly of cartridges, where petroleum tanks caught fire and exploded in 1889. The disaster spread to encompass all of the Amerika Dock in the north of Antwerp. Photograph by H. Colon, in Anonym., ca.1889. From La catastrophe d'Anvers: September 6, 1889 (Antwerp: Jos. Maes).

Numerous survivors were severely injured (Figure 13.5). To add to the horror of this extreme event, newspapers offered unusually realistic, if not crude, depictions of wounded and mutilated but still living victims, charred corpses, parts of bodies and limbs, many of them thrown into the air by the explosions and scattered over a wide perimeter in the north of Antwerp.[22]

FIGURE 13.5 Views of the exploded workplace, cartridges, and a destroyed house situated 500 m from the explosion and the site of the oil company Rieth and Co., in 1889. The image underneath shows the rescue of numerous victims. Page with drawings by Paul Destez, in L'Univers illustré, September 14, 1889.

Petroleum Zuid

The catastrophe of 1889 forced the city of Antwerp to take action and protect the inhabitants from the risk of fire and explosion. It was decided that all industrial oil activities would be concentrated on 54 ha in the polders south of the city, in Kiel. The new industrial site was developed under the name Petroleum Zuid on the grounds between the new southern quarters of the city of Antwerp and the Hoboken polder.[23] In 1898, the city expropriated the ground, while the state had to develop the railway infrastructure. The city was also responsible for the plant, the buildings, and installations. Petroleum Zuid, the first petroleum port in Belgium, was one of the most important petroleum port areas in Europe. The new location was completely isolated from the residential city.

To serve Petroleum Zuid, the long quay in front of the city, built at the end of the nineteenth century, was prolonged southwards. The new port of Petroleum Zuid conformed not only to higher security norms but also provided modern infrastructure. By means of a jetty in the Scheldt—the so-called "petroleum bridge"—the ships did not have to moor along the quay and could stay at a distance from the river bank (Figure 13.6). By 1901, all petroleum activities in the port of Antwerp were moved to Petroleum Zuid. In mid-August 1904, the first oil tanker moored.

Oil companies could obtain a concession in the vast allotted area of Petroleum Zuid. The companies had to equip the long narrow parcels with railways and all the necessary installations and buildings. Some 233 petroleum tanks were situated in Petroleum Zuid. The crude oil was easily emptied by means of pipelines from tankers moored at the jetty on the Scheldt to the refineries and directly to the tanks. Refined petroleum returned by the same path. From the jetty, it was shipped to other regions. For use inside the country, it was transported from the tanks by trucks, trains, or in barrels. A new station and a huge railway yard were built northeast of Petroleum Zuid, and in 1904 a tram connection with the city was established.

The location of Petroleum Zuid was chosen to afford more security to the city and other harbor installations. Newly patented security valves were used to avoid tanks exploding and letting the gas come out and burn like gigantic torches. Despite these measures, on August 6, 1904, the plant of Petroleum Zuid was struck by an immense fire. It started with one of the tanks exploding with an enormous blast and bursting open under the pressure of gas.

FIGURE 13.6 New terminal at Petroleum Zuid laid out by Antwerp Oil Wharves on the site of a former brickyard. The chimney of the brickworks is still standing. The jetty, clearly visible in the picture, enables ships to moor at a distance from the quay. *Source*: Tank Storage Verbeke nv.

FIGURE 13.7 Petroleum tanks on fire in Petroleum Zuid in 1904. Some of the burning petrol spread toward the Scheldt, but was stopped by the digging of canals. City Archives, Antwerp.

Liquid oil flew from the tank and was ignited by a mobile forge operating quite a distance away. Almost 100,000 m^3 of petroleum burned and spread the fire to forty other oil tanks, several of them exploding (Figure 13.7). The horizon was completely darkened by smoke and put the city of Antwerp in obscurity. A strong eastern wind drove the flames to the petroleum railway station where wagons caught fire and exploded. Thanks to trenches hastily dug by the firemen, the burning oil was kept from spreading. A pyrotechnic factory and a series of gasoline reservoirs were preserved (Figure 13.8). Six workers were killed on the spot.[24] However, in the minds of the inhabitants of Antwerp, the accident demonstrated that the new location was disaster-proof. Locating Petroleum Zuid at a distance, in the south of the city, had prevented the devastation of business and residential areas and other port installations.

During World War I, access to oil was a major strategic factor. The British-Belgian army ordered the petroleum tanks of Petroleum Zuid to be set on fire, although, when the German army conquered Antwerp in October 1914, they were still able to recover a significant amount of petroleum. Later, during the war, two fires and several bombardments caused enormous damage to Petroleum Zuid. On an international level, oil became rare. Following a request from Britain to prevent oil wells from falling into German hands, Romania destroyed more than 1,500 of its national wells. Romanian production dropped from 1,783,947 tons in 1914 to 518,460 tons in 1917 during the German occupation.[25] The outbreak of the Russian Revolution interrupted trade relations with Russia and eastern Europe.

After the depression of World War I, the port revived. The installations of Petroleum Zuid were rebuilt in one year.[26] A new extension was added to the area. The oil industry in Antwerp continued to develop successfully in the years prior to World War II. Petroleum

526 *REVUE UNIVERSELLE*

Le cadavre carbonisé d'un ouvrier.

L'incendie des docks de pétrole d'Hoboken, près d'Anvers.

L'Incendie des docks de pétrole d'Anvers. — Le 26 août, vers 10 heures du matin, un terrible incendie a éclaté à Hoboken, à quelques kilomètres d'Anvers, dans les réservoirs à pétrole de MM. Speth et Eiffe et de l'American Petroleum Company. En un instant ce fut un spectacle terrifiant; un nuage épais de fumée noire, chassé par un vent de sud-ouest, s'étendit sur la ville, obscurcissant l'atmosphère; les détonations se succédaient sans interruption, au fur et à mesure qu'éclataient les réservoirs, dont le contenu venait alimenter le feu.

Les pompiers d'Anvers, immédiatement accourus, se rendirent compte que toute tentative serait vaine pour arrêter la violence de cette fournaise; tous leurs efforts durent se borner à protéger des flammes les établissements voisins et notamment un atelier de pyrotechnie. Les soldats du génie s'employèrent à creuser des tranchées pour détourner le cours de ce fleuve incandescent, qu'il fallait empêcher, à tout prix, d'atteindre l'Escaut, où il eut incendié tous les navires à quai. On y parvint non sans peine, car l'incendie s'était étendu sur plus de 3 hectares. Sur quarante « tanks » (réservoirs) élevés en cet endroit, trente-huit ont été détruits.

La catastrophe serait due à une cause très minime : un filet de liquide, s'écoulant par une fissure de l'un des tanks de l'American Petroleum Company, s'enflamma au contact du feu d'une forge portative, installée près des réservoirs. Les dégâts ont été évalués à 7 ou 8 millions au moins.

On retrouva dans les décombres sept corps carbonisés, dont l'identité ne put être établie; mais le nombre des victimes doit être plus élevé.

L'Incendie de l'arsenal de Toulon. — Le 7 août, un incendie d'une violence inouïe s'est déclaré à Toulon, dans l'arsenal principal de la marine, gagnant bientôt les ateliers des mouvements généraux, la section du charpentage, la section de garniture. A droite et à gauche de ce vaste bâtiment se trouvent deux cales où précisément des torpilleurs vedettes étaient en réparation; deux ont pu être mis à l'eau, un troisième a été légèrement endommagé.

Les bâtiments ravagés par le feu ne for-

La cale de réparation des torpilleurs.

L'incendie de l'arsenal de Toulon. — Phot. Marius Bar, Toulon.

FIGURE 13.8 The tanks of the oil companies of MM Speth and Eiffe and of the American Petroleum Company caught fire on August 26, 1904, and quickly spread over the whole area of Petroleum Zuid. The photograph in the center shows the carbonized body of one of the victims, an oil worker. *Source*: Revue universelle, August 26, 1904.

Zuid was further extended into the Hoboken polder to accommodate the construction of a refinery, a new type of activity for the Port of Antwerp. A new jetty was built. Unfortunately, Petroleum Zuid was damaged during World War II and it never fully recovered. Also, the oil industry moved to the Marshall Dock—inaugurated in 1951 and initially called Petroleum Dock—in the new port areas in the north of Antwerp. Petroleum Zuid was abandoned little by little. In 1953, 4.3 million tons of oil were processed in Antwerp of which only 405,000 tons were handled at Petroleum Zuid.[27] The last concession in Petroleum Zuid will end in 2035.

Environment, Pollution, and Heritage

Images of oil slicks on the surface of oceans and rivers, burning oil wells in the desert as a result of war action, petroleum disasters caused by train accidents, and accidental explosions of petroleum storage tanks not only evoke ancestral fears of collective cataclysms, but also amplify worries about contemporary worldwide ecological disasters. Images do not need to be explicit to evoke the hazards inherent in oil. Consider the famous and fashionable photographer Ed Burtynsky on his recent photographic series, "Oil":

> The car that I drove cross-country began to represent not only freedom, but also something much more conflicted. I began to think about oil itself: as both the source of energy that makes everything possible, and as a source of dread, for its ongoing endangerment of our habitat.[28]

Despite all the benefits oil can produce for the immediate comfort and pleasure of individuals and society, it is increasingly considered a menace to a habitable environment.

With oil's dark side becoming more visible, two specific concerns came up when Petroleum Zuid was abandoned: the burden of its environmental pollution and the possible loss of its industrial heritage. These concerns represent separate ecological and historical values, but seen in a wider context, they require a form of intellectual integrity and integrative thinking, as the care for our environment and for the historical remains of what has preceded our society expresses our effort to preserve the future. Our history, let it be environmental, industrial, urban or architectural, is part of our future. It is challenging, but not surprising, to see these concerns in a highly industrial area as Petroleum Zuid combined as aspects of the same question.

Petroleum Zuid should be considered a place with an industrial memory. The former petroleum site with its machinery, storehouses, jetties, tanks, and pipelines is now abandoned but represents a certain idea of a petroleumscape. All this equipment and infrastructure form an ensemble of early twentieth-century petroleum infrastructure and architecture. Today, the site includes several historical elements which have recently been protected by law.[29] Indeed, the public action of the industrial archaeology group VVIA[30] has resulted in stopping the planned demolition of some of Petroleum Zuid's oldest buildings.[31] These buildings were the only ones that survived the fire of 1904. One of them was a unique, early concrete structure. In December 2014, the activist group succeeded in having some buildings listed as valuable and protected, including the Avia company refectory with the concierge's house and several petroleum tanks with masonry stairs and talus as well as a pump house of the same oil company, along with a concrete warehouse, the concrete jetty on the Scheldt, and a portion of the pipelines above ground.

On the other hand, how Petroleum Zuid is remembered also concerns the environment and its deterioration over a century of intensive industrial use. When the site started to function in the early twentieth century, oil was pumped through underground pipes from ships at the jetty to the tanks. Multiple leaks occurred and large quantities of oil infiltrated the soil, resulting in severe pollution of the area.[32] At the end of the 1930s, after long negotiations, the leaking underground pipes were replaced by conduits above ground. Nevertheless, the soil of Petroleum Zuid remains heavily polluted and requires remediation. In terms of intellectual integrity, industrial archaeology should also consider the ecological consequences of the industrialization of oil as part of an entire systemic complex. If the residual pollution left over from the industrial activities is perceived as a violation of nature and as an obstacle to the cultural valorization of the site, it recovers also a completely different meaning. Oil pollution on a site like Petroleum Zuid should be considered an integral part of the industrial system: the machines, installations, and buildings are witnesses of highly technological innovations and functional designs of an industrial system that involved also smoke, spilled oil, filthy soil as well as enormous societal costs in terms of human health and well-being. All these systemic consequences need a thorough didactic explanation besides the historical and archaeological documentation of oil industry technology.

To mark the difference with the previous period, by March 2016, Petroleum Zuid had been transformed into a project area of about 60 ha with the name Blue Gate Antwerp, a name that evokes renewal, cleanliness, openness, immateriality, and airiness. For the public partners of the project (the city of Antwerp and the Flemish region), "blue" is a symbol for the green economy and an ecological and economic future, and "gate" refers to the role of a doorway imagined as played by both city and river. In the mindset of both partners, the project is intended as an investment in sustainability and eco-efficiency based on "cradle-to-cradle" organization: it is a form of closed cyclic thinking where every kind of waste is nonexistent because it is avoided or recycled. It is designed to improve the negative imagery of oil in terms of pollution, risk, and the waste of precious soil and space. The ancient petroleum plant will be recycled and transformed into an activity zone with green areas, new clean industries, start-ups, leisure areas, and parks. Today, with the foresight of the Blue Gate Antwerp project in the south of Antwerp, the epic cycle of the oil industry in an urban context—transport in barrels and storage in reservoirs, shipping in tankers, pumping through pipelines, treatment in petrochemical complexes—is now ending.

Epilogue: A Carbon Age

The Antwerp port, city, and river have been privileged witnesses of the dramatic ascent of oil in modern society. In the wake of the historic trajectory that petroleum has traced in the port city of Antwerp, and in view of future projects like Blue Gate, something more than a postindustrial turn could open a window on the twenty-first century. The story of the petroleumscape is after all a mixture of utopia and hubris; its arrival as a basic industrial substance, its rise as an economic value, its abstraction through petrochemical processes, went along with the fabulous imagery of wealth and power and the risks and fears of fire, disaster, and pollution. At first, the oil industry was one of the fundamental pillars of an "affluent society."[33] But today, this cycle of production is to a large extent compromised. All the fears that have been adding up since oil's first appearance in the nineteenth century have culminated in its recent rejection. Carbon is the fourth most abundant element in the universe. Because of the unlimited use of nonrenewable energy resources, carbon accumulated by geological processes over millions of years has been released back into the atmosphere. Oil's immense carbon footprint is one of the main causes of climate change. If humanity's presence on Earth has created the Anthropocene, this has occurred as a result of what Eric Roston has called the Carbon Age.[34]

In this context, the case of Antwerp is even more intriguing. Oil and diamond, substances that historically engendered the city's wealth, are both carbon-based. Diamond is a solid form of carbon. Crude oil is a mixture of comparatively volatile liquid hydrocarbons that releases carbon when burned.[35] The carbon cycle, made of the different paths of carbon in the environment—between ocean, soils, vegetation, air, and so on—is one of the main dynamic elements of Earth's climate. Diamonds and oil symbolize inorganic and organic processes on different time scales. The materials are associated with opposing qualities: eternity and immanence with the one, and unbridled consumption, immediate effect, and exhaustibility with the other.

A reflection on the Carbon Age could offer Antwerp a new context for its carbon past and future, leading to a new narrative and the creation of museums, exhibits, courses, and educational activities, heritage tours and visits, marketing topics for business companies and industries, but also to the exploration and experimental observation of natural environments. Besides the urban and architectural dimensions of the petroleumscape, the mental landscapes shaped by the desires and fears initiated by oil could lead to new research about specific imagery, discourse, legends, dreams, and nightmares generated in the Carbon Age of Antwerp and other port cities.

Notes

1 Port of Antwerp, "1800–1930 Industrial Revolution," https://www.portofantwerp.com/en/1800-1930-industrial-revolution, accessed 23 November, 2018.
2 A. J. Hendrix, *Petroleum* (Antwerp: Kiliaan, 1920), 30–31.
3 Total, "50 Years of the Antwerp Platform," https://wiki.total/en/theme/50-years-antwerp-platform, accessed 20 July, 2020.
4 Frans Claes, *Inauguration de la Raffinerie du Kruisschans, Anvers* (Antwerp: Société Industrielle Belge de Pétroles, 1952). Quotation translated by author.
5 Ibid.
6 Ibid.
7 Port of Antwerp, "Expansion to the Left Bank," https://www.portofantwerp.com/en/1970-2000-expansion-left-bank, accessed 23 November, 2018.
8 Nationale Bank van België, "Het economisch belang van de chemie- en petroleumsector in de regio Antwerpen, 1990–1998." Unpublished report.
9 Louis Nauwelaerts, *Petroleum* (Tilburg: Het Nederlandsche Boekhuis, 1936); *Het groene goud* (Antwerp: Standaard, 1939). Nauwelaerts (1904–?) also wrote popular literary works, for example, *Verkenning op den vooravond* (Ghent: Snoeck-Ducaju, 1942) and *De brug van licht en liefde* (Leuven: Davidsfonds, 1944).
10 Nauwelaerts, *Petroleum*, 6.
11 Nauwelaerts, *Het groene goud*, 32–33.
12 Carola Hein, "Oil Spaces: The Global Petroleumscape in the Rotterdam/The Hague Area," *Journal of Urban History* 44, no. 5 (2018): 887–929.

13 On the history of the port of Antwerp, see Louis de Kesel, "Structurele ontwikkeling van de haven (1803–1914)," in *Bouwstoffen voor de geschiedenis van Antwerpen in de XIXde eeuw*, ed. Genootschap voor Antwerpse Geschiedenis (Antwerp: Algemene Drukkerijen LLoyd Anversois, 1964), 124–68; Karel Jeuninckx, "De havenbeweging in de Franse en Hollandse période," in *Bouwstoffen voor de geschiedenis van Antwerpen in de XIXde eeuw*, ed. Genootschap voor Antwerpse Geschiedenis (Antwerp: Algemene Drukkerijen LLoyd Anversois, 1964), 93–123; Michael Ryckewaert, "The Ten-Year Plan for the Port of Antwerp (1956–1965): A Linear City along the River," *Planning Perspectives* 25, no. 3 (2010): 303–22, 10.1080/02665433.2010.481179; Port of Antwerp, "History of the Port," http://www.portofantwerp.com/en/history-port; World Port Source, "Port of Antwerp," http://www.worldportsource.com/ports/review/BEL_Port_of_Antwerp_25.php.

14 Quoted from *Journal des Nations*, January 15, 1936, as mentioned in Nauwelaerts, *Het groene goud*, 37.

15 Congress of the United States, Joint Economic Committee, *Soviet Economy in a New Perspective: A Compendium of Papers* (Washington, DC: US Government Printing Office, 1976), 90.

16 Ibid.

17 Benno Barnard, "Een vaudeville met horizontaal vuurwerk; Antwerpen 93 voor Hollanders verklaard," NRC (31 December, 1993), https://www.nrc.nl/nieuws/1993/12/31/een-vaudeville-met-horizontaal-vuurwerk-antwerpen-7208704-a338653, accessed 14 January, 2019.

18 Guido de Brabander, *Na-kaarten over Antwerpen* (Brugge: Uitgeverij Marc Van De Wiele, 1988), 102.

19 *La Lanterne* (29 August 1882).

20 "La catastrophe d'Anvers," *L'Argus* 13, no. 499 (15 September 1889).

21 Ibid. Most articles give the same information, for instance *L'Univers illustré* (21 September 1889); *La Justice* (August 8, 1889); *La Lanterne* (9 and 10 September 1889).

22 See, e.g., detailed descriptions in *La Lanterne* (10 September 1889).

23 On the history of Petroleum Zuid, see, for example, Louis de Kesel, "Les travaux du nouveau port pétrolier d'Anvers. (Introduction à la visite du chantier)," *Bulletin de la Société royale belge des ingénieurs et des industriels* 6 (1950): 286–300.

24 Concerning these dramatic events, see *Armée et Marine–Armes et Sports* (September 8, 1904); *L'Univers* (August 29, 1904); *La Croix* (August 29, 1904); *La Lanterne* (August 28 and 29, 1904); *Le Matin* (27 August 1904).

25 https://sp.lyellcollection.org/content/465/1/1.1#sec-4, accessed 23 July, 2020.

26 Mario Janssens, *Expeditie Blue Gate Antwerp. Een wandelzoektocht in het voormalige "Petroleum Zuid"* (Antwerp: Stad Antwerpen/Tom Meeuws, 2011), 11.

27 UrbEx.nl, "Petroleum Zuid," https://www.urbex.nl/petroleum-zuid/, accessed February 11, 2019.

28 Edward Burtynsky, "Oil," https://www.edwardburtynsky.com/projects/photographs/oil, accessed February 28, 2019.

29 Hooft 2011; Adriaan Linters, *Antwerpen Petroleum Zuid: urgentie-onderzoek, samenvattend verslag* (Wevelgem: Kleio, 2001); Moermans 2011.

30 VVIA stands for Flemish Association for Industrial Archaeology.

31 Elise Hooft, Inventaris van het bouwkundig erfgoed, Provincie Antwerpen, Stad Antwerpen: Petroleum Zuid. Bouwen door de eeuwen heen in Vlaanderen ANT4 (unpublished document); Linters, *Antwerpen Petroleum Zuid*; K. Moermans, "Antwerpen Petroleum Zuid, Visienota Onroerend – discipline bouwkundig en industrieel erfgoed," n.d.

32 See, on these topics, Erfgoed and Visie, "Onderzoek industrieel erfgoed Petroleum-Zuid Antwerpen" unpublished study report; Janssens, *Expeditie Blue Gate Antwerp*; Linters, *Antwerpen Petroleum Zuid.*

33 John Kenneth Galbraith, *The Affluent Society* (New York: Houghton Mifflin, 1958).

34 Eric Roston, *The Carbon Age: How Life's Core Element has become Civilization's Greatest Threat* (New York: Walker & Co, 2008).

35 Encyclopædia Britannica, "Crude Oil," https://www.britannica.com/science/crude-oil, accessed 3 July, 2020.

14

POWER STATIONS AND PETROLEUM HERITAGE IN ITALY

The Case of Porto Tolle

Chiara Geroldi and Gloria Pessina

In Italy, as in other parts of the world, thermoelectric power plants fueled by oil are being decommissioned and transformed, raising questions about the materiality of the energy transition and the role of power stations in the global petroleumscape. Making good decisions about what to do with these plants and the surrounding areas requires research on the historical emergence of this petroleum-related infrastructure as well as its location, spatial form, and connection to ports and consumption sites. We began to see the need for the research we present here while participating in a consultancy undertaken by the Politecnico di Milano[1] for Enel, a formerly state-run energy company founded in 1962, privatized in 1999, and now a multinational energy company. With its Futur-e project, launched in 2015, Enel asked Politecnico di Milano to analyze the territories of ten power stations (out of twenty-three undergoing closure) in order to provide information to participants in competitions for designers (e.g., in a call for ideas) as well as to entrepreneurs who might be interested in acquiring and redeveloping the sites.[2]

Intrigued by the material we collected during the consultancy and visits to several power plants undergoing closure, we started the research presented in this chapter. Through the lens of the literature on the global petroleumscape, we began analyzing several oil-fueled power plants as elements that direct and redirect flows. Afterward, we decided to focus on a particularly interesting case, the power station of Porto Tolle, showing its role in the global petroleumscape and highlighting the specific historical, geographical, environmental, and sociopolitical factors that characterize its territory and may influence its post-oil future. Our research seeks to fill a gap in the design and planning literature, in which the landscape of (oil-fueled) thermoelectric power stations is little explored, thus offering a necessary addition to the emerging research on the petroleumscape.[3]

Power Stations in Remote Areas as Part of the Global Petroleumscape

The power stations involved in the Futur-e project are spread out across Italy and vary in size, from only a few hectares to over 100 ha. Most are located in marginal sites, often near the

DOI: 10.4324/9780367816049-18

sea, and far from major urban centers. A few are located in ports or industrial areas. The large oil-fueled power station of Porto Tolle, which once produced 10 percent of Italy's energy, lies in the particularly remote and fragile territory of the Po River delta. It closed in 2015.

Although not physically close to cities, power stations such as Porto Tolle can be considered "urban" using the concept of *planetary urbanization* proposed by Neil Brenner and Christian Schmid.[4] In particular, they can be considered part of extended urbanization or what Brenner defines as an "operational landscape."[5] Brenner and Schmid argue that spaces usually considered part of the "non-urban realm," such as infrastructures, pipelines, "nature parks" as well as oceans, deserts, or even the atmosphere, have now become part of "the worldwide urban fabric."[6] The global petroleumscape is a perfect expression of planetary urbanization, as it connects places of extraction, often located in areas difficult to reach, with consumers, who are mainly located in agglomerations. Brenner further develops the concept of planetary urbanization, stressing that urbanization contains both concentration and extension and noting that urban theory has focused its attention primarily on processes of concentration,[7] ignoring the wider areas and sites that support these processes, such as oil production sites or power stations. Agglomerations must be connected to large-scale processes of territorial reorganization and resource extraction that encompass the world.[8]

Porto Tolle power station lies in the protected area of the Po River delta, which in 2015 was recognized as a Biosphere Reserve within the UNESCO Man and Biosphere Programme. Although the plant might appear remote, it was once tightly connected to global flows of oil and national flows of electrical energy serving residential and industrial areas. As such, the power plant can be seen as part of the urban, being an "operational landscape" that supports agglomerations through energy production. Such a remote site also exemplifies important transformations, particularly those related to the energy sector at the Italian national level and current European policies for energy transition. Our analysis of this site reveals the large landscape of oil to which the power station is connected and identifies the oil-related elements in this and other oil-fueled power stations in the Futur-e project.

Global Landscapes of Oil, Architecture, and Heritage

As Carola Hein writes, even though "the development of petroleum-based technologies and infrastructure has produced significant changes in the built environment,"[9] such oil-related changes in cities and landscape have been largely ignored by architecture and urban studies. Various approaches to oil have been promoted in other disciplines, including political ecology, energy humanities, economic geography, planning history, cultural history, and the history of the built environment.[10]

Studies in the field of political ecology were among the first attempts to highlight the contradictions and conflicts emerging in the territories where oil is extracted and energy is produced.[11] Moreover, political ecology scholars started analyzing elements through which oil circulates,[12] how oil is embedded in sociopolitical relations and how it affects territories.[13] Concerned with the materiality of oil and building on previous studies on the flows of various types of "matter" (e.g., water, waste, fuels) in the field of political ecology,[14] Gavin Bridge lists both the specific "biophysical characteristics" and the "material forms of oil," identifying three types of extractive spaces: (1) a "geography of holes" such as a mine shaft or an oil well, (2) the structures that drive the flows of fuels, and (3) waste and waste spaces produced from extractive industries.[15]

While such approaches unveiled the complexity of oil-affected areas, with an in-depth overview of the interrelations between matter, space, power, and society, the landscape of lines, axes, and nodes was merely listed and localized by early political ecology scholars, who aimed at identifying emerging environmental conflicts, unequal exposure to possible sources of pollution, and imbalances of power. Architects, designers, and planners have been contributing to the debate on the global petroleumscape through an in-depth analysis of the material forms of spaces, attempting to identify which parts of the landscapes of oil could be kept as heritage and how they could be reimagined for future uses.[16] After having studied the topic for the past decade,[17] Hein has introduced the concept of a "palimpsestic global petroleumscape," which includes the physical forms of the built environment, its representations and values as well as its uses.[18]

Moreover, Hein emphasized that oil networks are "powerful producers of new landscape" and that the global oil industries have generated an "international 'landscape of oil,'" which also influences our understanding of territories and movements.[19] Building on this literature, this chapter considers the landscapes in which oil is transformed into electricity, highlighting flows of oil, connected infrastructure, emerging conflicts as well as related architectural elements. While extensive work has been conducted on the conflicts as well as the flows of matter and power related to sites of electricity generation, especially nuclear power stations[20] and hydropower,[21] less has been written on thermoelectric power plants in terms of conflicts and flows[22] as well as of landscape and built environment.[23] Therefore, we introduce the topic of the contemporary landscapes of oil and electricity in Italy as an addition to the existing international literature on the petroleumscape and on the political ecology of energy. Then, we focus on a specific plant and the surrounding territory to show how an in-depth description of the spaces of the petroleumscape can be used to identify valuable elements for post-oil landscapes.

Contemporary Landscapes of Oil and Electricity in Italy: Four Examples

In the global context of the energy transition, Italy is adopting international policies aimed at reducing the use of fossil fuels for energy production. In 2015, when Enel launched the Futur-e project for twenty-three thermoelectric plants with a significant total gross capacity of 13 GW, the company explained that these large plants running on fossil fuels have become marginal in Italian electricity production, due to diminishing industrial demand[24] and an increase in small renewable energy plants throughout the territory.[25] The production capacity of Italian thermoelectric plants now exceeds demand.[26] Several architectural and other elements of thermoelectric power station landscapes—physical structures that may remain after the energy transition—can be interpreted as the heritage of the future and as objects with the potential for new uses.

In what follows, we analyze four oil-fueled power stations showing the variety of oil networks in which the power stations were located: Porto Tolle, Montalto di Castro, Rossano, and Bari. Emphasizing the materiality and the diverse environments of the stations, we explain the various ways oil can reach the power stations and explain how oil-fueled power stations can be reconverted to other fuels. Calls for a manifestation of interest in buying and redeveloping them have been made for all four stations. At the time of writing (2020), the most advanced in the transformation process was Porto Tolle, for which a preliminary sale agreement has been signed. Three of the plants—Porto Tolle, Montalto di Castro, and Rossano—are located quite far from major urban centers in agricultural areas, in the north, center, and south of Italy, respectively. They face the sea and use sea water for plant cooling (Porto Tolle drew water from both the sea and the Po River). The Bari station is located in the city of Bari, in a transition zone between the city center and an industrial area, adjacent to the dismantled Stanic oil refinery (owned by the Eni oil and gas company). These sites have not yet been sold and their regeneration has not started yet.

The sites vary in terms of size and gross capacity: Porto Tolle occupies 380 ha and had a gross capacity of 2.640 MW; Montalto di Castro extends over 220 ha and had a gross capacity of 3.600 MW; Rossano occupies 50 ha with a gross capacity of 1.738 MW, while Bari extends over 6.8 ha and had a gross capacity of 205 MW. Some power stations were fueled by oil throughout their life span, while others underwent conversions or exploited both oil and gas (Figure 14.1). Porto Tolle has been always fueled by oil; its four steam-powered electric sections began operating between 1980 and 1984. Enel proposed a conversion to coal in 2005, but Italy's Ministry of the Environment, Land, and Sea ultimately rejected it in 2013. The thermoelectric plant of Rossano began operating in the mid-1970s. At first, it was fueled by oil; later, gas was preferred as it was considered to have less of an impact on nearby agricultural fields.[27] In the mid-1990s, Enel added several turbogas repowering units, which are fueled by gas. The Bari plant, constructed in the late 1950s, was initially powered by coal, later by natural gas and oil, and after 2008, exclusively by natural gas. In the late 1970s, a nuclear power station was partially constructed on the Montalto di Castro site, but it never became operational due to the 1987 referendum rejecting nuclear power across Italy. In the 1990s, Enel constructed eight turbogas groups and a thermoelectric plant at the site, fueled by oil and gas.[28]

Oil reached these power stations in various ways, after long journeys involving global systems of infrastructure and logistics that transport oil from extraction sites to power stations, which converted the oil to electrical energy. The Porto Tolle power station received oil mainly from a pipeline of 92 km, which started at a depot in the industrial area of the port of Ravenna. The station of Montalto di Castro was serviced by a 35 km oil pipeline in the sea, starting at the Enel power plant of Torrevaldaliga Nord, near the port of Civitavecchia in the metropolitan area of Rome. The Torrevaldaliga Nord plant obtains oil from a pipeline of ca. 3 km extending from a single point mooring tower in the sea.[29] The Rossano plant had its oil delivered by tanker trucks coming from the depots of Taranto and Crotone, at a distance of

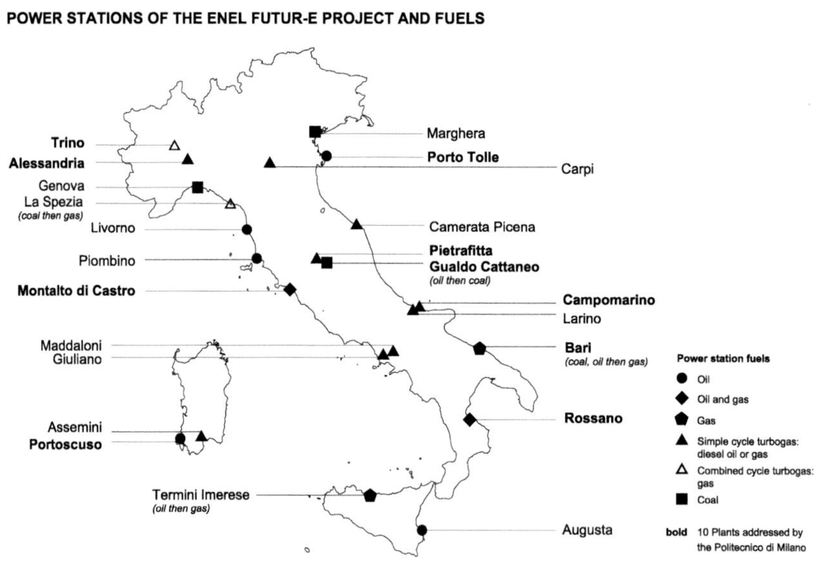

FIGURE 14.1 Thermoelectric power stations involved in the Enel Futur-e project. *Sources*: Chiara Geroldi and Enel S.p.A.

150 and 95 km, respectively.[30] Enel had planned an oil pipeline to connect the Rossano plant to the nearby port of Corigliano Calabro, but this project was highly contested and never implemented. Some oil was also delivered from this port via tanker trucks (Figure 14.2).[31]

This brief look at oil flows within the Italian territory reveals connections between oil-fueled power stations and the port areas, where oil is delivered and where refineries and oil depots are often present. The pipelines connecting some power stations are not just "lines"; indeed, they are often a source of conflict. As an alternative or in addition to pipelines, oil was also carried by tanker trucks, or in the case of Porto Tolle, also by barges. An even broader view of flows of oil shows the high connectivity of these power stations on a global scale. Although these plants are quite disconnected from their immediate surroundings, they are highly connected to a global network of fuel flows, in addition to the Italian energy network.

Although each plant has some unique features, there are many common elements. Thermoelectric plants have large machinery halls, high chimneys, and other features that depend on the type of fuel used (coal, oil, oil and gas, or gas alone) such as coal yards, oil tanks, oil pipelines, and methane pipelines. The ones located far from major water sources also contain cooling towers. Turbogas simple-cycle plants (also involved in the Futur-e project) are very different and much smaller as they are meant to operate only during periods of high demand or emergencies.[32] They were fueled by diesel oil or gas and served by tanker trucks and methane pipelines. All the power stations in the Futur-e project are equipped with oil tanks or they have them in their proximity, an example of thermoelectric plants' connection to the petroleumscape. Oil tanks vary in terms of number (e.g., from two in turbogas plants to nine at Porto Tolle), dimension (e.g., from 7,500 to 20,000 m^3 at Bari and from 50,000 to 100,000 m^3 at Porto Tolle and

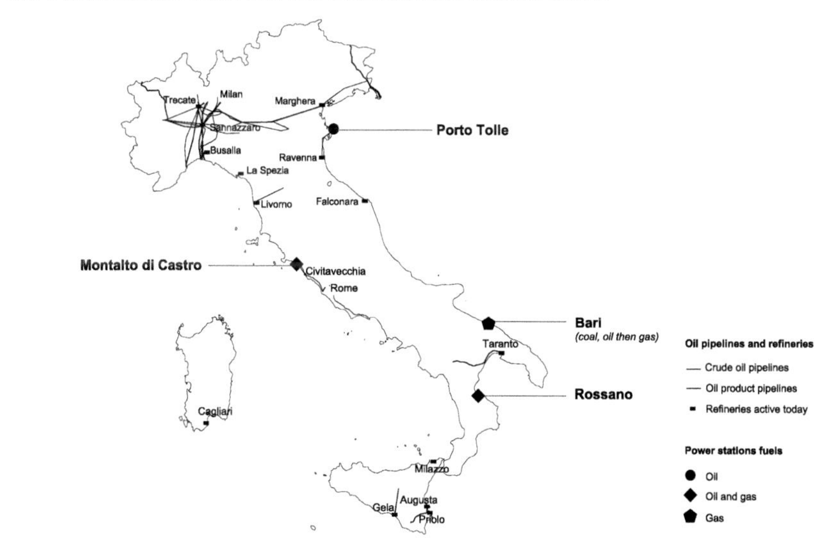

FIGURE 14.2 Oil pipelines, active refineries, and the power stations of Porto Tolle, Montalto di Castro, Rossano, and Bari. *Source*: Chiara Geroldi. Pipelines base map information: © PLATTS for the underlying grids for electricity, gas, and oil, 2016 © European Union, 2019.

Montalto di Castro), and total area occupied. They are surrounded by earth embankments or by concrete walls that create basins to contain oil in case of leaks.

From a design reuse perspective, several elements of the plants are of interest, including the large and impressive machinery halls, the high chimneys that serve as local landmarks,

FIGURE 14.3 The power station of Porto Tolle, located in the Po River delta. *Source*: Enel Produzione S.p.A.

FIGURE 14.4 The power station of Montalto di Castro, located in a former Etruscan territory on the Tyrrhenian Sea. *Source*: Enel Produzione S.p.A.

the hyperbolic cooling towers, and the oil tanks.[33] Specific elements of the power stations' context, environments, and history can also be particularly relevant in terms of "heritage" to be enhanced. Montalto di Castro, for instance, contains large and thick unused concrete structures intended for the nuclear plant and Etruscan tombs (the site is in a former Etruscan territory). A wide strip of protected Mediterranean landscape is located along its border, close to the beach. Rossano, which has a particularly luminous machinery hall, is located in front of the beach and has olive trees on the property. Porto Tolle, which is particularly large, includes fragile environments, rich in biodiversity, both on land and in the water, typical of the Po River delta (Figures 14.3–14.6).

FIGURE 14.5 The power station of Rossano, located on the Ionian Sea. *Source*: Enel Produzione S.p.A.

FIGURE 14.6 The power station of Bari, located close to the dismantled Stanic Refinery. *Source*: Enel Produzione S.p.A.

The Po River Delta: A Fragile Equilibrium Between Land, Water, Oil, and Gas

The Porto Tolle power station is located in the territory of Polesine, part of the Po River delta in the Veneto region (Figures 14.7–14.8). Geographers and geologists have highlighted rare features of the area that result from the accumulation of centuries of detritus from the rivers Adige and Po.[34] Our description of the context elaborates on the existing literature on the territorial and environmental fragility[35] of the Po River delta[36] produced by a wide range of disciplines, including geology, geography, sociology, and urban, rural, and heritage studies.

Human interventions, climatic and alluvial factors have been shaping the Po River delta throughout its history, and since the end of the 1880s, large-scale drainage works were promoted by land reclamation consortia.[37] Despite many attempts to control the water, several

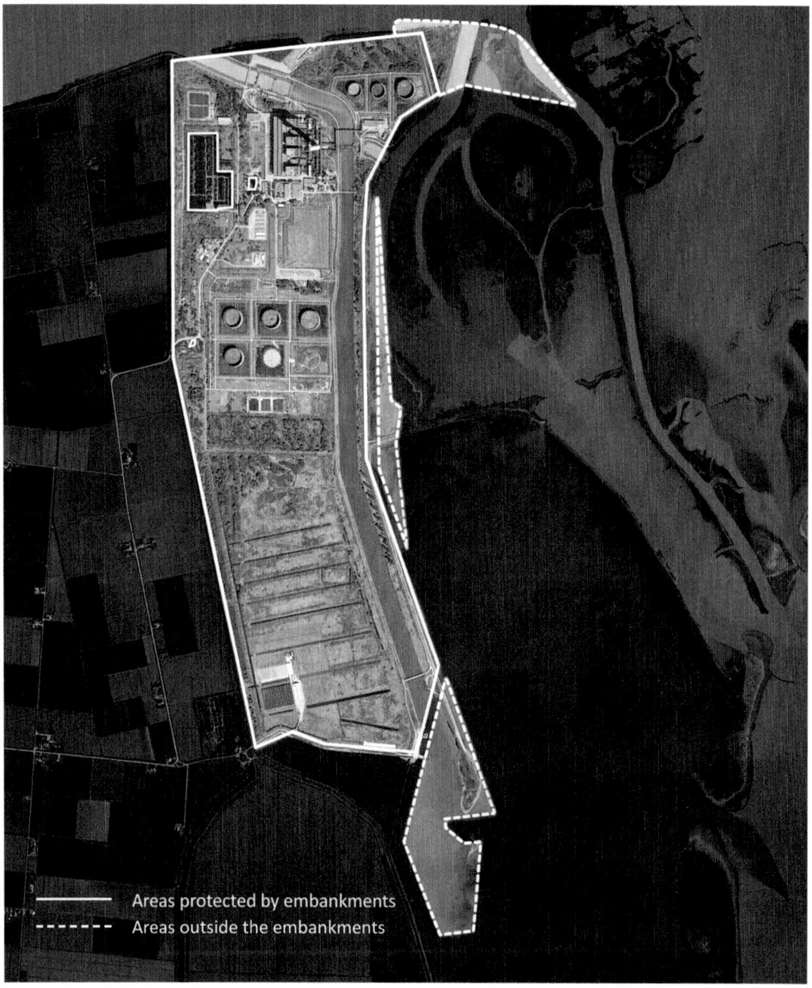

FIGURE 14.7 The power station of Porto Tolle. *Source*: Google Earth. Image © 2016 TerraMetrics – Data SIO, NOAA, US Navy, NGA, GEBCO modified by the authors.

FIGURE 14.8 The power station of Porto Tolle in the context of Polesine (Po River delta). *Source*: Contains modified Copernicus Sentinel data (2020) (texts added by the authors).

FIGURE 14.9 Abandoned methane gas extraction wells in Ca' Pisani (Porto Viro, RO). *Source*: Moreno Bonifacio.

tragic inundations have occurred in Polesine. According to geologists,[38] the frequency and the severity of the floods increased between the late 1930s and early 1960s, due to the extraction of underground water rich in methane gas (Figure 14.9), which accelerated land subsidence in the area.[39] In 1951 and 1966, two major floods devastated Polesine, causing the deaths of several hundred people and destroying thousands of properties.

The relative poverty of the area—where until the 1950s livelihoods centered on fishing and minor agricultural activities—and the increasing exposure to hydrogeological risks made Polesine an "area to escape from," according to sociologists who studied the substantial outmigration from the area to the industrial cities of Northern Italy in the 1950s–1960s.[40] In an effort to address the poverty, various local economic development programs have been promoted since the 1970s.[41] The placement of the power station of Porto Tolle in such an area was part of a political effort to bring employment opportunities to a place where there were few.

At the time of the conception of the Porto Tolle power station, the surrounding territory between Porto Marghera in the Veneto region and Ravenna in the Emilia Romagna region had already become established as part of the so-called "petrochemical quadrilateral" of the Po River valley, the multipolar industrial area extending also to Mantua in the Lombardy region and Ferrara in the Emilia Romagna region.[42] The petrochemical industry, which mostly specialized in the production of plastics and fertilizers for agriculture, depended on the refining of crude oil for energy and raw material. The "petrochemical quadrilateral" was highly interconnected by land and sea infrastructures (pipelines, roads, railways, barges, etc.) intended to facilitate the flow of oil and chemical products. The location of the Porto Tolle oil-fueled power station was chosen by Enel for a variety of reasons, including the possibilities for the oil to be stored, refined, and transported by the infrastructures of the "petrochemical quadrilateral."

Since the conception of the power station of Porto Tolle and the introduction of intensive agricultural activities based on the use of chemical fertilizers in the early 1970s, the landscape of water, land, oil, and gas in the Po River delta has been the object of a contentious process of economic development and environmental protection. As documented on film, recorded at the time of its construction, the population of Porto Tolle was aware of the environmental threats caused by a thermoelectric power station fueled by oil, but many accepted it because of the promise of jobs.[43] As a man declared in an interview filmed in 1976, "It's better to die from smog in our land, in our Delta, rather than in Milan."[44]

At the time of the construction of the power station (Figure 14.10), only a few groups contested the possible impacts of the new development measures on the fragile and marshy landscape of the Po Delta, which was home to a wide variety of animal species.[45] Local environmentalists advocated for environmental protection and obtained the creation of the Bocche di Po regional natural reserve (1977), still considered one of the areas of greatest avifaunal interest in Italy.[46] Despite the environmental concerns, economic and logistical reasons led the transformation of the area. In addition to generating employment, the power

FIGURE 14.10 Photographs taken during the construction of the Porto Tolle power station: (a) transport of sand to the southern part of the site, aimed at preparing the ground of the future oil depots; (b) an oil depot in the northern area of the site and the Po River entering the Adriatic Sea; (c) the machinery hall and the boilers; and (d) the chimney, oil depots, and piling rigs. *Source*: Vinicio Zanardi.

station's location was convenient for the Enel company, since the site had access to both sea and river water to cool the plants and it could be easily connected to oil depots in Ravenna or in Porto Marghera.[47] Moreover, the Porto Tolle power station could use 800,000 tons of oil that Ravenna's refinery (owned by Sarom) would produce over the following decade, at a time when it was at risk of closure.[48]

Environmental conflicts intensified at the end of the 1980s, when the emissions of the power station attracted the attention of some local residents. Following various struggles, the process of environmental protection that started at the end of the 1970s continued with the creation of two Po River delta regional parks (in Emilia Romagna in 1988 and in Veneto in 1997), one of which was included in 1999 on the list of UNESCO sites, and several areas received recognition as Sites of Community Importance (SIC) and Special Protection Areas (ZPS) in the framework of the EU's Natura 2000.[49] At that time, the protected areas were becoming part of a natural heritage, from which any form of economic exploitation of water, land, oil, or gas was prohibited.

The beginning of the new millennium saw a simultaneous resurgence of environmental activism and economic development initiatives related to oil and gas in Polesine. While environmentalists were working to protect local ecosystems from various threats,[50] Enel was proposing to convert the oil-fueled Porto Tolle power station to coal. Edison was creating a new power energy plant fueled by gas[51] a few kilometers away from the Enel power station, and a joint venture by Exxon Mobil, Qatar Terminal Limited, and Snam was starting the first Italian offshore LNG regasification plant in the Adriatic Sea.[52] Following the release of data about the exceptional rate of cancer, respiratory, and cardiovascular diseases registered in Polesine,[53] the struggle against Enel's proposal to convert to coal the Porto Tolle plant intensified. In 2013, the project was rejected by local and national authorities, and in 2014 the former chief executive officers of Enel were sentenced and later released for the crime of "threat of sanitary disaster."[54]

A new phase in the life of the Po River delta started in 2015, after the closure of the power station, when the delta was listed as a UNESCO Biosphere Reserve. In the same year, a team coordinated by the Consortium for the Land Reclamation of the Po River Delta (Consorzio di Bonifica Delta del Po) started working for the inclusion of Polesine among the marginal areas eligible to receive funds from the National Strategy for the Development of Inner Peripheries (SNAI, Strategia Nazionale per le Aree Interne 2014–2020).[55] In 2018, SNAI funded a local development program (Contratto di Foce Delta del Po) working to enhance local heritage related to water and to promote slow tourism in the area.[56] After several contentious decades, the program seems to be trying to reconcile environmental protection and economic development in Polesine, even though the environmental impact of the gas industry in the area is still being researched.

The Porto Tolle Power Station as Heritage

The power station of Porto Tolle is one of several large thermoelectric plants constructed in Italy in the final decades of the twentieth century. Earlier hydroelectric and thermoelectric plants had often been designed by well-known architects and engineers. But Porto Tolle was built during a period when, as Rosario Pavia notes, less attention was paid to design and plants were increasingly imagined as detached from their territories.[57] The plants built at this time nevertheless include noteworthy spaces, architecture, and environments that can be profitably enhanced through regeneration projects. Thermoelectric plants, including oil-fueled ones, can represent heritage for the future, a sign of recent history and fossil fuel-dependent energy production. Porto Tolle, formerly one of the most important plants nationally and located in a unique and fragile environment, can become part of the petroleumscape heritage of tomorrow.

In this section, we consider how this power plant might find a new role in connection with its territory and which elements might be profitably enhanced in a process of regeneration. We also consider how the petroleumscape it belonged to could be made apparent, thereby cultivating knowledge and awareness.

One way the Porto Tolle power plant has the potential to find new meaning is in relation to its biodiverse territory. The Po River delta offers new possible relationships in terms of sustainable tourism and slow mobility. Indeed, the site could belong to a network of bike paths and boat tours exploring the delta and its nature, avifauna, aquaculture, and fishing activities as well as its historic buildings like palaces or *casoni* and the decommissioned drainage pump structure that today hosts the Museo della Bonifica di Ca' Vendramin (the drainage museum, part of a global network of water museums), but also structures like the abandoned methane wells. A regeneration project could take advantage of the unspoiled beaches, marshes, and bodies of water located within or right next to the power stations. The Enel site hosts not only the power station (the plant's surface covers about 200 ha out of 380 ha) but also a plant nursery, wetlands, a canal, a small woodland as well as areas overtaken by spontaneous vegetation. These could become part of a connected system, integrated with the surroundings. Several oil-related structures could also be enhanced and reused, avoiding a tabula rasa approach and educating visitors about the wider petroleumscape.

The petroleumscape of the power station, which ran on heavy fuel oil, includes the aforementioned 92 km-long oil pipeline that linked the plant to the Integrated Fuel and Oil Pipelines Plant depot (IICO—Impianto Integrato Combustibili e Oleodotti) in Ravenna. Fuel destined for Porto Tolle was carried to the IICO depot via an oil pipeline connected to a sea terminal, to which Panamax delivered fuel imported mainly from Libya and to a lesser extent, from other Middle Eastern countries as well as from the North Sea, Russia, Brazil, and the US.[58] These connections remind us of the global relationships behind power stations and extraction sites. In case of emergency, the oil would instead be carried by tanker trucks or barges traveling along the Po River from the power stations of Sermide and Ostiglia (formerly owned by Enel),[59] in the province of Mantua, about 120–130 km from Porto Tolle. In the past, before the pipeline had begun operating, oil was also delivered from Marghera via barges or tanker trucks (Figures 14.11–14.12). These oil infrastructure and transport systems connecting to the Porto Tolle plant are mainly out of sight, underground, under or on water, far from inhabited areas.

Despite being in a remote area, the power station has a strong physical presence. A visitor would notice the 250 m high chimney from afar, as it stands out on the horizon of the Po River delta, as well as the electric trellis and cables, both of which are material signs of the conversion from oil to electricity. Drawing nearer, one notices the large machinery hall, its four high boilers, and to a lesser extent, the oil tanks (present in two areas) (Figure 14.13). Once inside the Enel site, one sees the electric stations (Figure 14.14), depots, complementary plants (e.g., for the treatment of process waters and of wastewater), pipes, offices, an information area, the "natural" areas, and elements of the plant cooling system, particularly two canals. The water was collected both from the river (Po di Pila) via a canal and in

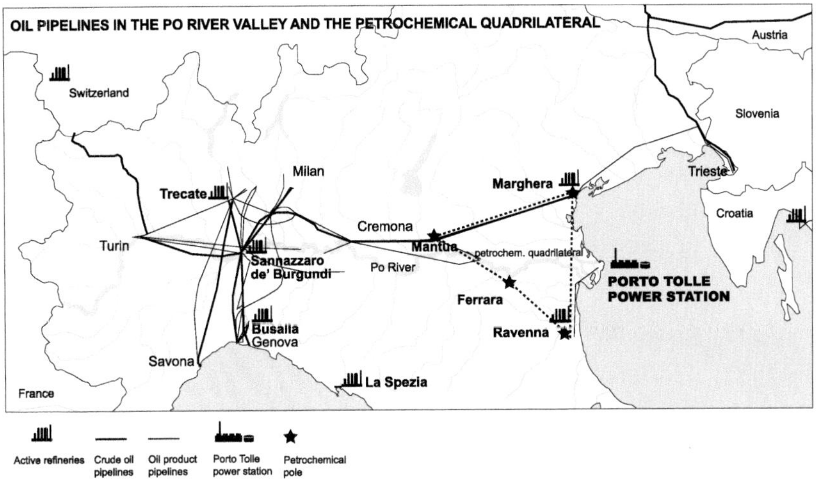

FIGURE 14.11 Oil pipelines in the Po River valley, the "petrochemical quadrilateral," and the location of the Porto Tolle power station. *Source*: Chiara Geroldi. Pipelines base map information: © PLATTS for the underlying grids for electricity, gas, and oil, 2016 © European Union, 2019.

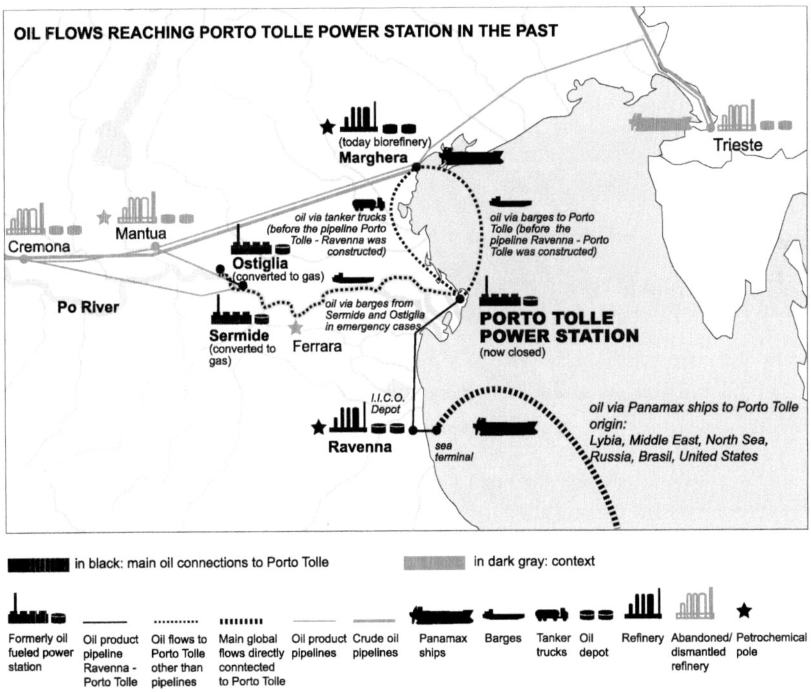

FIGURE 14.12 Oil flows reaching the Porto Tolle power station in the past and the pipeline and refinery system present in the territory. *Source*: Chiara Geroldi. Pipelines base map information: © PLATTS for the underlying grids for electricity, gas, and oil, 2016 © European Union, 2019.

FIGURE 14.13 Porto Tolle power station. *Source*: Chiara Geroldi.

FIGURE 14.14 Electric station and system of pylons extending from the power station. *Source*: Chiara Geroldi and Gloria Pessina.

periods of low water, from the sea via a canal in the Sacca del Canarin. After the cooling, water was then returned through a gate system with two canal branches leading back to its origin (river or sea). The plants are built on a raised pile foundation (3 m a.s.l.) with embankments for additional flood protection (4.5 m a.s.l.).[60]

Specific elements of the plant are part of the fuel system. The oil tanks, surrounded by earthen embankments and located at the north and south sides of the plant area (two of 50,000 m^3 and one of 100,000 m^3 in the north, six of 100,000 m^3 in the south, two of which have been dismantled), occupy, respectively, about 8 and 23 ha. The capacity of these tanks ensured steady production at the power station in case of oil delivery problems.[61] The tanks have a floating roof and external stairs to access the top and to walk on the upper rim (Figure 14.15). Beyond the tanks and other main parts of the plant (e.g., machinery hall and boilers), other elements related to the fuel system include the oil pipeline input, the dock (Figure 14.16), an

FIGURE 14.15 Oil tanks and view of the two canals used for plant cooling. *Source*: Chiara Geroldi.

FIGURE 14.16 The dock for barge unloading. *Source*: Chiara Geroldi.

alternate pier for unloading barges, a tanker truck unloading area, a pumping station, a system of aboveground oil pipes, two basins for collecting ashes, and a fire prevention system.

This inventory and description of oil-related elements and more general artifacts of thermoelectric power stations (Figure 14.17) indicates the variety of components one might find in a landscape of oil as well as the architectural/spatial value of the different elements. Such information is useful when discerning what is worth maintaining as heritage. By tracing the

FIGURE 14.17 Intricate system of tubes and pipes in the proximity of the chimney. *Source*: Chiara Geroldi and Gloria Pessina.

flows of fuel, we can better connect this site to both worldwide and more local networks of oil, that is, to the petroleumscape. This history of connectedness should be taken into consideration, conveyed, and enhanced. One possibility is to collect and display documentary evidence, photographs, and archival materials, and to construct maps of the previous flows and existing structures as well as to design informative devices or specific design installations related to this history to be located in the territory and at the plant.[62]

Our description of the oil flows and power station elements undertaken also represents a way to maintain and transmit the history of these structures and the roles they have played. The power station of Porto Tolle contains elements of the petroleumscape that could be maintained and reused—as architecture and to convey the history of the site. In particular, the concrete chimney could become a very high lookout point, and the large machinery hall and the boiler structures could easily host a variety of uses. Some of the oil tanks could be reused, for instance, for sports. The docks for unloading barges could be used by boat tours. The artificial canals constructed to take and return water represent human interventions that impacted the local ecology and will need to be reinterpreted. The embankments of the power station represent a specific part of the history of the anthropogenic operations that have constructed the modern delta. They could be enhanced as slow mobility paths, with connections between the site and the broader territory. The terrestrial and aquatic "natural" areas, the embankments, canal system, and structures could thus find new meanings and uses in a close relationship with the unique territory of the Po River delta.

Conclusion

As research consultants for Enel, when we started to analyze the territories and the architecture of ten thermoelectric power stations in Italy included in the Futur-e project, we entered an area given little attention in urban, landscape, and architecture scholarship. Focusing on the requests of the company, we initially strived to provide detailed descriptions of the plants and their contexts, in order to give adequate information to those who might respond to the project's call for ideas and manifestations of interest. In the course of the work, we realized that our studies and our observations of the plants' territories provided a helpful starting point for further research. As a result, we started to explore various theoretical approaches that might help us interpret the phenomena, the territories, the landscapes, and the architecture. Starting by addressing the topic from our previous knowledge of planetary urbanization, landscape architecture, and political ecology, we encountered the literature on the petroleumscape, which is represented in its variety and complexity in the book you are currently reading.

This chapter considers oil-fueled thermoelectric power stations as among the "global landscapes of oil." It could be easy to overlook these power stations' role in the petroleumscape, because they produced electricity. But like so much else, they were fueled by oil and would not have existed without it. The station we have been discussing in the most detail, Porto Tolle, is located in a particularly unpopulated and environmentally fragile landscape characterized by interaction between land, water, oil, and gas and by contentious processes of development and heritigization that started in the 1970s. In the course of time, the Porto Tolle power station became integrated in the network of infrastructures of the "petrochemical quadrilateral"[63] and firmly attached to the destiny of the whole territory, despite its seemingly marginal position. Hence, while reflecting on the past and on possible futures for this tiny example of the "global landscapes of oil," we have had to consider the destiny of the larger territory.

Today, the four poles of the quadrilateral lie in a state of partial abandonment and severe pollution, still characterized by the some of the activities of the main refineries (Marghera and Ravenna), in line with the claim that "old refineries rarely die."[64] The related chemical plants are also partially active, but several oil-related infrastructures have ceased their operations. Although the industrial poles of Marghera, Ravenna, Mantua, and Ferrara have been classified as contaminated Sites of National Interest,[65] eligible for special national depollution plans, the major petrochemical companies often did not comply with their obligation to decontaminate the sites[66] and managed to sell them to large companies active in the field of logistics.[67]

Ironically, such highly polluted compounds are located at the outskirts of four of the most important UNESCO sites of Italy: "Venice and its lagoon," "Ferrara, city of the Renaissance and its Po Delta," "Mantua and Sabbioneta," and "Early Christian Monuments of Ravenna."[68] Located in a less populated area, surrounded by a landscape characterized by exceptional environmental qualities, and recognized as a UNESCO Biosphere Reserve, the Porto Tolle power station allows us to observe in depth the main features of a portion of the Po River valley as part of a "palimpsestic global petroleumscape."[69] It also enables us to imagine a future in which we can rebalance the broken equilibrium between land, water, and oil and overcome outdated and harmful economic development models.

Contributions: The two authors jointly conceived the structure of this chapter and finalized the entire piece. Chiara Geroldi authored the first, third, and fifth sections and Gloria Pessina authored the second, fourth, and sixth sections. The authors wrote the introduction together. The work of Chiara Geroldi is supported by the Department of Architecture and Urban Studies (DAStU) at Politecnico di Milano, while Gloria Pessina's research is supported by the Excellence Project Fragilità Territoriali/Territorial Fragilities (2018–2022; financed by the Italian Ministry of University and Research L.232/2016) of the DAStU/Politecnico di Milano.

Notes

1 The Politecnico di Milano consultancy research group was coordinated by G. Azzone and A. Balducci. It has included, in different time periods, M. A. Bruzzese, C. Geroldi, G. Pessina, V. Galli (DAStU Dept.), M. Arena, I. Bengo, P. Colzani, L. Dell'Agostino, and S. Ratti (DIG Dept.).

2 For more information on the plants, see Chiara Geroldi and Gloria Pessina, "Le centrali elettriche coinvolte nel progetto Futur-e," *Territorio* 86, no. 3 (2018): 72–84, DOI:10.3280/TR2018-086011. For the Futur-e project, see https://corporate.enel.it/en/futur-e.
3 Some literature exists on the spatiality of hydroelectric plants and of a variety of energy production plants, for example, Rosario Pavia, ed. *Paesaggi elettrici: territori, architetture, culture* (Venezia: Enel, Marsilio, 1998); Michael Jakob, ed. *Architecture and Energy*, 2G n. 18 (Barcelona: Editorial Gustavo Gili, 2001). On thermoelectric plants, see Chiara Geroldi, "Landscapes and Architecture of Thermoelectric Power Stations in Italy," *Territorio* 86 (2018), DOI:10.3280/TR2018-086013. On the petroleumscape, see Carola Hein, "Oil Spaces: The Global Petroleumscape in the Rotterdam/The Hague Area," *Journal of Urban History* 44, no. 5 (February 2018): 887–929, DOI10.1177/0096144217752460.
4 Neil Brenner and Christian Schmid, "Planetary Urbanization," in *Urban Constellations*, ed. Matthew Gandy (Berlin: Jovis, 2012), 11–13.
5 Neil Brenner, "Urban Theory Without an Outside," in *Implosions/Explosions: Towards a Study of Planetary Urbanization* (Berlin: JOVIS, 2014), 14–35.
6 Brenner and Schmid, "Planetary Urbanization."
7 Neil Brenner, "Eight Theses on the Urbanization Question: Introducing the Urban Theory Lab/GSD," in *Instigations Engaging Architecture Landscape and the City*, eds. Mohsen Mostafavi and Peter Christensen (Baden: Lars Müller Publishers, 2012), 346–63.
8 Ibid.
9 Carola Hein, "Global Landscapes of Oil," in *New Geographies 2: Landscape of Energy*, ed. Rania Ghosn (Cambridge: Harvard University Press, 2009).
10 Hein, "Oil Spaces."
11 Matt Huber, "Energy and Social Power. From Political Ecology to the Ecology of Politics," in *The Routledge Handbook of Political Ecology*, eds. Tom Perreault, Gavin Bridge, and James MacCarthy (Oxon: Routledge, 2015).
12 Michael Watts, "Crude Politics: Life and Death on the Nigerian Oil Fields," *Niger Delta Economies of Violence Working Papers*, no. 25 (2009): 1–27; Michael Watts, "Antinomies of Community: Some Thoughts on Geography, Resources and Empire," *Transactions of the Institute of British Geographers* 29, no. 2 (2004); Michael Watts, "Oil-City: Petro-Landscapes and Sustainable Futures," in *Ecological Urbanism*, eds. Gareth Doherty and Mohsen Mostafavi (Baden: Lars Müller Publishers, 2010), 96.
13 Watts, "Oil-City," 97.
14 Karen Bakker and Gavin Bridge, "Material Worlds? Resource Geographies and the 'Matter of Nature'," *Progress in Human Geography* 30, no. 1 (2006): 5–27.
15 On the material forms of oil, see Gavin Bridge, "Past Peak Oil: Political Economy of Energy Crises," in *Global Political Ecology*, eds. Richard Peet, Paul Robbins, and Michael J. Watts (London: Routledge, 2011), 316. On extractive spaces, see Gavin Bridge, "The Hole World: Scales and Spaces of Extraction," in *New Geographies 2: Landscape of Energy*, ed. Rania Ghosn (Cambridge: Harvard University Press, 2009), 46–47.
16 Hein, "Global Landscapes of Oil."
17 Carola Hein, "Port Cities: Nodes in the Global Petroleumscape between Sea and Land," *Technosphere Magazine*, 2017 [online publication].
18 Hein, "Oil Spaces," 888.
19 Ibid., 33, 42.
20 Stu Daultrey, "The Political Ecology of Nuclear Power in Ireland or, How Nuclear Wastes," *Antipode* 12, no. 1 (June 1980): 108–11, https://doi.org/10.1111/j.1467-8330.1980.tb00310.x
21 Karen Bakker, "The Politics of Hydropower: Developing the Mekong," *Political Geography* 18, no. 2 (February 1999): 209–32; Maria Kaika, "Landscapes of Energy: Hydro-Power from Techno-Natures to Retro-Natures," in *New Geographies 2: Landscape of Energy*, ed. Rania Ghosn (Cambridge: Harvard University Press, 2009), 103–10; Michael Webber, "The Political Economy of the Three Gorges Project," *Geographical Research* 50, no. 2 (May 2012): 154–165, https://doi.org/10.1111/j.1745-5871.2011.00725.x
22 See Gloria Pessina, "Nodi da sciogliere nello spazio dei flussi. Vita nei territori delle centrali elettriche del progetto Futur-e," *Territorio*, 86 (2018), DOI:10.3280/TR2018-086014, for an initial contribution on these topics.
23 Geroldi, "Thermoelectric Power Stations."
24 Enel S.p.A. b, "The Project," https://www.enel.it/en/futur-e/a201611-the-project.html, accessed April 28, 2017.

25 Giuseppe D'Orsi, "La riqualificazione delle centrali Enel: l'esempio del Progetto Futur-e," *Energia, ambiente e innovazione*, no. 2 (April–June 2018): 30–34, DOI:10.12910/EAI2018-031.
26 Enel S.p.A. b, "The Project."
27 Pessina, "Nodi da sciogliere nello spazio dei flussi. Vita nei territori delle centrali elettriche del progetto Futur-e," 106.
28 For data and information on the four plants, see Geroldi and Pessina, "Le centrali elettriche coinvolte nel progetto Futur-e."
29 Comune di Civitavecchia, *Piano di Emergenza Comunale*, 2016, 28.
30 MWH S.p.A., "Centrale Termoelettrica di Rossano Relazione di Riferimento ai sensi del DM 272/2014 Enel Produzione S.p.A," December 2015, 10, 11.
31 Ibid.
32 For more physical descriptions, see Geroldi, "Thermoelectric Power Stations."
33 For more design reflections, see Ibid.
34 Elio Migliorini and Mario Brunetti, "Polesine," in *Enciclopedia Italiana di Scienze, Lettere e Arti* (Roma: Treccani, 1935), http://www.treccani.it/enciclopedia/polesine_%28Enciclopedia-Italiana%29/.
35 Catherine Dezio, Bruna Vendemmia, Giulia Setti, Domenico D'Uva, Fabio Lepratto, Lavinia Dondi, Nicole De Togni, et al. "Territorial Fragilities in Italy. Defining a Common Lexicon," *Territorio*, 91 (2019).
36 Giorgio Osti, "Un'economia leggera per aree fragili: criteri per la sostenibilità ambientale del Nord Italia," *Sviluppo Locale* 27 (2004–2005): 9–31; Enrico Anguillari, "Abitare territori fragili. Il caso del delta del Po descritto dagli amministratori locali," *Archivio di Studi Urbani e Regionali*, 97–98 (2010): 261–81, DOI: 10.3280/ASUR2010-097015; Marina Bertoncin, *Logiche di terra e acqua. Le geografie incerte del delta del Po* (Verona: Cierre, 2004).
37 Tiziana Quaglia, Francesco Musco, and Graziano Caramori, eds. *Aree protette e sviluppo sostenibile. Un piano per la conservazione della biodiversità e lo sviluppo dell'economia locale nel delta del Po* (Venezia: Regione Veneto, 2011): 29–30.
38 Pietro Caloi, "How Nature Reacts to Human Interventions. Responsibilities of Those Who Cause and Who Interpret Such Reactions," *Gli Annali di Geofisica*, Istituto Nazionale di Geofisica e Vulcanologia, no. 23 (1970): 283–305.
39 Caloi, "How Nature Reacts."
40 Achille Ardigò, "Il Polesine, un'area di fuga nell'Italia settentrionale," in *Il Polesine*, ed. Centro Nazionale di Prevenzione e Difesa Sociale (Milano: Edizioni di Comunità, 1964), 97–142.
41 Ministro dell'Interno, "Decreto Costitutivo del Consorzio per lo sviluppo economico e sociale del Polesine," n. 2050.15.100.128, February 19, 1963; Consiglio Regionale del Veneto, "Interventi a favore dei consorzi tra enti locali per la realizzazione dei piani di cui all'art. 27 della legge 22 ottobre 1971 n. 865," *Regional Law*, April 22, 1977, n. 3 (BUR n.18/1977).
42 Diego Marazza, "Il quadrilatero del petrolchimico in Pianura Padana," *Ecoscienza* 4 (2016): 58.
43 Vittorio Marchetti, "Città e campagna. Venezia, Marghera e il Delta del Po," video documentary, 43:33 (1976), http://www.teche.rai.it/1976/03/citta-e-campagna-venezia-marghera-e-il-delta-del-po/.
44 Marchetti, "Città e campagna."
45 Quaglia, Musco, and Caramori, *Aree protette*, 19.
46 PPRA, *Relazione Illustrativa. Piano Paesaggistico Regionale d'Ambito Arco Costiero Adriatico Laguna di Venezia Delta del Po* (Venezia: Regione Veneto e MiBACT, 2015): 131.
47 Roberto Bianchin, "Si fermerà tra una settimana la centrale delle polemiche," *La Repubblica*, October 18, 1984, http://ricerca.repubblica.it/repubblica/archivio/repubblica/1984/10/18/si-fermera-tra-una-settimana-la-centrale.html.
48 "Polemiche a Ravenna sulla raffineria AGIP," *La Repubblica* (February 27, 1985), http://ricerca.repubblica.it/repubblica/archivio/repubblica/1985/02/27/polemiche-ravenna-sulla-raffineria-agip.html.
49 Quaglia, Musco, and Caromori, *Aree protette*, 20.
50 Giovanni Carrosio, "Ingiustizia ambientale nel bacino del Po: il conflitto tra il Polesine e la città di Milano per l'inquinamento delle acque," *Partecipazione e conflitto* 6, no. 1 (2013), https://doi.org/10.3280/PACO2013-001005.
51 Giorgio Osti, "Dalla terra al mare. Esperti e istituzioni nella realizzazione di una grande infrastruttura Energetica," in *I conflitti ambientali. Esperti, politica, istituzioni nelle controversie ecologiche*, ed. Luigi Pellizzoni (Bologna: il Mulino, 2011).
52 Osti, "Dalla terra al mare."

53 Paola Zambon, ed., "Analisi dell'incidenza dei tumori nell'ASL 19 di Adria," Istituto Oncologico Veneto, IRCCS Registro Tumori del Veneto (Padova: Regione del Veneto, 2008).
54 Alexander H. Bell, "Il processo alla centrale termoelettrica di Porto Tolle: gli ex Amministratori Delegati di Enel condannati per pericolo di disastro sanitario," Tribunale di Rovigo, March 31, 2014, https://archiviodpc.dirittopenaleuomo.org/d/3355-il-processo-alla-centrale-termoelettrica-di-porto-tolle-gli-ex-amministratori-delegati-di-enel-cond; Giuseppe Fornari and Enrico di Fiorino, "Emissioni nocive dalla centrale di Porto Tolle: la cassazione mette la parola fine," Cassazione Penale, Sez. I, January 19, 2018 (ud. January 10, 2018), n. 2209, https://www.giurisprudenzapenale.com/2018/04/04/emissioni-nocive-dalla-centrale-di-porto-tolle-la-cassazione-mette-la-parola-fine/.
55 https://www.agenziacoesione.gov.it/lacoesione/le-politiche-di-coesione-in-italia-2014-2020/strategie-delle-politiche-di-coesione/strategia-nazionale-per-le-aree-interne/.
56 http://contrattodifocedeltadelpo.it.
57 Pavia, *Paesaggi elettrici*.
58 Enel S.p.A., written communication with the authors, May 4, 2017.
59 Ibid.
60 Enel S.p.A. a. *Centrale Termoelettrica di Porto Tolle*, printed booklet retrieved at the Porto Tolle plant.
61 Ibid.
62 For similar arguments in a context of a coal-fuel power station see Geroldi, "Landscapes and Architecture of Thermoelectric Power Stations in Italy."
63 Marazza, "Il quadrilatero del petrolchimico in Pianura Padana," 58–59.
64 Carola Hein, "'Old Refineries Rarely Die': Port City Refineries as Key Nodes in the Global Petroleumscape," *Canadian Journal of History/Annales canadiennes d'histoire*, 53, no. 3 (2018): 450–79, https://doi.org/10.3138/cjh.ach.53.3.05.
65 "Contaminated sites of national interest (SIN)," ISPRA, n.d., https://www.isprambiente.gov.it/en/activities/soil-and-territory/copy_of_contaminated-sites/contaminated-sites-of-national-interest-sin.
66 Commissione Parlamentare d'Inchiesta sulle attività illecite connesse al ciclo di rifiuti e su illeciti ambientali ad esse correlati, "Relazione sulla situazione delle bonifiche dei poli chimici: il «Quadrilatero del Nord» (Venezia-Porto Marghera, Mantova, Ferrara, Ravenna)," Doc. XXIII, n. 11, Camera dei Deputati, Senato della Repubblica Italiana.
67 Gilda Zazzara, "I cento anni di Porto Marghera (1917–2017)".
68 UNESCO – World Heritage Centre, "Properties inscribed in World Heritage List," http://whc.unesco.org/en/statesparties/IT.
69 Hein, "Palimpsestic Petroleumscape."

15

PETROLEUMSCAPE AS HERITAGE LANDSCAPE

The Case of the Dunkirk Port City Region

Carola Hein, Christine Stroobandt, and Stephan Hauser

In 2009, Jean-Francois Vereecke, an economist with the Urbanism Agency of Flanders-Dunkirk (Agence d'urbanisme de la région Flandre-Dunkerque [AGUR]), developed the Toile Industrielle (Industrial Canvas®), a visualization of the interindustrial network of Dunkirk, an industrial city in northern France (Figure 15.1).[1] The illustration situated the economic network of Dunkirk in the context of international markets, subcontractors, and competing ports. The accompanying analysis revealed the economic impact of the proposed closure of the Total refinery, particularly on the port of Dunkirk.[2] This helped local actors negotiate with the national government, which agreed to the construction of a liquefied natural gas (LNG) terminal in the city to strengthen the local industrial ecosystem. Furthermore, in 2016, when the Total refinery finally closed most of its activities in Dunkirk, the company kept part of the refinery running as a training center called "Oleum"—the first in Europe—therefore retaining some of the local oil-related knowledge in the city.[3]

While the Toile Industrielle visualizes the economic presence of oil in Dunkirk, it does not capture oil's spatial impact—inscribed in the pollution of its soils—or its impact on the narrative of the city. There is a need to further explore the physical reality of a place where the petroleum economy is coming (or must come) to an end and where the oil industry has vacated large sites, upsetting the long-standing economic, spatial, and social structure of the city. Since the 1860s, the life and death of petroleum sites in port cities have occurred with a lack of complete information regarding the places and dates where related activities took place. This legacy has left patches of pollution in port cities, knowledge about their existence is sometimes lost, and most of them have never been cleaned up. This chapter explores the petroleum history of Dunkirk and the importance of the role of port city regions in the establishment, continuity, and imaginary of the petroleumscape. It reveals the challenges involved in addressing the heritage embedded in former petroleum sites that have not been fully documented, even though they form part of the urban and social fabric and the imaginary of generations of inhabitants.

The history of Dunkirk is a prime example of how private and public oil interests have shaped port city regions for more than 150 years and have led to the population's dependence

DOI: 10.4324/9780367816049-19

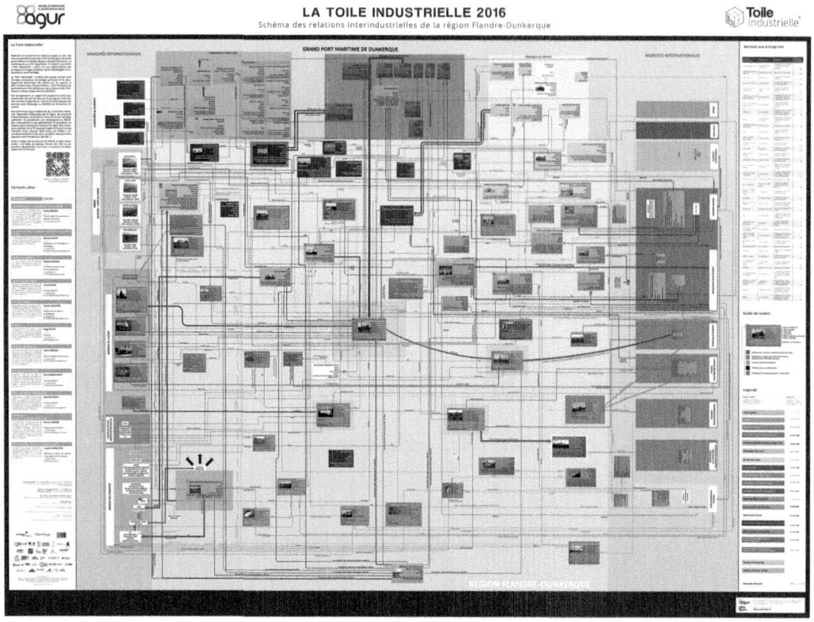

FIGURE 15.1 The Toile Industrielle in 2016. *Source*: Jean-François Vereecke. Source: https://fr.wikipedia.org/wiki/Toile_industrielle#/media/Fichier:AGUR_Toile_industrielle_2016.jpg

on petroleum as a provider of jobs and a creator of spaces. Global and local actors have built industrial sites—refineries, storage tanks, and port facilities—and other oil-related sites, such as administrative, housing, and leisure facilities throughout the city. Petroleum shaped the city and its structures remain, even if the petroleum has largely left. The closure of oil facilities, including the refinery, requires acknowledgment not only of the economic but also of the spatial impact of the end of the petroleumscape.[4] With the refinery closing, Dunkirk has moved to the forefront in the development of post-oil economies and landscapes. Planning for the future of Dunkirk demonstrates the challenges and opportunities of transforming an area that has long been dedicated to the transshipment and refining of oil in line with new energy practices and the social and spatial conditions they require. The transformation is not limited to industrial spaces, as exemplified by the demolition of the Cité des Ingénieurs in 2018, a historic housing district for oil company employees designed by Max Wenders in 1931.

Through the lens of Dunkirk's spatial transformation, and particularly the impact of two major oil companies—BP and Total— on the city's spatial development, government, and culture, this chapter considers how various actors in Dunkirk's history have adapted to changing economic and spatial conditions. What kinds of oil spaces have they created and through which means? The chapter also analyzes recent design proposals for a post-oil Dunkirk that were inspired by the concept of the petroleumscape. Understanding the long-term impact of oil on governance and the built environment in Dunkirk can help us prepare for the transformation of other (port) cities in northern Europe and around the world.

The Role of Ports in the Global Petroleumscape

Petroleum is a natural product that was used sparingly for thousands of years as it bubbled out of the ground; it only gained its current ubiquity with industrial drilling, refining, and scientific innovations that increased the product's usability (see Hein, introduction to this volume). Throughout the nineteenth century, inventors, businessmen, and chemists worked to create an efficient lamp and to refine petroleum (replacing whale oil, which was expensive, and other oils that could be used as food).[5] In the twentieth century, improvements in refining technology and chemistry provided the foundation for transforming petroleum into a rapidly increasing number of new products, including well-known ones, such as vinyl flooring, paint, and plastics, and also many that are less familiar, including petroleum-based fibers such as nylon, acrylic, or polyester as well as microplastics like microbeads used in toothpaste and skincare products. Oil companies, through their refineries, have provided an endless number of products to suit particular needs and interests.

With the rapid rise of petroleum consumption in places where few oil sources existed, places of oil extraction, refining, administration, and retail had to be connected around the world, including across seas. Producers located storage sites and refineries along the flows of oil. In the early years of petroleum production in Pennsylvania, Baku, and Gallicia—to name just a few early production sites—producers located refineries and storage sites close to extraction sites. It took several years for shipping and trading companies to find locations for storage and refining facilities that could handle the specific environmental and fire hazards of petroleum.[6] Private entrepreneurs ran much of the business in the mid-nineteenth century, each controlling a part of the industry on their own land. Over time, local traders were pushed out of the market by a handful of transnational oil companies, including Exxon Mobil, BP, and Shell, which emerged in the late nineteenth century. As nation-states came to realize the importance of petroleum as a fuel, including for military purposes, they collaborated with petroleum companies in the discovery, extraction, and resale of petroleum. Over time, ports, cities and their regions began to thrive on the global petroleum trade as shipping linked diverse places of production and consumption around the world and as port cities emerged as transportation hubs and sites of petroleum refining.

Port cities were important as both export and import hubs. The industrial nations of Western Europe had only a small number of local petroleum sources and depended on imports. Port cities on the North Sea, including Dunkirk, Antwerp, Rotterdam, Hamburg, and Wilhelmshaven, became nodes in the oil distribution system, first for lighting oil, then for motor fuel, and later for oil used in manufacturing other products. The story of petroleum refineries in European ports really started in the 1860s, when small companies began to import petroleum from the US. The first barrels of oil were transported by sailing ships, arriving in ports that were equipped to store and transship a range of other, less dangerous, goods. Soon afterwards, companies built specialized storage sites and refineries along the northwest European coast to hold and process imported petroleum. As national entrance gates for petroleum flows from abroad—including from colonies or former colonies—port cities became an important asset for their respective governments. They served as access points of petroleum storage and refining, supplying crude and refined products to the hinterland, including the large inland capitals. As a country with limited petroleum within its borders and in its colonies, France, like the UK, became a big petroleum importer in the mid-nineteenth century and developed its own petroleum policy through the creation of the company Total. Dunkirk was one of the main hubs that kept petroleum flowing in France. Its urban form has been marked by oil transport and transformation, by the appearance and disappearance of petroleum sites, with its soil retaining a 150-year legacy of pollution.

Dunkirk in the French Oil System

Like most of Western Europe, Dunkirk's industrial petroleumscape was largely shaped by private business with political support by both local and national institutions. Occasionally, the same individuals represented private business and local or national institutions. Entrepreneurs were among the first to establish a petroleum refinery on the northwest European coast. In Dunkirk, a refinery may even have preceded the arrival of foreign petroleum. In 1861, the Raffinerie Trystram et Crujeot, in Petite-Synthe, opened along the canal of Mardyck (today part of Dunkirk). The first documented arrival of petroleum from Pennsylvania did not take place until 1863.[7] The early construction of the refinery may have given Dunkirk a head start in the petroleum business.

The further development of refineries was the result of political interventions. In 1863, the French government issued a tax decree—the first appearance of oil in the French customs tariff system[8]—that put taxes on the import of refined products to ensure that the refining of petroleum took place on French soil. To avoid these taxes, oil companies opted to build refineries in France and erected the necessary infrastructures for its transport. In 1874, the Raffinerie Clère et Boilet opened in Coudekerque-Branche, south of Dunkirk, with a river connection to the harbor, and a few years later, the refinery received dedicated access to the newly built railway station (financed in large part by Clère).[9] This refinery was also refining American crude oil, which it then sent via canal and railway to the hinterland, particularly to Paris.

This and other petroleum installations appeared in the city with little attention to their environmental impact or the fire hazard they posed (Figure 15.2). The refinery in Coudekerque-Branche burned down in 1888 and again in 1891, the second event receiving much media attention, and perhaps even a third time.[10] Nonetheless, the facility was again rebuilt on the same site and only closed in 1906. But refineries are rarely replaced once established. Without a strong incentive, it is more efficient for companies to keep an old site and benefit from the existing infrastructural and economic networks than to demolish a retired refinery. Because refineries are tied to a specific physical location and local infrastructure, they are hard to move; and because they are hard to move, in time they become even more tightly linked to their location. As the international professional services company Ernst and Young put it: "Old refineries rarely die."[11]

Starting in the nineteenth century, economic and political interests surrounding oil were closely connected. A number of petroleum companies emerged in France after the 1850s; they would go on to become the core of the French petroleum trade and their owners would wield significant political power. Edmond Paix established his refinery in 1863 in Saint-Amand near Douai; Henri Desmarais set up shop in 1878 in Le Havre; Alexander Deutsch (de la Meurthe) established his first industrial oil facility in 1862 in Pantin; and Fenaille, Châtillon et Despeaux began its petroleum refining in 1868 in Aubervilliers. Desmarais frères, Deutsch de la Meurthe, and Fenaille et Despeaux alone refined about 90 percent of French oil imports in 1895. Seven other companies started importing after 1895: G. Lesieur et fils, Compagnie générale des pétroles, Compagnie Industielle des pétroles,

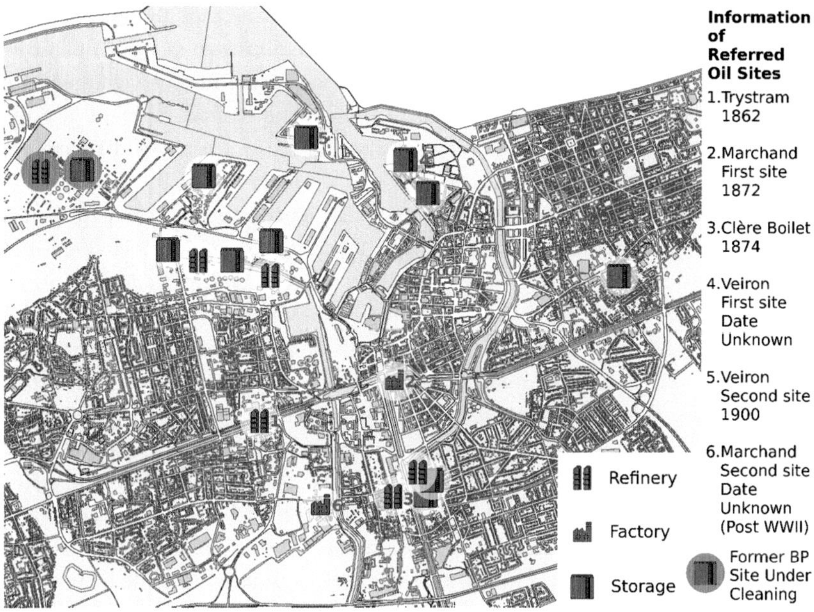

FIGURE 15.2 Oil-related sites (refineries, storage sites, and factories in Dunkirk). Map by Stephan Hauser, based on recent geographic information about the city provided by the Learning Center of Dunkirk.

Raffinerie du Midi, Société Lille-Bonnières-Colombes, Raffinerie de pétrole du Nord, and Paix et Cie.[12] All these companies, which became known as the Cartel of Ten, bought petroleum from Standard Oil of New Jersey (Rockefeller), giving the American company lobbying power with the French state. In 1903, the Parliament (Chambre des Députés) pushed the government to propose a law giving the French state a monopoly over the refining of petroleum.[13] As the French government imposed taxes on the import of refined products in 1863, refining in France became more lucrative. In 1903, the government introduced another tax, this time on crude imports reaching French refineries, to balance what it considered an excessive profit from refining petroleum companies.[14] Taken up again in 1909, the new taxation project remained without effect as Standard Oil threatened to cease oil deliveries if its economic interests were challenged—at least according to right-wing sources.[15] Already in the days of lighting oil, before cars became major petroleum consumers, the predecessor of Exxon Mobil had gained enough global power to sway national politics and decision-making.

Other private companies connected to national projects formed in those years would similarly lay claim to entire parts of the globe, strategically deploying their national hinterland or colonial reach as well as the political and economic alliances of their owners. The Russian connection, for example, was strong in Dunkirk. In 1891, on Dock 4 of the Dunkirk port, the André oil company built storage for mineral oils arriving from Baku in Azerbaijian and from Batumi in Georgia on the Black Sea, selling them for Branobel, run by brothers Robert and Ludvig Nobel (whose brother Alfred Nobel founded the Nobel Prize). Branobel collaborated with Shell and with Mazout (a daughter company of Bnito owned by the Rothschild family) to keep Standard Oil, an American company, out of the Russian markets. In 1912, the company G. M. Lianozov Sons (also known as G. M. Lianosoff Fils), the third-largest financial and industrial group in Russia, administered by Stepan Lianozov (or Lianosoff)—called the Russian Rockefeller—joined the Russian General Oil Corporation, a group of over twenty investment banks and French, Russian, and Belgian funds as well as Russian oil industry leaders.

The new oil industry reshaped the industrial petroleumscape in port cities throughout Europe and beyond as oil storage and refining territories expanded. In 1912, Lianosoff opened an industrial site in France's St. Pol s/Mer for storing and packaging Russian lubricants and other oil products that it then transported elsewhere. The same year, he founded the company of G. M. Lianozov Sons with branches in London, Antwerp, and Paris.[16] These branches served as entrance points for the development of European sales of oil in competition with the Nobel and Rothschild companies until 1916, when Emmanuel Nobel, Ludvig Nobel's eldest son, took over G. M. Lianozov. After the 1917 Russian Revolution cut off the Nobel procurements, the André Company started importing oil from the US and the Dutch Indies. In 1920, Branobel was nationalized, but the Nobels managed to sell 50 percent of their stock to the Standard Oil Company of New Jersey. Together, Lianozov and Emmanuel Nobel participated in anti-Bolshevik activities with Russian immigrants in Paris in 1920 (they were trying to defend their ownership of extraction and production sites).[17] Private petroleum actors had become unable to efficiently push the government to protect their interests.

With the emergence of the car and other vehicles with oil-burning engines (including ships) and increasing petroleum consumption in the early 1900s, energy consumption in Europe increased. The major American, Russian, English, and Dutch companies started to compete for access to European markets, and thus for land and influence in the main port cities. Some local companies expanded, and petroleum-related industries formed new organizational clusters. By 1908, Georges Lesieur had separated from Desmarais and took over the Dunkirk Oil Refinery in exchange for land ceded to the previous owners: Fenaille and Despeaux, Paix, Desmarais, Deutsch, who had established the Raffinerie de Pétrole de Dunkerque in 1896.[18] In 1909, Dunkirk's Chamber of Commerce granted permission for a pipeline to bring refined oil from Dock 4 in Dunkirk to a petroleum refinery (Raffinerie de Dunkerque [ex-Trystram]) and warehouse in Petite-Synthe and to build storage.[19] Many of petroleum's dangers were known but mostly ignored by decision-makers.

While the early years of petroleum were characterized by private entrepreneurs, the business of oil developed quickly into a matter of national security for countries around the world, including France. Petroleum extraction in colonies was key to the development of many European countries. World War I made it clear that national dependence on foreign products was not a long-term option, and within a few years a French national company would emerge.[20] In the 1920s, the French state took control of oil importing and took over former German installations in the Middle East. The role of port cities as nodes in the transportation and refining of petroleum became even more important. Between 1920 and 1930, another seven industrial petroleum sites opened, inserting Dunkirk and other French ports into larger petroleum networks. Georges Lesieur—a partner in the Cartel of Ten—and Paix established the Compagnie Occidentale des Produits du Pétrole (COPP), built storage in the Dunkirk port, and developed the Courchelettes refinery to refine petroleum from Abadan, a refinery and oil city in southern Iran.[21] In 1920 and 1921, Paix-Lesieur[22] signed an agreement with the Anglo Persian (later Iranian) Oil Company (APOC; a predecessor to BP).

Then, in 1920, five more entities—APOC, SNO (Société Navale de l'Ouest), La Banque de la Seine, and Sir Basil Zaharoff—established a refining company with headquarters in Paris, the Société Générale des Huiles de Pétrole (SGHP), and relaunched the activities of the Courchelettes refinery.[23] The refineries in Dunkirk were thus within national control and firmly embedded in international petroleum flows.

The city government of Dunkirk began to promote its growing role as distribution and production center. At the time, the infrastructure of oil was a celebrated part of the public imaginary and perceived as a sign of economic growth. The Chamber of Commerce in Dunkirk found the changes brought by oil so important that it commissioned Hugo d'Alesi to document them in a monumental painting for the World Exhibition in Paris (Figure 15.3). The painting shows Dunkirk from the waterside with the various docks and storage facilities. In the front, to the right of the entrance channel, the petroleum port was closed off by a floating barrier when a petroleum ship was in port. Another monumental painting made between 1923 and 1930 appears to have been inspired even more strongly by the pride in oil storage and handling. Viewing the scene from the sea, this anonymous painting again puts port activity at center stage (Figure 15.4). Former fortifications had been transformed into a green belt that appears ready to be populated with (industrial) activities: multiple oil storage tanks visible on both sides of the river give a vision of the port's industrial future after World War I. The painting shows the tanks lit by the sun, and the new structures are gleaming white in the front right of the image. This imagined future would come almost immediately to Dunkirk with the installation of large new refineries and storage sites. Dunkirk's Chamber of Commerce may have commissioned the painting; they displayed it at industrial and commercial exhibitions in France. Petroleum promised a better life and was not yet seen as an environmental threat or a cause of war.

France was keen on controlling petroleum for national purposes, and the ports played an important role in plans to achieve this goal. In order to make France less dependent on foreign petroleum companies, in 1924, French Prime Minister Raymond Poincaré created the Compagnie Française des Pétroles (CFP), the first national oil company, a predecessor

FIGURE 15.3 Panorama of Dunkirk port around 1900, painted by Hugo d'Alesi. Collection Musée portuaire Dunkerque. Cliché Studio Mallevaey.

FIGURE 15.4 View of the City of Dunkirk around 1923. Collection of the Port Museum of Dunkirk.

of Total. Before World War II, France and Italy were the only countries that encouraged companies to site refineries within their territories so that they no longer depended on foreign deliveries of refined products.[24] This company opened up new markets and sources (including in France), and started to import crude oil from Mesopotamia and Venezuela. The new oil company chose to route the oil to Paris by way of Le Havre and Marseilles.[25] In Dunkirk, foreign countries, including the UK, had a major stake in the oil business. Foreign and French companies collaborated to buy up existing petroleum sites in Dunkirk to process their foreign-purchased crude oil. In 1920 and 1921, the future BP (then Anglo-Iranian Oil Company through the French company SGHP) took control of the refineries in Dunkirk (Lesieur) and Courchelettes (Paix Cie) and their depots (Cie Occidentale des Produits Pétroliers). In 1929, the French state and investors (including Desmarais frères and Lille Bonnières et Colombes et ses filiales) helped establish the refining company CFR (Compagnie Française de Raffinage).[26] PétroFina France then built the new Raffinerie de Pétrole du Nord in Dunkirk in 1932, on the site of what would become the BP refinery.[27] All these intricate economic maneuvers ultimately had an impact on the spatial development of Dunkirk, on the industrial, infrastructural, or ancillary spaces of the local petroleumscape.

Like in many other port cities, petroleum sites in Dunkirk have been the target of military action, but they have also been rapidly rebuilt. In May of 1940, the refinery in Dunkirk was bombed and dismantled by the Germans. By the end of World War II, PetroFina lacked funding to rebuild and struck an agreement with SGHP-BP, the French arm of BP, in exchange for petroleum delivery, at a time when companies were hungry for access to crude oil. In 1947, the Monnet Plan (1947–1953) aimed at modernizing and developing basic and strategic industries (energy, steel, transport) as a way of relaunching economic activity and taking advantage of the Marshall Plan, which allowed SGHP-BP to rebuild the refinery in Dunkirk in the location occupied by the Raffinerie de Pétrole du Nord (and later Purfina) before the war. In 1948, the SGHP-BP started to build the new refinery on the western side of the port of Dunkirk in the area of Saint-Pol-sur Mer.[28] In 1950, the refinery at Courchelettes was dismantled and moved to Dunkirk with a capacity of 350,000 tons per year. At the end of 1951, the first part of the refinery was built; in 1952, the refinery was completed and began operating.[29] In agreement with the French government, British

Petroleum PLC rebuilt and expanded the site of the former Raffinerie de Pétrole du Nord, which took the name of SGHP-BP in 1952, SF-BP in 1954, and later became BP.

The presence of the foreign petroleum companies in Dunkirk extended from the industrial petroleumscape to ancillary elements. The BP refinery, described by the French writer André Maurois as "the factory on the dunes," is an important part of the history of Dunkirk and exemplifies the economic strength of the region and the nation.[30] In the 1950s, BP added storage of 500,000 m^3 capacity and transported oil by rail, tank truck, and through the canals to the hinterland.[31] To facilitate oil production in the war-destroyed city of Dunkirk, SGHP-BP needed housing for its employees. The company rebuilt the housing known as the "Cité des Ingénieurs," which was located next to the refinery (Figure 15.5). The neighborhood consisted of twenty-three houses with accommodations for the refinery's director, engineers, and foremen. With the housing development and its park-like setting, the company expressed both the early twentieth-century garden city concept and the 1930s modernist idea of connecting places of working and living. There was also a so-called "Cercle," a place for parties that also served as a restaurant. Public spaces were provided where families could meet and children could play.

The Cité des Ingénieurs exemplifies the way petroleum was perceived as an agent of welfare and a better life. In addition to the Cité des Ingénieurs, additional housing was needed for workers. The Cité Bayard in Saint-Pol-sur-Mer consisted of 160 houses and 333 apartments 2 km from the refinery. The BP also built twelve apartments in Rosendaël. Furthermore, the BP helped employees to build houses in Petite-Synthe, in an attempt to demonstrate that the company cared about the welfare of its employees.[28] The lack of equipment and available houses pushed employees to also build their own homes. This movement of collaborative auto-construction called Castors was supported by funds from the refinery in the city of Dunkirk. Eventually, 200 houses were constructed.

FIGURE 15.5 Raffinerie de Pétrole de Dunkerque, 1955. Archives de Dunkerque CMUA, 14FI63, copyright Lapie.

Even in the early postwar period, petroleum appeared to companies, workers, and citizens alike as a unique opportunity rather than an environmental challenge. It would take several decades for the environmental and health hazards of the emerging technology to become evident. In the 1930s, when the Cité des Ingénieurs was built next to the refinery, it was considered a desirable place to live. Equipped with all the modern comfort available, the houses were, however, within the industrial territory of the port of Dunkirk. In the 1970s, the Cité des Ingénieurs began to empty, especially after the construction of the steel site USINOR, less than a kilometer away. The 1976 chemical incident of Seveso, in Italy, triggered a European classification of dangerous industries, with distance required between industry sites and residences. Refining sites were included in the dangerous classification and became subject to the distance requirement. In addition to the risk of accidents, everyday life for the inhabitants of the Cité des Ingénieurs included noise, odor, and pollution. Incidents occurred regularly in the refinery and the increasing pollution pushed the residents to leave. By 1995, it was a decaying neighborhood in the industrial port, deserted and walled. Instead of relocating the petroleum industry sites, it was the residents who had to move.

Extension of the Port to Anchor the Petroleumscape

The port and region of Dunkirk, like many other port city regions, thrived on petroleum into the late twentieth century. Accommodating it led local development. The refining activity in Dunkirk attracted the Pechiney aluminum industry, and the establishment of the USINOR steel industry in 1963 indicated the French government's commitment to create a coherent policy for the management of ports of national interest. In 1965, a law made ports autonomous—rather than controlled by the French state—as public institutions with financial autonomy.[32] In 1966, the Port Authority of Dunkirk was established to improve the efficiency of the port and facilitate the settlements of economic actors. By the end of the 1960s, the port of Dunkirk was the third-most important French port in terms of trade—the major items being steel, petroleum, and textiles.[33] Through the inclusion of local actors in the decision-making of the port authority, extensions and management of the port received additional support. As local authorities gained voices in port's strategy, more and more industries settled in Dunkirk and the expanding port redefined the industrial landscape of the city. The increase in shipping traffic and the size of ships in these years entailed significant changes, including moving the port away from the city. Deeper water allowed the berthing of container ships and oil tankers of 200,000 tons. The oil trade, which increased after 1972, when imports from North Sea exploration rose to a level similar to those from the Middle East, needed more space and deeper water.[34] The BP refinery, then the only one in operation in Dunkirk, and the refineries in other parts of France were not able to satisfy domestic demand, so imports of refined oil from the Netherlands, Belgium, and Great Britain increased considerably.[35]

The oil economy and its industries remained strong in the 1970s. In 1974, the French oil company Total opened the new Flanders refinery in the municipality of Mardyck, the industrial port area of Dunkirk, close to existing heavy steel and metallurgical and petrochemical industries (including Copenor from Qatar). The new refinery, the last to be built in France, then occupied 230 ha and was located east of the port of Gravelines, along the grand canal Dunkirk-Denain. Pipelines connect it to the Flanders jetty, built in 1975 in the new western outer port, 8 km from Gravelines. The latter allows the berthing of oil tankers of 300,000 tons. Seeing its market share decline, the authorities of the BP refinery located in the older part of the industrial port with limited access capacity to 33,000 tons ships, abandoned the manufacture of fuels, and in 1981, turned to the production of lubricants. As early as 1976, it had benefited from the possibility of installing new storage tanks next to those of the Raffinerie des Flandres and using pipelines to connect the tanks to its production site.[36] With such adjustments, until the 1980s, the oil industry in Dunkirk was able to continue making a positive impact on the urban economy and provided good housing, leaving an image that today many locals recall fondly.

In the 1980s, the tide turned for the oil industry and its unchecked growth came to an end. The 1979 oil crisis and the increasing price of oil had a strong impact on the rivalry between BP and Total. BP finally ended its oil refining operations in 1982 and dedicated its refinery exclusively to the refining of lubricants, waxes, and bitumen. In 2010, the refineries of Flanders (Dunkirk), Reichstett, Berre, and Petit-Couronne began to close, leaving behind all at once large sites to clean and transform in cities and leading to an abandonment of the old housing complex (Figure 15.6). The commercial activities of Total's refinery ceased in 2016, but the oil industry in Dunkirk still has an impact on the area through the refinery training center under the name Oleum.[37] Together with Le Havre and Marseille, Dunkirk is the only port in France that can accommodate the largest container ships and large oil tankers. This is one of the reasons why the port authority intends to continue using the port as a hub for oil and gas transport.[38]

Today, the former Cité des Ingénieurs is a polluted and abandoned brownfield site in Dunkirk, exemplifying the relationships between global and local players and the impact of

FIGURE 15.6 Abandoned building of the Comité d'Entreprise of the Société de la Raffinerie de Dunkerque. *Source*: Carola Hein.

overcapacity in the petroleum sector on local urban development.[39] It raises the question of how to reshape the spatial petroleumscape, the growth of which has shaped cities for some 150 years, and what message to send to future citizens about the oil heritage, which is an important part of the represented petroleumscape. In addition to providing educational training sites for the education of oil workers, Dunkirk now contains an experimental area for biofuels. Some sites are being dismantled and decontaminated and plans are being considered for an asphalt warehouse. Dunkirk now faces the challenge of reinventing itself by repairing and recycling its port, its port industries, its port heritage buildings, and the abandoned and unused areas between the port and the city. There are also opportunities to provide examples of experiments with transitions in energy, technology, and society and their spatial impact. All stakeholders, public and private, professional and lay, have an opportunity to consider the issues of the future of former oil-based industrial cities.[40]

With the support of the metropolitan area of Dunkirk, the Hauts-de-France region, and European funds and programs, several actors have begun urban and architectural renewal. Since 2010, the Port Authority of Dunkirk has implemented a comprehensive strategy for protecting its natural heritage, a Natural Heritage Master Plan (Schéma Directeur du Patrimoine Naturel [SDPN]).[41] This tool has given rise to a series of initiatives to conserve and enhance nature within the port, such as one developing the former St Georges rail line between Bourbourg and Gravelines as an eco-landscape corridor. In addition, the port authority has opened the port to the city through visits, projects, and activities in schools. In addition, several actors aim to connect port and city. The Learning Center Halle aux Sucres educates the local community on issues of sustainability. A new Port Center aims to communicate with the citizens and to encourage collaboration between the port authority, the city of Dunkirk, and the port museum. These activities made Dunkirk one of the five finalists for the European Sea Port Organization Award in 2016 with the theme "Nature in Ports," for ports preserving natural heritage and connecting ports with the public.[42]

Spatial and Social Development Beyond Oil

In France and elsewhere, port cities like Dunkirk were shaped by global industry and national interests. The end of refining spells uncertainty for the future of flows through the port and for the landscape. Since the refinery ceased its commercial activities in 2016, the petroleum areas have become places for the development of new projects, visions, and practices. One important concern is that the oil industry not only affected the spatial planning of the area but also polluted the soil. Industry locations were often not recorded, and the short lifespan of many facilities presents problems for both authorities and citizens.[43] Successive public authorities transformed these former industrial landscapes into urban areas, now with schools, housing, and parks. Many abandoned or transformed sites may still contain invisible pollution in the soil. This is a potential environmental and health catastrophe. Regular incidents around these industrial sites impacted air quality and are still affecting residents with potential limitations on planting vegetable gardens and use of surface water. In the case of Dunkirk, most of these sites are also in flood zones, which increase the risk of pollution spreading to a wider area. These soils need to be identified, studied, and the hazards addressed to avoid endangering the health of the inhabitants and users—and to fully address the heritage of oil. The narrative of the petroleum heritage of Dunkirk in all its facets—positive and negative—needs to be carefully studied and discussed as part of the green energy transition.

The Learning Center's activities are examples of efforts to educate citizens on the impact of oil on the built environment and the challenges and opportunities of overcoming petroleum dependency. Exhibits have included "Or Noir" (October 2018 to June 2019), which explored spatial approaches for Dunkirk beyond oil.[44] Developed by students from Delft University of Technology, it considered the future of post-oil cities, focusing on Dunkirk and its mix of hazardous industries, refineries, huge steel plants, petroleum tanks, and other industrial structures. Students developed approaches and methods based on an analytical reading of maps, historical documents, and materials from the archives in Dunkirk. They then translated their research into individual approaches, and created visual representations of their findings as well as new projects for the future of the former refinery sites.[45] Students considered the cultural meanings of oil in a post-oil world and how to provide opportunities for future generations without relying heavily on oil or other fossil fuels. *What does post-oil mean?* With this question in mind, they reflected on how to reshape the relationships between global and local, private and public, economy and ecology, people and nature, new and old cultures, energy and infrastructure, land and sea.

The students have explored energy narratives, examined the political and economic drivers of the transition, developed alternative infrastructure and food systems, proposed new construction materials and systems, and studied ways to turn polluted industrial sites into educational tools and creatively designed spaces for people and nature. They devised possible developments that tie in with the history of the sites while also giving the sites a completely new future, one that makes industrial spaces of oil available for new practices, that uses historical narratives to pave the way for postindustrial futures, and that relegates petroleum to a non-nostalgic past. Some of the students worked on how to reclaim polluted soil, others on how to realize new kinds of industries in the industrial area of the port and nearby urban districts, how to reuse oil industries in a creative way, or on redesigning entire pieces of coastline and finding new narratives and cultural values to once again link the port with the city.

Rashid Ayoubi provided perhaps the most critical and dystopian project. He imagined a past where four "companies"—Oil Arch, GreenLeaf, Every Drop Matters, MADInc.—created a giant mountain over an abandoned refinery to produce the last drops of oil (Figure 15.7). While the refinery satisfies the society's needs for petroleum as a component in medicine, this hidden back end serves the people of Dunkirk, who use the global container trade and additional green energies to devote themselves to consumption and play.

Transition strategies were key to several projects. For more than 150 years, the petroleum industry has polluted water, soil, and air around its installations. Toxic material has seeped deeply into the ground. Clean-up, when not ignored, is often left partly or totally to local public partners. New approaches, technologies, and practices can help make clean-up a process that generates money, promotes innovation, and responds to collective needs. Ege Cakir proposed large autonomous "animals" that will roam the site of the Total refinery to clean up the soil and turn the remediated landscape into a recreational park. Select oil structures—refinery elements and storage tanks—remain as sculptures, a reminder of the industrial petroleumscape in its heyday and its negative impact on the environment and health.

Other students engaged concretely with new production and consumption patterns post-oil. Gemma Galeno suggested using materials with low amounts of embodied energy that

FIGURE 15.7 Rashid Ayoubi's dystopian vision of post-oil Dunkirk, Beyond Oil Studio Spring 2018.

are fabricated without the use of oil and do not contain oil (Figure 15.8). She proposed a new joint system for bamboo, a bamboo plantation, and a bamboo research center high-rise for Dunkirk. These projects represent just a few possible future developments and are early attempts at conceptualizing the impact of transitions involving energy, technology, and ways of life on our future built environment. Understanding the possible impact of new materials and new technologies as well as of new lifestyles and new utopian (and dystopian) narratives can help nurture a necessary conversation about a future beyond oil.

New actors and actor constellations are needed to transform the petroleumscape into a post-oil landscape. It is now up to the local population, policy makers and new industries to find and support a sustainable and meaningful use of areas facing a major change in the wake of new industrial energy landscapes. They require political action and legal tools to overcome the spatial, economic, and social legacies of oil. This process will need to involve a complete rethinking of the current cultural and economic model and it will require commitment by the government, private sector, and local community.[46] Dunkirk's Toile Industrielle is an important instrument to measure economic flows; now, it is time to also map the spatial impact of petroleum over time and to confront its heritage—partly undocumented

FIGURE 15.8 Gemma Galeno, proposal for a bamboo-based post-oil system for Dunkirk, Beyond Oil Studio Spring 2018.

in its soil. Dunkirk and other port cities are not only historical hubs of the petroleumscape but must play a key role in transforming it.

Throughout the petroleum era, corporate actors—often very large ones—have shaped local territories in collaboration with national and local actors, leaving spatial and representational traces that have tied regions such as Dunkirk to petroleum. This long tradition has created path dependencies as defined in the political sciences. Such path development is self-reinforcing, in part because the "embeddedness" of new decisions in already established dynamics implies significant costs to changing strategy.[47] In making complicated decisions, it is often easier to rely on familiar, proven strategies. Paths thus follow an immanent logic[48] that make course changes more difficult. Such path development is self-reinforcing, in part because the "embeddedness" of new decisions in already established dynamics implies significant costs to changing strategy.[49] Path dependence theory emphasizes the role of critical junctures in institution formation, particularly as moments which privilege some pathways over others. The closure of the Dunkirk refinery can be such a critical juncture.

In northwest European port cities, the closure of refineries and ultimately most of the oil-related infrastructure is not a question anymore; time is the only unknown factor. Local authorities of ports must anticipate this outcome and plan for a post-oil future. To achieve such conversion, the EU supports an increasing number of local actors through programs, funding, rules, and policies that support sustainable development and energy transition. The European Green Deal of the new European Commission, which offers a wide range of climate policies and environmental measures, is a first step toward achieving a carbon-neutral Europe by 2050.[50] As key economic and industrial facilities for the EU's competitiveness, port cities must be at the forefront of the "effective and fair transition" that the commission supports.

These objectives and the effects of the 2020 COVID-19 pandemic on prices of oil barrels have pushed European oil companies to rethink their strategies, both to protect the environment and their survival.[51] This unique context must incite all public and private actors to acknowledge and address the petroleumscape and to design the post-oil future in port cities.

Notes

1 AGUR Dunkerque, "La Toile industrielle ® Schéma des relations inter industrielles du Dunkerquois," http://www.agur-dunkerque.org/publications/publication45, accessed June 12, 2020.
2 M. F. "Total. La 'Toile Industrielle' Endommagée," *L'Humanité*, April 16, 2010. https://www.humanite.fr/node/436604, accessed June 12, 2020; "France's Dunkirk Lng Terminal Receives First Cargo," *Offshore Energy* (July 11, 2016), https://www.offshore-energy.biz/frances-dunkirk-lng-terminal-receives-first-cargo/, accessed June 12, 2020.
3 "Société de la raffinerie de Dunkerque: un pas de plus vers la fermeture définitive de l'usine," *La Voix du Nord* (April 6, 2016). http://www.lavoixdunord.fr/archive/recup%3A%252Fregion%252Fsociete-de-la-raffinerie-de-dunkerque-un-pas-de-plus-ia17b47588n3430564, accessed June 10, 2020; Total, "Oleum," https://www.oleum.total.com/en, accessed June 10, 2020.
4 Carola Hein, "Oil Spaces: The Global Petroleumscape in the Rotterdam/The Hague Area," *Journal of Urban History* 44, no. 5 (September 2018): 887–929, DOI:10.1177/0096144217752460.
5 R. J. Forbes, *More Studies in Early Petroleum History 1860–1880* (Leiden: E. J. Brill, 1959).
6 Ernst & Young, *The Oil Downstream: Vertically Challenged?* (Ernst & Young, 2012), 2, petroleumclub.ro/downloads/links/The_oil_downstream_vertically_challenged.pdf, accessed September 4, 2018; Carola Hein, "'Old Refineries Rarely Die': Port City Refineries as Key Nodes in the Global Petroleumscape," *Canadian Journal of History* 53, no. 3 (2018): 450–79.
7 Chambre de Commerce de Dunkerque, "Port de Dunkerque, Exposition universelle de 1878: Recueil des notices etc. produits à l'exposition spéciale des ports de commerce," *Bulletin de la Société de géographie commerciale de Bordeaux*, no. 5 (March 8, 1884): 134; *Plan de Dunkerque et des environs avec indication des nouveaux travaux en cours d'exécution ou en projet, dressé par DURAND, conducteur spécial des waeteringues et Delamare, graveur*, 1864 (Paris : imprimerie Janson). See Henri Durin, *Dunkerque sous le Second Empire in Dunkerque à travers les siècles*, Vol. 2, 1899.
8 Marcel Amphoux, "Une nouvelle industrie française: le raffinage du pétrole," *Annales de géographie* 44, no. 251 (1935): 509–33.
9 Géo Risques, "Fiche détaillée Basias," http://fiches-risques.brgm.fr/georisques/basias-detaillee/NPC5909510, accessed June 10, 2020.
10 "Dunkerque. Incendie De L'usine a Petrole De Coudekerque-Branche." *L'Illustration*, no. 2519 (June 6, 1891).
11 Hein, "'Old Refineries'," 450–79.
12 Michael Stephen Smith, *The Emergence of Modern Business Enterprise in France, 1800–1930* (Cambridge, MA: Harvard University Press, 2006).
13 Michel Bruguière, *Administration et contrôle de l'économie, 1800–1914* (Paris: Librairie Droz, 1985), 23.
14 Amphoux, "Une nouvelle industrie," 509–33.
15 Pierre Fontaine, "Pierre Fontaine. L'Aventure du pétrole français," *Les Sept Couleurs* (1967): 30.
16 Le Capitaliste, *Journal de la banque parisienne*, no. 12 (March 20, 1913): 196.
17 Jonathan D. Smele, *Historical Dictionary of the Russian Civil Wars, 1916–1926* (Lanham, MD: Rowman & Littlefield, 2015), 678
18 "Le trait d'union," *Société Française des Pétroles BP*, no. 43 (Noël 1958): 57.
19 Recueil des procès-verbaux des séances de la Chambre de Commerce, 1909, 430–31.
20 Gabriel Chalandon, "La Politique Francaise Du Petrole," *La Revue hebdomadaire*, no. 15 (1937); Smith, *Modern Business*.
21 Recueil des procès-verbaux des séances de la Chambre de Commerce, 1920, XXXI.
22 Paix-Lesieur signed the agreement with the participation of the SNO, the Société Estrine et Cie, and the Banque de La Seine.

23 Amphoux, "Une nouvelle industrie," 509–33.
24 Alexander Melamid, "Geographical Distribution of Petroleum Refining Capacities: A Study of the European Refining Program," *Economic Geography* 31, no. 2 (1955): 168–78.
25 Amphoux, "Une nouvelle industrie," 509–33.
26 Total, "Total, a Pioneering Spirit," http://www.total.com/en/our-group/thumbnail/total-leading-energy-operator-almost-century, accessed June 10, 2020.
27 Dubessy, "Colas met en vente sa raffinerie de Dunkerque," Econostrum, January 15, 2016. https://www.econostrum.info/Colas-met-en-vente-sa-raffinerie-de-Dunkerque_a21330.html, accessed 15 January, 2021.
28 André Maurois, J. Blancard, J. Huré, and X. Normand, "La société générale des huiles de pétrole BP présente sa nouvelle raffinerie de Dunkerque 1952," *Société Générale des Huiles de Pétrole BP*, 1952.
29 Ibid.
30 Ibid.
31 Ibid.; *Le trait d'union de la Société Générale des Huiles de Pétrole*, no. 18 (December 26, 1952).
32 Stephan Hauser, "The Interplay of Economic Development and Environmental Protection: Dunkirk and the Search for Balance," *PORTUSplus* 8 (November, 2019), https://portusplus.org/index.php/pp/article/view/176.
33 Organe de la technique française des travaux publics et du ciment armé, "Travaux," *Editions Science et Industrie*, 1960.
34 Fabien Guggemos, E. Misak, and D. Mombel, "Importations, exportations, facture énergétique", in *Panorama énergies-climat*, edited by Ministère de l'Environnement, de l'Énergie et de la Mer, 2016. https://www.ecologie.gouv.fr/sites/default/files/dgec_panorama_energie_climat_16.pdf.
35 Dominique Lejeune, *La France des trente glorieuses: 1945–1974* (Paris: Armand Colin, 2015).
36 Joseph, Yvon. "Ports Maritimes", in the opinions presented to the French Senate, Tome VI Equipment et logement, on the 19 November, 1970. https://www.senat.fr/rap/1970-1971/i1970_1971_0056_06.pdf.
37 Total. "Total Announces Plan to Repurpose its Dunkirk Refinery Site," March 8, 2010. http://www.total.com/en/media/news/press-releases/total-annonce-le-projet-de-mutation-industrielle-de-son-etablissement-de-dunkerque, last accessed September 15, 2017.
38 Caroline Britz, "Interview: Stéphane Raison, Président du Directoire du port de Dunkerque," *Mer et Marine*, July 7, 2015, accessed 22 February, 2019. https://www.meretmarine.com/fr/content/interview-stephane-raison-president-du-directoire-du-port-de-dunkerque.
39 British Petroleum, "BP Statistical Review of World Energy 2017," *BP Statistical Review of World Energy*, June 2017; Saudi Aramco, "Saudi Aramco and Total Launch Engineering Studies to Build Giant Petrochemical Complex in Jubail," October 8, 2018, https://www.saudiaramco.com/en/news-media/news/2018/saudi-aramco-and-total-launch-engineering-studies-to-build-giant-petrochemical-complex-in-jubail.
40 Carola Hein, "How the Fourth Industrial Revolution Will Change the Energy Landscape," *The Beam*, no. 4 (August 17, 2017).
41 Dunkerque Port, « Conserver la biodiversité sur le territoire portuaire de Dunkerque-port," *Grand Port Maritime de Dunkerque*, 2010, http://www.dunkerque-port.fr/publicmedia/original/131/25/fr/Dunkerque-Port_Conserver_Biodiversite.pdf, accessed June 26, 2020.
42 ESPO award 2016, https://www.espo.be/news/bremenports-wins-espo-award-2016.
43 Stephan J. Hauser, "Long Live the Heritage of Petroleum—Discoveries of Former Oil Sites in the Port City of Dunkirk," *Urban Science* 4, no. 2 (2020): 22.
44 The course was developed by Carola Hein, who co-taught it with Henri van Bennekom, with the assistance of Paolo De Martino.
45 Hein, "How the Fourth Industrial Revolution Will Change"; Carola Hein and Paolo De Martino, "Designing Post-Carbon Dunkirk with the Students from Tu Delft," *The Beam*, no.5 (February 26, 2018).
46 Almut Reichel, Mieke De Schoenmakere, J. Gillabel, Jeroen Martin, and Ybele Hoogeveen, "Circular Economy in Europe: Developing the Knowledge Base," *European Environment Agency*, 2016, https://eco.nomia.pt/contents/documentacao/thal16002enn-002.pdf.
47 Scott E. Page, "Path Dependence," *Quarterly Journal of Political Science* 1 (2006): 87–115.

48 The German sociologists Helmuth Berking and Martina Löw introduced the term "Eigenlogik" in 2005, focusing on the peculiarities and unique selling points of cities. Helmuth Berking and Martina Löw, *Die Eigenlogik der Städte. Neue Wege für die Stadtforschung* (Frankfurt am Main: Campus 2005).
49 Page, "Path Dependence."
50 European Commission, "COM(2019) 640: The European Green Deal," December 11, 2019, accessed January 16, 2020, https://ec.europa.eu/info/sites/info/files/european-green-deal-communication_en.pdf.
51 R. Bousso and Shadia Nasralla, "Coronavirus Widens Climate Rift between European and U.S. Oil Majors," *Reuters* (May 18, 2020).

BIOGRAPHIES

Nancy Couling is an associate professor at Bergen School of Architecture in Norway and was a Marie Curie Fellow at TU Delft in 2017–2019. She also collaborates with the ETH Zürich After architecture education in New Zealand and international practice that included cofounding cet-0/cet-01 in Berlin. She completed her PhD on ocean urbanization at École Polytechnique Fédérale de Lausanne.

Ben de Vries is a historian and an expert on Indonesian history and culture. He is a program director at the Dutch Cultural Heritage Agency. His coauthored books include *Post-War Reconstruction in the Netherlands 1945–1965* (2016), *Monumenten van de Wederopbouw in Nederland 1940–1965* (2013), and *66 Kanjermonumenten* (2006).

Nelida Fuccaro is a professor of Middle Eastern history at NYU Abu Dhabi. She is the author of *Histories of City and State in the Persian Gulf* (2009) and the editor of "Histories of Oil and Urban Modernity in the Middle East" (2013) and *Violence and the City in the Modern Middle East* (2016).

Chiara Geroldi is a postdoctoral researcher and adjunct professor in landscape architecture at Politecnico di Milano. Her research interests include the uses of discarded fill (PhD thesis: "Designed Landscapes of Discarded Fill," 2015) and landscapes of electricity. Her articles have appeared in *JoLA*, *Extractive Industries and Society*, and *Territorio*.

Stephan Hauser is a PhD candidate in the Chair of History of Architecture and Urban Planning at Delft University of Technology. His research focuses on the impact of oil companies on the development of port cities and on the creation and application of regulations linked to spatial planning, health, and the environment.

Carola Hein is professor of history of architecture and urban planning at Delft University of Technology. Her authored and (co-) edited books include *Adaptive Strategies for Water Heritage* (2019), *The Routledge Handbook of Planning History* (2018), *Port Cities* (2011), *Cities, Autonomy and Decentralisation in Japan* (2006), *The Capital of Europe* (2004), and *Rebuilding Urban Japan after 1945* (2003).

Laura Hindelang is a postdoctoral researcher at the University of Bern, Institute of Art History. Her research focuses on visual cultures of modern architecture and urban planning and petro-modernity and oil culture in Kuwait and the Persian Gulf region. Her monograph on petro-modernity and visual culture in mid-twentieth-century Kuwait will be published by De Gruyter (2021).

Li Hou is a professor of urban planning (Tongji University). Her books include *Building for Oil: Daqing and the Formation of the Chinese Socialist State* (2018), *The History of Chinese Urban and Rural Planning Discipline* (2018), and *Richard Paulick in Shanghai, 1933–1949: The Postwar Planning and Reconstruction of a Modern Chinese Metropolis* (2016).

Alan Lessoff is a university professor of history at Illinois State University in Normal, IL, USA. A specialist in US and comparative urban history, he is the author, coauthor, or editor of six books, most recently *Where Texas Meets the Sea: Corpus Christi and Its History* (2015).

Gloria Pessina is a postdoctoral researcher ("Territorial Fragilities" Excellence Project – DAStU/Politecnico di Milano) and an adjunct professor in urban and territorial analysis (AUIC/Politecnico di Milano). Her articles have appeared in *Urban Research and Practice*, *CPCL*, *PACO*, and *Territorio*. She is coeditor of *Ricomporre i divari* (2021), a book on spatial inequalities in Italy.

Drielli Peyerl is a young investigator of the energy transition, history of technology, and decarbonization (University of São Paulo). Her books include *The Oil of Brazil* (2019), *History, Exploration & Exploitation of Oil and Gas* (2019), and *Opportunities and Challenges of Natural Gas and Liquefied Natural Gas in Brazil* (2020).

Stephen J. Ramos is an associate professor of urban planning and design at the University of Georgia. He is the author of *Dubai Amplified: The Engineering of a Port Geography* (2016, 2010) and coeditor of *Infrastructure Sustainability and Design* (2012) and *New Geographies 01: After Zero* (2009).

Giulia Scotto is an architect and a PhD candidate in urban studies at the University of Basel, Switzerland. Her dissertation, "Postcolonial Logistics: ENI's 'Disegno Africano,'" investigates the infrastructural operations of the Italian national hydrocarbon agency (Ente Nazionale Idrocarburi or ENI) in postcolonial Tanzania.

Christine Stroobandt is responsible for exhibitions, projects, and research at the Musée Portuaire in Dunkirk. Her research focuses on the transformation of ports as ecosystems and witnesses of societal evolution over time.

Christoph Strupp is a research fellow at the Research Centre for Contemporary History in Hamburg (Forschungsstelle für Zeitgeschichte in Hamburg). His publications include *Johan Huizinga: Geschichtswissenschaft als Kulturgeschichte* (2000), *Nahverkehr und Nationalsozialismus: Die Hamburger Hochbahn AG im "Dritten Reich"* (2010), and (coedited with Malte Thießen) *Moderne Stadtgeschichte*, no. 1 (2019).

Imre Szeman is a university research professor of environmental communication at the University of Waterloo. His main areas of research are in energy and environmental studies, social and political philosophy, and critical theory and cultural studies. In 2020, he was named a fellow of the Royal Society of Canada.

Pieter Uyttenhove is a full professor of history and theory of urbanism at Ghent University. He is the author of *Marcel Lods: Action, Architecture, Histoire* (2009), coauthor of *Labo S Works 2004–2014: A Landscape Perspective on Landscape* (2015), and editor of *Recollecting Landscapes: Rephotography, Memory and Transformation 1904–1980–2004–2014* (*2018*, with B. Notteboom).

Caleb Wellum holds a PhD in modern US history from the University of Toronto. His research explores the intertwined histories of energy and crisis. He is currently completing work on *Energizing the Right: The 1970s Energy Crisis and the Making of Neoliberal America*, which will be published by the Johns Hopkins University Press.

INDEX

Note: **Bold** page numbers refer to tables and *italic* page numbers refer to figures.

For Product Safety Concerns and Information please contact our EU representative GPSR@taylorandfrancis.com Taylor & Francis Verlag GmbH, Kaufingerstraße 24, 80331 München, Germany

Batch number: 10722134

Printed by Printforce, the Netherlands